Communications in Computer and Information Science 2532

Series Editors

Rationale
The CCIS series is devoted to the publication of proceedings of computer science conferences. Its aim is to efficiently disseminate original research results in informatics in printed and electronic form. While the focus is on publication of peer-reviewed full papers presenting mature work, inclusion of reviewed short papers reporting on work in progress is welcome, too. Besides globally relevant meetings with internationally representative program committees guaranteeing a strict peer-reviewing and paper selection process, conferences run by societies or of high regional or national relevance are also considered for publication.

Topics
The topical scope of CCIS spans the entire spectrum of informatics ranging from foundational topics in the theory of computing to information and communications science and technology and a broad variety of interdisciplinary application fields.

Information for Volume Editors and Authors
Publication in CCIS is free of charge. No royalties are paid, however, we offer registered conference participants temporary free access to the online version of the conference proceedings on SpringerLink (http://link.springer.com) by means of an http referrer from the conference website and/or a number of complimentary printed copies, as specified in the official acceptance email of the event.

CCIS proceedings can be published in time for distribution at conferences or as post-proceedings, and delivered in the form of printed books and/or electronically as USBs and/or e-content licenses for accessing proceedings at SpringerLink. Furthermore, CCIS proceedings are included in the CCIS electronic book series hosted in the SpringerLink digital library at http://link.springer.com/bookseries/7899. Conferences publishing in CCIS are allowed to use Online Conference Service (OCS) for managing the whole proceedings lifecycle (from submission and reviewing to preparing for publication) free of charge.

Publication process
The language of publication is exclusively English. Authors publishing in CCIS have to sign the Springer CCIS copyright transfer form, however, they are free to use their material published in CCIS for substantially changed, more elaborate subsequent publications elsewhere. For the preparation of the camera-ready papers/files, authors have to strictly adhere to the Springer CCIS Authors' Instructions and are strongly encouraged to use the CCIS LaTeX style files or templates.

Abstracting/Indexing
CCIS is abstracted/indexed in DBLP, Google Scholar, EI-Compendex, Mathematical Reviews, SCImago, Scopus. CCIS volumes are also submitted for the inclusion in ISI Proceedings.

How to start
To start the evaluation of your proposal for inclusion in the CCIS series, please send an e-mail to ccis@springer.com

Timoteo Carletti · Thierry-Sainclair Njougouo ·
Elio Tuci
Editors

Artificial Life and Evolutionary Computation

18th Italian Workshop, WIVACE 2024
Namur, Belgium, September 11–13, 2024
Revised Selected Papers

Springer

Editors
Timoteo Carletti
University of Namur
Namur, Belgium

Thierry-Sainclair Njougouo
IMT School for Advanced Studies Lucca
Lucca, Italy

Elio Tuci
University of Namur
Namur, Belgium

ISSN 1865-0929 ISSN 1865-0937 (electronic)
Communications in Computer and Information Science
ISBN 978-3-031-93630-2 ISBN 978-3-031-93631-9 (eBook)
https://doi.org/10.1007/978-3-031-93631-9

This Springer imprint is published by the registered company Springer Nature Switzerland AG
The registered company address is: Gewerbestrasse 11, 6330 Cham, Switzerland

Preface

This volume of Communications in Computer and Information Science contains the proceedings of WIVACE 2024, the XVIII Workshop on Artificial Life and Evolutionary Computation. The event was successfully held on the Campus of the University of Namur, Belgium, during September 11–13, 2024. This is the second edition of the WIVACE conference series outside Italy; indeed after a long series of Italian locations, Samperi (2007), Venice (2008), Naples (2009), Parma (2012), Milan (2013), Vietri sul Mare (2014), Bari (2015), Salerno (2016), Venice (2017), Parma (2018), Rende (2019), Gaeta (2022), and Venice (2023), WIVACE has been organized in Winterthur (Switzerland) in 2021 and now in Namur (Belgium), strengthening in this way the international dimension of the workshop.

The International Workshop on Artificial Life and Evolutionary Computation aims at bringing together researchers to present and share their results and ideas in a multidisciplinary context. The workshop provides a forum for the discussion of new research directions and applications in Artificial Life, Evolutionary Computation and related fields, where different disciplines and research areas could effectively meet. In particular, the workshop brings together computer scientists, mathematicians, biologists, psychologists and cognitive scientists to discuss issues related to the origins of life, evolution and adaption, collective and social behaviours and other topics related to the development of technological solutions inspired by biological principles.

The workshop was organized on two and half days, based on single-track sessions, with oral (22) and poster presentations (7). The workshop was structured using a format specifically designed to allow doctoral students and early career researchers to meet in order to discuss their ideas and to disseminate their work in a collaborative and constructive environment. We managed to have five keynote speakers of international renown; we are thus deeply indebted for their participation to Monika Scholz (Max Planck Institute for Neurobiology and Center of Advanced European Studies and Research (caesar), Germany), Yara Khaluf (Wageningen University, The Netherland), Pietro Liò (University of Cambridge, UK) and Aldo Genovesio (Università del Piemonte Orientale "A. Avogadro", Italy). A special thanks goes to the origami artist Dewi Brunet (Dewi Origami, Belgium), who showed the links between art and science.

Overall, we received 42 submissions, out of which 23 were selected for this volume of proceedings after a two-round single-blind review process performed by Program Committee members for an acceptance rate of about 54%. We would like to warmly thank the PC members for their help in the selection process. We take the opportunity to also thank the Local Organizing Committee composed of Nemanja Antonic, Aymeric Vellinger, Martin Moriamé and Jean-François de Kemmeter from the Namur Institute of Complex Systems, naXys, for their valuable help. Last but not least, we thank various members of the staff of the University of Namur and the host institution for making the infrastructure available. A special thanks to all authors for having contributed to the success of this workshop.

Let us finally acknowledge the precious advice of the staff at Springer, who provided their professional support through all the phases that led to this volume.

February 2025

Timoteo Carletti
Thierry-Sainclair Njougouo
Elio Tuci

Organization

General Chairs

Timoteo Carletti	University of Namur, Belgium
Elio Tuci	University of Namur, Belgium
Thierry Njougou	IMT School for Advanced Studies Lucca, Italy

Program Committee

Marco Baioletti	Università degli Studi di Perugia, Italy
Mauro Birattari	Université Libre de Bruxelles, Belgium
Nicolas Bredeche	Sorbonne Université, France
Stefano Cagnoni	University of Parma, Italy
Alexandre Campo	Université Libre de Bruxelles, Belgium
Giulio Caravagna	Università di Trieste, Italy
Antonio Chella	Università di Palermo, Italy
Franco Cicirelli	Università della Calabria, Italy
Cilia Nicole Dalia	Università degli studi di Enna Kore, Italy
Giulia De Masi	Technology Innovation Institute, United Arab Emirates
Claudio De Stefano	Università di Cassino e del Lazio meridionale, Italy
Luca Di Gaspero	Università di Udine, Italy
Federico Divina	Universidad Pablo de Olavide, Spain
Marco Dorigo	Université Libre de Bruxelles, Belgium
Gusz Eiben	VU Amsterdam, the Netherlands
Alessia Faggian	Università degli studi di Trento, Italy
Harold Fellermann	Newcastle University, UK
Eliseo Ferrante	VU Amsterdam, the Netherlands
Alessandro Filisetti	Officinae Bio S.r.l., Italy
Francesco Fontanella	Università di Cassino e del Lazio meridionale, Italy
Mattia Frasca	Università di Catania, Italy
Rudolf Marcel Füchslin	Zürcher Hochschule für Angewandte Wissenschaften, Switzerland
Mario Giacobini	Università di Torino, Italy
Alexandros Giagkos	Aston University, UK

Onofrio Gigliotta	Università di Napoli Federico II, Italy
Alex Graudenzi	IBFM-CNR, Italy
Antonio Guerrieri	ICAR-CNR, Italy
Alvaro Gutierrez	Universidad Politécnica de Madrid, Italy
Giovanni Iacca	University of Trento, Italy
Ignazio Infantino	CNR, Italy
Tanja Katharina Kaiser	TU Dresden, Germany
Fred Labrosse	Aberystwyth University, UK
Antonio Liotta	Free University of Bozen-Bolzano, Italy
Gianfranco Lombardo	Università di Parma, Italy
Luca Manzoni	Università di Trieste, Italy
Roberto Marangoni	Università di Pisa, Italy
Davide Marocco	Università di Napoli Federico II, Italy
Giancarlo Mauri	Università di Milano-Bicocca, Italy
Alexandre Mauroy	University of Namur, Belgium
Eric Medvet	Università di Trieste, Italy
Giorgia Nadizar	Università di Trieste, Italy
Stefano Nolfi	ISTC CNR, Italy
Dimitri Ognibene	Università di Milano-Bicocca, Italy
Mario Pavone	Università di Catania, Italy
Riccardo Pecori	Universitas Mercatorum, Italy
Anthony Pipe	University of Bristol, UK
Andreagiovanni Reina	University of Konstanz, Germany
Simone Righi	Università di Venezia, Italy
Riccardo Righi	European Commission, Belgium
Andrea Roli	Università di Bologna, Italy
Federico Rossi	Università di Siena, Italy
Stephan Scheidegger	Zürcher Hochschule für Angewandte Wissenschaften, Switzerland
Yaroslav Sergeyev	Università della Calabria, Italy
Roberto Serra	Università degli Studi di Modena e Reggio Emilia, Italy
Debora Slanzi	European Centre for Living Technology, Italy
Annalisa Socievole	CNR ICAR, Italy
Pasquale Stano	Università del Salento, Italy
Thomas Stützle	Université Libre de Bruxelles, Belgium
Pietro Terna	Università di Torino, Italy
Dari Trendafilov	Université de Namur, Belgium
Vito Trianni	ISTC CNR, Italy
Nicolas Vandewalle	Université de Liège, Belgium
Marco Villani	Università degli Studi di Modena e Reggio Emilia, Italy

Andrea Vinci	ICAR CNR , Italy
Mathias Weyland	Zürcher Hochschule für Angewandte Wissenschaften, Switzerland
Myra Wilson	Aberystwyth University, UK
Olli Yli-Harja	Tampere University of Technology, Finland

Additional Reviewers

Nemanja Antonic
Aymeric Vellinger
Martin Moriamé
Jean-François de Kemmeter

Contents

Emergent Dynamics in Heterogeneous Life-Like Cellular Automata 1
Aarati Shrestha, Felix Reimers, Sanyam Jain, Paolo Baldini, Michele Braccini, Andrea Roli, and Stefano Nichele

Autonomous Traffic Management: Integrating Vissim Traffic Model with a Swarm of Drones 16
Davoud Alahvirdi, Julien Pietquin, Alexandre Mauroy, and Elio Tuci

Analysing Opinion Dynamics via a Cognitive Model of Structured Beliefs 28
Eleni Michaelidou, Eirini Ioannou, Keren Tapper, Benjamin D. Goddard, and Guillermo Romero Moreno

When is Reservoir Computing with Cellular Automata Beneficial? 42
Tom Eivind Glover, Evgeny Osipov, and Stefano Nichele

Chemotactic Swarming of *C. elegans* as a Collective Decision-Making Process 56
Nemanja Antonic, Aymeric Vellinger, and Elio Tuci

Factors Impacting Landscape Ruggedness in Control Problems: A Case Study 70
Michel El Saliby, Eric Medvet, Giorgia Nadizar, Erica Salvato, and Sarah L. Thomson

Steps to Chemically Embodied Cognition 84
Sergio Rubin, Pasquale Stano, Andrea Roli, and Luisa Damiano

A Comparative Study on Decision-Making Mechanisms in a Site Selection Task 98
Ahmed Almansoori, Dari Trendafilov, Muhanad Alkilabi, and Elio Tuci

Emergent Naming System in an Unknown Dynamic Environment: A Collective Perception Scenario 111
Dat Nguyen Tien Hoang and Nicolas Cambier

Mixed-Variable Surrogate-Based Optimization Using Affinity Features to Improve the Distance for Categorical Variables 125
Charlotte Beauthier, Rajan Filomeno Coelho, Sylverio Pool Marquez, Caroline Sainvitu, and Annick Sartenaer

Using Pseudo-attractors of Genetic Regulatory Networks to Analyze Biological Data 137
Gianluca D'Addese, Carlo Crovetti, Roberto Serra, and Marco Villani

Costs of Information Gathering Strategies and Their Impact on Animal Dispersal 148
Antoine Sion, Matteo Marcantonio, Stefano Masier, and Elio Tuci

Cooperation in Evolved Modular Soft Robots 160
Paolo Baldini, Michele Braccini, Filippo Benvenuti, Eric Medvet, Andrea Roli, and Francesco Rusin

Energy-Aware Neural Architecture Search: Leveraging Genetic Algorithms for Balancing Performance and Consumption 173
Antoine Gratia, Paul Temple, Pierre-Yves Schobbens, and Gilles Perrouin

Explainable Multi-label Classification for Predictive Maintenance 186
Antoine Hubermont, Aymeric Vellinger, and Elio Tuci

Cross-Entropy Regularization with Mutual Information in Training CNNs 198
Ch. Zaugg, R. Ingold, D. Trendafilov, and A. Fischer

Implementation of the Response to Synchronization in E-Puck2 Robots 208
Cinzia Tomaselli, Gonzalo Marcelo Ramírez-Ávila, Lucia Valentina Gambuzza, Mattia Frasca, Elio Tuci, and Timoteo Carletti

Attractive-Repulsive Challenge in Swarmalators with Time-Dependent Speed 223
Steve J. Kongni, Thierry Njougouo, Gaël R. Simo, Patrick Louodop, Robert Tchitnga, and Hilda A. Cerdeira

A "Plug and Play" Table for Composable Molecular Neural Networks Built with DNA 236
Ethan Collinson and Harold Fellermann

Educational Robotics: From Computational to Complexity Thinking 250
Onofrio Gigliotta, Michela Ponticorvo, Erica Chinzer, Clara Nobile, Diana Giorgini, and Alessandra Vitanza

An Ecological Model of Cancer Development and Control 261
Elpida Tzafestas

Stability of Input Representations in Biological Neural Network Reservoirs 276
Trym A. E. Lindell, Ola H. Ramstad, Ioanna Sandvig, Axel Sandvig, and Stefano Nichele

Latent Agency: A Proto-Cognitive Property 292
Fernando Rodriguez and Phil Husbands

Author Index 307

Emergent Dynamics in Heterogeneous Life-Like Cellular Automata

Aarati Shrestha[1], Felix Reimers[1], Sanyam Jain[1], Paolo Baldini[2], Michele Braccini[2], Andrea Roli[2], and Stefano Nichele[1,3](✉)

[1] Østfold University College, Halden, Norway
{felix.s.reimers,stefano.nichele}@hiof.no
[2] University of Bologna, Cesena, Italy
[3] Oslo Metropolitan University, Oslo, Norway

Abstract. The Game of Life (GoL), one well known 2D cellular automaton, does not typically ensure *interesting* long-term phenotypic dynamics. Therefore, while being Turing complete, GoL cannot be said to be open-ended. In this work, we extend GoL with the opportunity for local mutations, thus enabling a heterogeneous life-like cellular automaton guided by an evolutionary inner loop. Additionally, we introduce the concept of cell ageing to ensure that cell aliveness (activated by inheritance with variation, and controlled by ageing) and actual cell computation (governed by life-like rules on local neighborhoods) are kept conceptually separated. We conduct an experimental campaign to identify suitable parameters that produce long-term phenotypic dynamics and favor genotypic innovations.

Keywords: Cellular Automata · Game of Life · Open-Ended Evolution

1 Introduction

Conway's Game of Life (GoL) [3,7] is a very well known cellular automaton (CA) that has been proven to be Turing complete [16]. Therefore, in principle, it can execute a universal constructor or any Turing machine. GoL has been widely used as a model system to study the emergence of complex behaviors from simple rules [1,21]. In fact, the exploration of complex moving structures, such as spaceships (complex gliders) that can carry information and encode computations, is an active area of research [5]. Additionally, gliders have been studied in relation to agency [4]. However, such computational structures are rather fragile in that even small perturbations will destroy the precisely handcrafted computations. Additionally, from a random initial configuration typically GoL settles into a rather boring and not particularly useful behavior, as most of the living cells die out besides a few persistent or oscillating structures. Therefore, while in principle GoL supports arbitrarily complex computations, in practice it does not appear to be open-ended. Open-endedness is a property of a system, observed for example in biological systems, that ensures never ending innovation and discovery of novel solutions, thus providing continuously increasing complexity [12].

T. Carletti et al. (Eds.): WIVACE 2024, CCIS 2532, pp. 1–15, 2025.
https://doi.org/10.1007/978-3-031-93631-9_1

In order to allow more interesting behaviors to emerge, McCaskill and Packard [9,13] investigated a GoL world where local mutations diversify GoL rules to other life-like rules, named genelife. Genelife created more diversity in the local dynamics and therefore allowed the emergence of different kinds of non-uniform gliders and structures. Other attempt at coupling life-like rules and evolution are Sprout Life [17] and evolife [24]. On the other hand, in such CA worlds there is no distinction between the aliveness of cells (agents) and their cellular state. In fact, aliveness and computation carry the same meaning. In other words, there is no distinction between the agent and the computation it carries, i.e. aliveness is the computation. However, it may be beneficial to differentiate the conditions for cells to become alive and survive (and therefore generate offspring through inheritance and variation) while keeping the actual computation based on local information conceptually separated.

Medernach et al. introduced the concept of cell age in their HetCA system [10,11], while using evolved genetic programming (GP) rules as mechanisms to update the states of cells based on local neighboroods. In their work, each cell possesses an age counter that increments at each time step. Cells can live only for a certain number of steps and during those steps they can update their states through their local rules. When the age limit is reached, cells transition to the decay state, i.e., a living state where cells do not update their state any longer (however their state is still available to their neighbors). After the decay phase, cells die and empty their locations. Empty locations that are neighbors of living cells may probabilistically become alive by inheriting the GP rules from a living neighbor through mutation. As such, HetCA produces interesting long term phenotypic dynamics without any sign of stagnation or repetitive behavior.

One open question is whether a more simplified rule-set, such as life-like rules (instead of intricate GP programs), can support similar long-term phenotypic and genotypic dynamics by incorporating the concept of ageing, together with an evolutionary inner loop. Additionally, the concept of age allows for incorporating environmental factors such as "energy", which may increase/decrease the lifespan of cells. This may be favourable for simulating different ecologies, as well as plugging-in tasks to be solved in the form of "rewards and penalties". One may imagine a substrate that initially supports long-term evolutionary dynamics (without a goal). Then energy is introduced and this life-like "solver" adapts to the environmental conditions. Finally, the task is removed and open-ended life continues. In this work, we experiment to find out the ideal conditions to support long-term dynamics in heterogeneous life-like CA with age constrains and local evolution.

2 Background

2.1 Game of Life Cellular Automaton

Conway's Game of Life (GoL) is a 2D Cellular Automaton where each cell in the grid is either in an alive or dead state. The state of a cell is determined by its eight surrounding neighbours, known as the Moore Neighbourhood, and

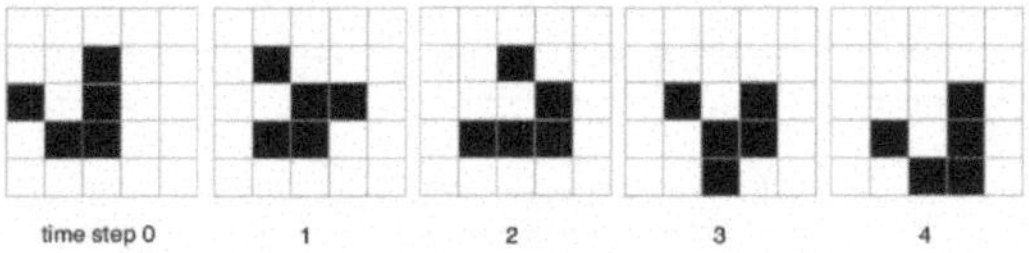

Fig. 1. States of a glider in Conway's Game of Life

follows a simple transition rule denoted as B3/S23. This notation signifies that a cell is born if it has precisely 3 live neighbours, and it survives if it has 2 or 3 live neighbours; otherwise, it dies. Figure 1 depicts the subsequent stages of the GoL's glider pattern, showcasing the intricate interplay of simple rules leading to emerging dynamics.

While extensive research has been conducted into the exploration of 2D GoL, it is crucial to acknowledge the parallel efforts dedicated to investigating higher dimensions. Notably, Bays [2] contributed significantly by extending GoL in three dimensions, introducing variations in the transition rules.

2.2 Life-Like Cellular Automata

Life-like CA (signifying that they are similar to the Game of Life) allow for variations of the GoL state transition function. In particular, for all CA in this family, the new state of a cell can be represented by a function of the number of neighbors that are currently alive (including the cell's own state). As for GoL, a typical notation of a life-like rule would consist in a string denoted by B (born) followed by the number of neighbors (from 0 to 9), and then a string denoted by S (survive), also followed by the number of neighbors (from 0 to 9) [23]. Therefore, there exist a total of $2^{18} = 262,144$ rules [6]. Only a few have been extensively explored[1]. For a recent exploration of complexity in life-like rules, please refer to [14].

2.3 Heterogeneous and Evolving CA

While typically the state transition function governing state updates of all cells in the CA is homogeneous, i.e., all cells are governed by the same rules, in heterogeneous CA each cell has its own (potentially unique) rule. Such rules may also vary over time. As such, the space of possible CA becomes rather vast and artificial evolution is often used to search for suitable rules [18–20,22].

Evolution may be used as an outer loop, where a genotype consists of the concatenation of the rules for all cells in the CA. In this case, once the CA phenotype (the actual CA execution for a certain number of steps) is evaluated by a fitness function, the genetic operators affect the overall (concatenated) genotype. Subsequently, a new CA execution is carried out and the results are further evaluated.

[1] https://conwaylife.com/wiki/List_of_Life-like_rules.

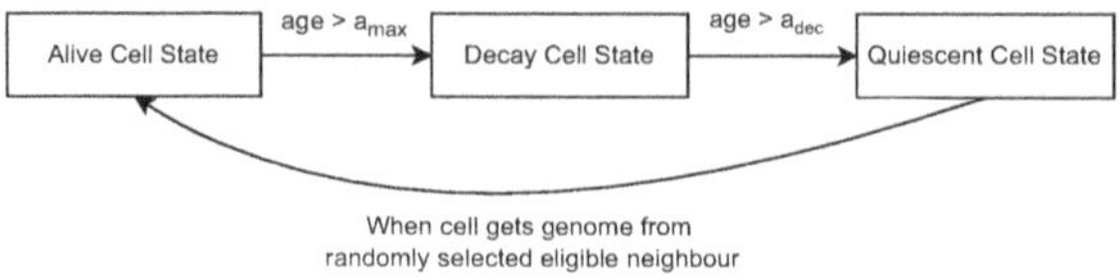

Fig. 2. Life cycle of a cell's *Cell State*.

In the case of an inner evolutionary loop, however, genomes encoding cell rules evolve locally and in an online fashion. This means that inheritance with variation is governed by local factors and happens from quiescent (dead) cells to alive cells while the CA computation is being executed. Examples of inner-loop CA evolution include [8–11,15].

3 Methodology

Our model, described in detail in this section, consists of a 2-dimensional heterogeneous cellular automaton where each cell is updated according to distinct life-like rules that operate in local neighbourhoods. Additionally, an inner evolutionary loop controls the local evolution of life-like rules by allowing cells to grow into empty nearby sites through inheritance with variation. At the end of this section, we describe how genotypic and phenotypic diversity are measured.

3.1 CA Substrate

The CA substrate consists of a 2D grid of cells with a tripartite *Cell State* life cycle, i.e., alive, decay, quiescent (inspired by [11]). The CA is initialized with cells either in an alive or quiescent state. The alive cells are those cells that actively apply their *Transition Rule* (genome) to change their Boolean *Grid State* based on a Moore's neighborhood. Quiescent cells are temporarily dead cells awaiting to become alive, therefore they do not have a genome. When initialized, all alive cells have identical Game-of-Life genomes. As the cellular automaton iterates through generations, the cell's *Cell State* undergoes a sequential transformation, transitioning cells from alive states to decay and ultimately to quiescent states. A cell cannot transition directly from decay to alive or from quiescent to decay (nor from alive to quiescent). Instead, these states follow an unidirectional progression, as illustrated in Fig. 2.

The *Cell State* of a given cell is governed by the age of the cell, specifically by two parameters: a_{max} and a_{dec}. The a_{max} represents the age threshold for remaining in an alive state. Once a cell's age crosses this threshold, it transitions into a decay state. Cells in a decay state are neither alive nor dead. They hold their genomes but do not apply it to change the *Grid State*. When the age of a decay cell crosses the a_{dec} age threshold, they become quiescent and their age is reset to 0. Quiescent cells may at some point become alive by inheriting a

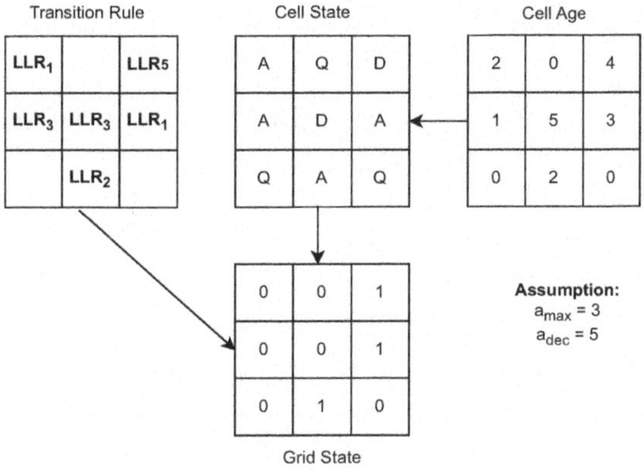

Fig. 3. Example of a 3×3 CA substrate, where the genotype is represented by the *Transition Rules* and the phenotype is composed by three components: *Cell Age* (a counter), *Cell State* (alive, decay, quiescent), and *Grid State* (Boolean cell state representing the ongoing computation).

mutated genome from a random alive neighbour. When a cell becomes alive, its age counter starts from 1 and is incremented by one at each CA iteration.

Further, alive cells get their genome when they are born and their genome remains unchanged throughout their life cycle. Cells in decay (and quiescent) state do not apply any transition rule and therefore do not update their *Grid State*. Cells in decay state retain an unaltered *Grid State* (either 0 or 1) throughout the duration of their decay period, while quiescent are considered to be in a *Grid State* 0.

Figure 3 illustrates the composition of the CA substrate on a small 3×3 grid. The genomes of each cell are represented by the *Transition Rule* layer, where each cell contains a possibly different Life-Like Rule (LLR). The *Cell State* layer represents whether cells are alive, decaying, or quiescent, while the *Cell Age* layer indicates how many iterations have passed since they became alive. Finally, the *Grid State* layer indicates the actual computation, which consists of a Boolean state that results from the execution of the LLR on the corresponding cell based on the local (Moore) neighborhood.

3.2 Evolutionary Inner Loop

Life-like rules may evolve over time by a mechanisms that allows alive cells to replicate their mutated genome into empty (i.e., quiescent) cells. This process is governed by an inheritance probability parameter that acts on all quiescent cells that have at least a living neighbor. A quiescent cell transitioning to an alive state inherits therefore the genome from a randomly selected alive neighbouring cell. The received genome, which consists of a life-like rule, may undergo mutation. A mutation consist of one of the following genome changes:

- A symbol is added, removed, or changed from the B section of the genome;

- A symbol is added, removed, or changed from the S section of the genome.

Legitimate life-like genomes range from empty $B \cdot S \cdot$ strings to $B0123456789$ $S0123456789$. This means that there are $2^{18} = 262,144$ possible combinations. Appropriate checks are carried out to avoid illegal genomes (for example duplicated symbols). As an example, a GoL genome represented by the string B2S23 (a cell is born when exactly 2 neighbors are alive, a cell survives when 2 or 3 neighbors are alive, otherwise the cell dies) can be mutated as follows:

- A random symbol is added to the B section: B2**9**S23
- A random symbol is removed from the B section: BS23
- A random symbol is changed in the B section: B**6**S23
- A random symbol is added to the S section: B2S23**9**
- A random symbol is removed from the S section: B2S3
- A random symbol is changed in the S section: B2S**1**3

It is worth to highlight that we have kept the same life-like notation, i.e., B = born and S = survive. However, in our substrate B and S rules **do not** governe the birth and survival of cells, since the *Cell State* (alive, decay, quiescent) is only controlled by the *Cell Age* progression. Therefore, B and S in fact control the computational state update represented by the CA *Grid State*. The actual meaning of B is whether a cell in *Grid State* 0 should transition to state 1, while S indicates whether a cell in *Grid State* 1 should remain in state 1.

3.3 Phenotypic and Genotypic Measures

We aim at measuring the variation of phenotypes and genotypes over time. The following Tables 1 and 2 give an overview of the qualitative and quantitative measures used, respectively, together with the corresponding color palettes.

The genotype of each cell is represented by its life-like rule. A qualitative overview of the current genotypic variability over the entire CA can be visually represented by a 2D-grid, where cells with identical genomes are represented by equal colors. Quiescent cells (a) do not possess any genome and are represented in white, while GoL genomes (b) are always represented in yellow. Other unique genomes (c) are represented by one unique color.

To quantitatively measure the overall genotypic variation, we count the total (cumulative) number of unique life-like rules discovered during each generation. This gives an indication on whether the CA substrate keeps discovering new genotypes over time.

The phenotype variation is qualitatively visualized with two 2D grids representing:

- the *Cell State* and age - quiescent (a) in white, decay (b) in orange, alive (c) in different green color shades indicating their current *Cell Age* counters,
- the actual *Grid States* (0, 1) - quiescent cells (a) are depicted in white, 1 (b) in green, and 0 (c) in red.

Table 1. Qualitative Phenotypic and Genotypic Measures

Measure	Grid representation
Genotypic variability	Color-coded genome 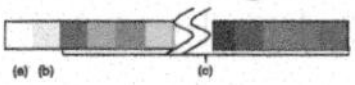
Cell state and age	Tri-color (color shades for age) 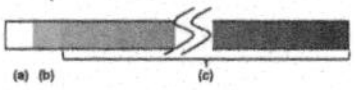
Grid state	Tri-color (quiescent in white) 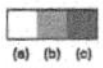

Table 2. Quantitative Phenotypic and Genotypic Measures

Measure
Cumulative number of discovered rules
Grid state fluctuation
Number of alive, decay, and quiescent cells
Number of cells in state 0, 1, and quiescent

To quantitatively assess the phenotypic variation over time, we are employing various metrics.

In order to quantitatively measure fluctuations of the Grid State, the number of cells that change state (either from 0 to 1, or from 1 to 0) between two consecutive iterations are measured. This indicates the phenotypic movement between two consecutive generations.

Another approach involves tallying the number of living cells in state 1 (living in this case includes both alive and decay cells), the number of alive cells in state 0, and dead (quiescent) cells, across each generation. This method offers insight into the overall computation in the substrate, by measuring the fluctuation of activities within the *Grid State*. Minimal variation across generations suggests a lack of significant activity within the system, while noticeable fluctuations indicate a more dynamic behaviour. Persistent and long-term fluctuations over time suggest long-term dynamics within the cellular automaton's *Grid State*.

Additionally, the count of cells in different *Cell States* is tracked for each generation, serving as a measurement of the distribution of cells across various states (alive, decay, quiescent), independent of the actual computation happening on the *Grid State*.

4 Experimental Setup

We conduct a series of experiments to identify suitable conditions that may support long-term dynamics in our heterogeneous life-like CA with age constrains

Table 3. CA Parameters

Parameter	Value
Eligible Cell State to Inherit Genomes	[Alive]
Probability Initial Alive Cell State	0.5
Probability Alive Cell State with Value 1	0.5
Inheritance Probability	0.125
Mutation Probability P_{mut}	0.2
a_{max}	10, 50
a_{dec}	15, 70
Initial Rule	B3S23

and local evolution. In this section, we provide details on the experimental setup and utilized parameters.

The used parameters are given in Table 3. In particular, the *Probability Initial Alive Cell State* determines the ratio of alive to quiescent cells in the first generation of the CA. In our experiments, the initial generation of a CA begins with expectedly half of the cells in a quiescent state, while the remaining half are alive. Initially, all alive cells start with a GoL genome, i.e., B3S23. Among the living cells, a stochastic allocation designates half with a phenotypic state of 0, while the remaining half is assigned a phenotypic state of 1, as dictated by the parameter *Probability Alive Cell State with Value 1.* The *Inheritance Probability*, set to 0.125, governs the likelihood of quiescent cells transitioning to an alive state, i.e. on average one out of the eight neighbors to a single alive cell is expected to become alive. Only alive cells are eligible for genome transmission to quiescent cells, determined by the parameter *Eligible Cell State to Inherit Genomes.* This means that decay cells cannot reproduce. The inherited genomes of new alive cells may undergo mutation with a probability indicated by *Mutation Probability* (P_{mut}). Suitable values for *Inheritance Probability* and *Mutation Probability* used in this study are determined experimentally.

Additionally, we experiment with two age/decay budgets, a_{max} and a_{dec}, namely 10/15 and 50/70.

Finally, we benchmark our results with two GoL versions:

- a GoL version with age constrains where $P_{mut} = 0.0$, however $a_{max} = 50$, $a_{dec} = 70$. All the other parameters are kept as in Table 3
- a classical GoL execution (no age and no mutation), where $P_{mut} = 0.0$ and a_{max} / a_{dec} are set to a value larger than the number of executed generations. In this case, all cells are alive in the initial state, either in Cell State 0 or 1 (with 0.5 probability).

The experiments are run on a 50 × 50 grid for 10,000 generations and on a 500 × 500 grid for 1,000 steps. The experiments were repeated 10 times and

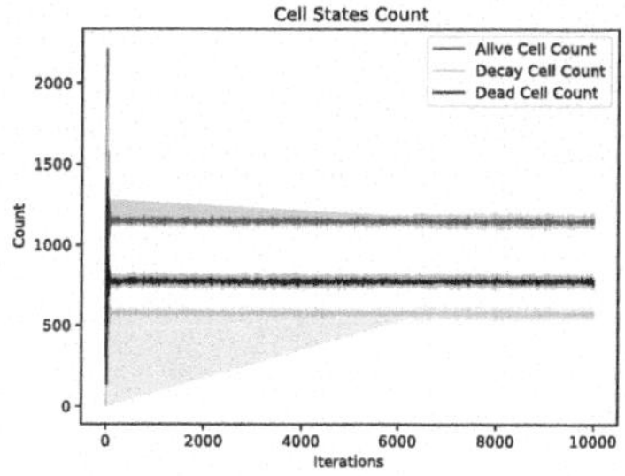

(a) *Cell States* count, a_{max} 10 and a_{dec} 15.

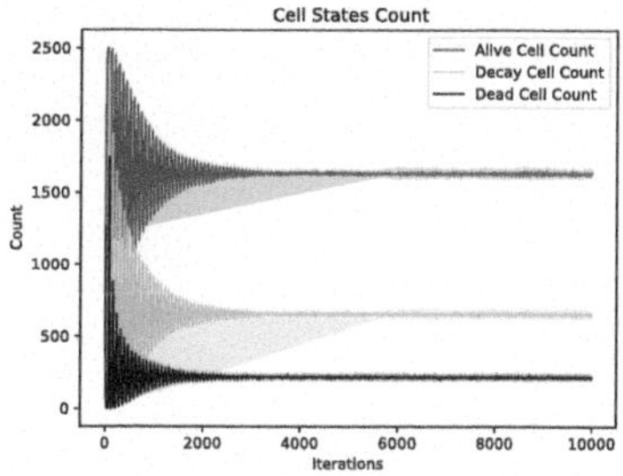

(b) *Cell States* count, a_{max} 50 and a_{dec} 70.

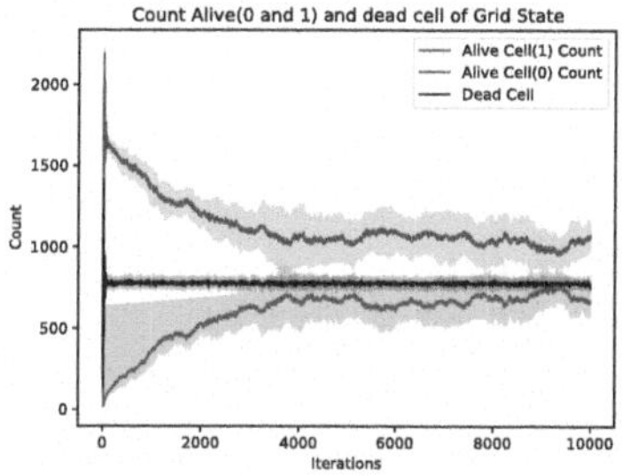

(c) *Grid States* count, a_{max} 10 and a_{dec} 15.

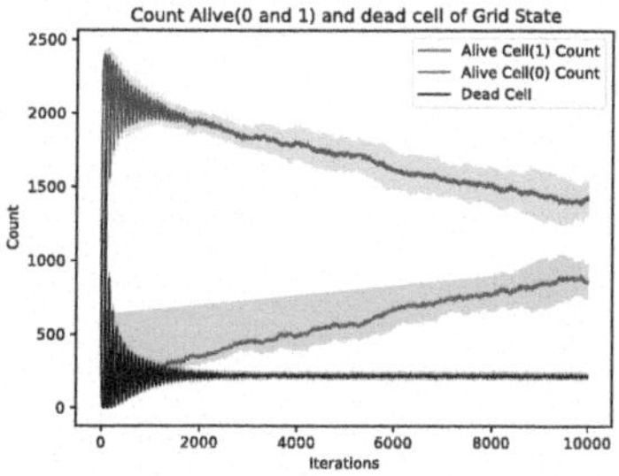

(d) *Grid States* count, a_{max} 50 and a_{dec} 70.

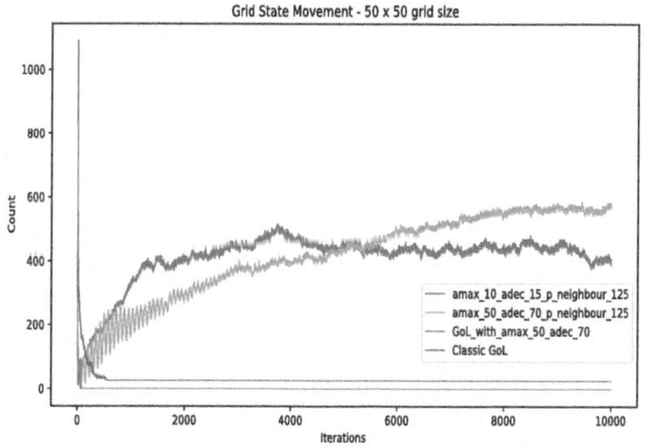

(e) Comparison of *Grid State* fluctuations.

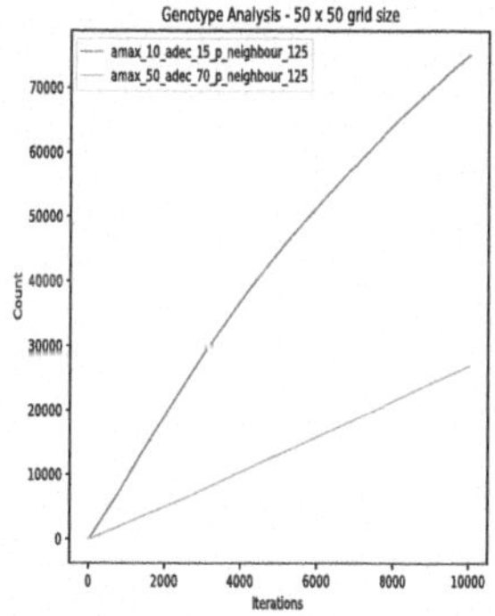

(f) Cumulative Genotype count.

Fig. 4. Grid Size of 50 × 50 for 10,000 generations. Averages over 10 runs.

for the quantitative metrics, results were averaged and the deviation calculated. Sample videos are available here: https://tinyurl.com/vrkfs63a.

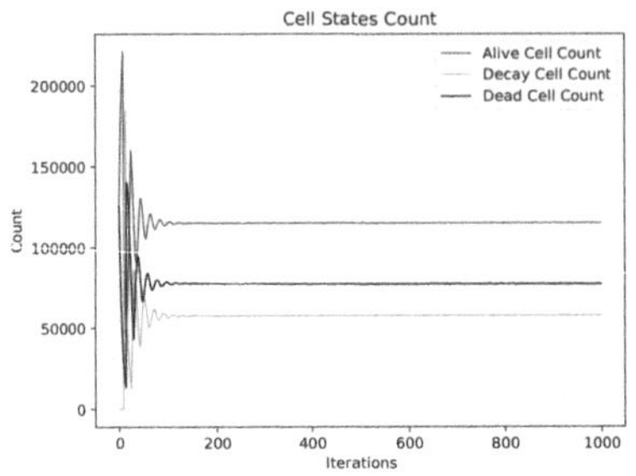

(a) *Cell States* count, a_{max} 10 and a_{dec} 15.

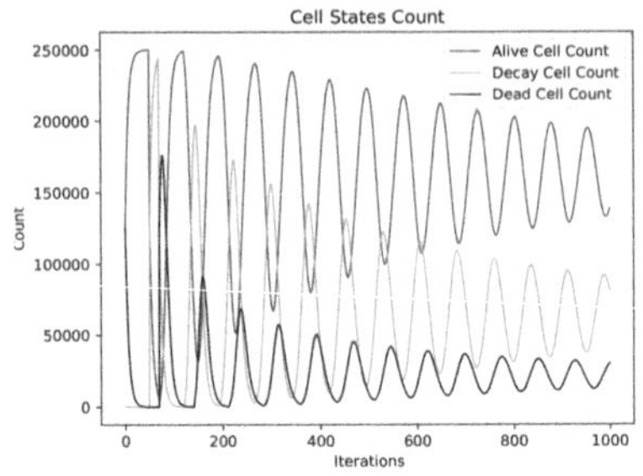

(b) *Cell States* Count, a_{max} 50 and a_{dec} 70.

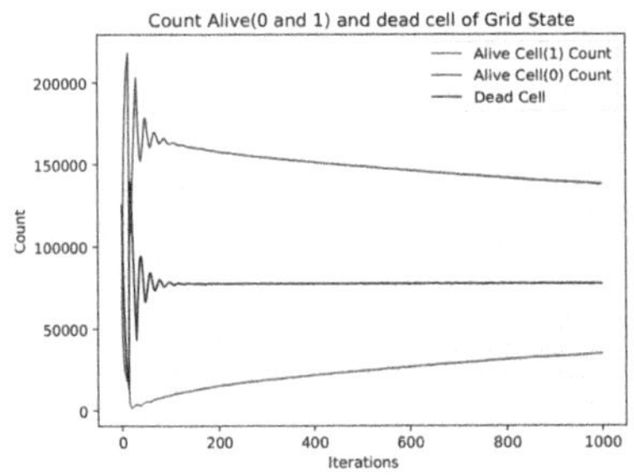

(c) *Grid States* count, a_{max} 10 and a_{dec} 15.

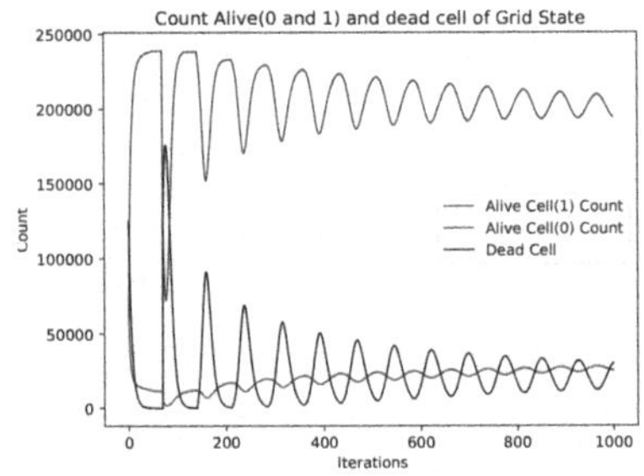

(d) *Grid States* count, a_{max} 50 and a_{dec} 70.

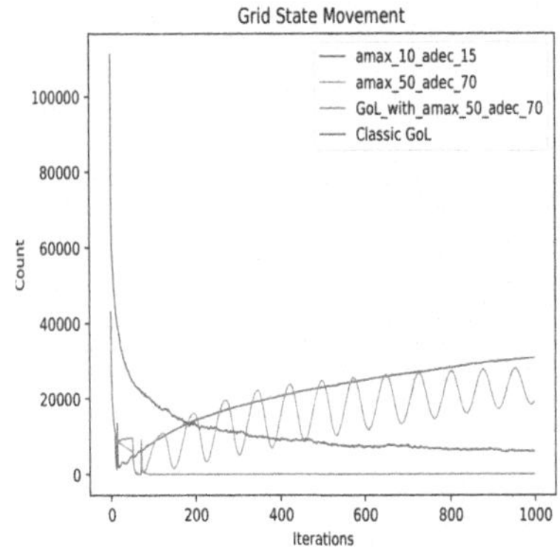

(e) Comparison of *Grid State* fluctuations.

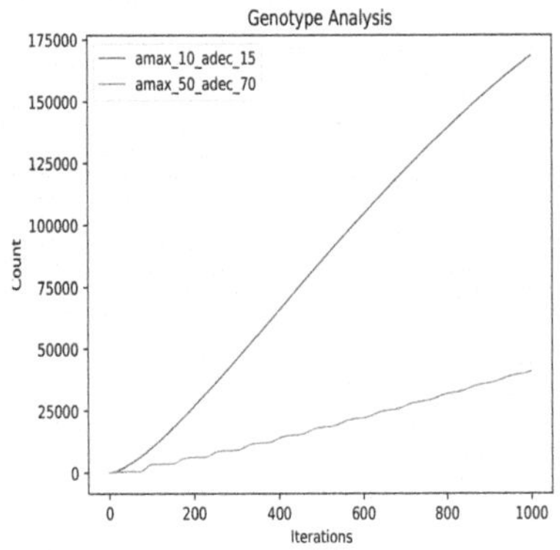

(f) Cumulative Genotype count

Fig. 5. Grid Size of 500 × 500 for 1,000 generations. Averages over 10 runs.

5 Results and Discussion

The results for the experiments on the 50 × 50 grid are depicted in Fig. 4 and for the larger grid size of 500 × 500 in Fig. 5. In 4a and 4c are the results for the shorter lifespan of $a_{max} = 10$ and $a_{dec} = 15$. Here, all counts have large variability for the first few generations. After this initial period, the *Cell State* counts seem to be almost constant with small deviations afterwards. In this equilibrium there are consistently more alive than dead and more dead than decaying cells. Considering the choice for a_{max} and a_{dec}, the ratio of alive to decaying cells seems natural in such a state. Whether the amount of dead cells can be simply explained by the inheritance and mutation probability needs to be investigated. For the *Grid States*, the count of dead cells is the same as in the other plot. The amount of cells with state 0 and 1 seem to slowly converge, which means that the alive cell with state 0 slowly become fewer and the ones with state 1 increase for roughly the first 4,000 iterations. There is quite some variation, both for the mean as well as the deviations between runs. The growing amount of cells with state 1 and its perseverance over 10,000 generations gives an indication of long-term phenotypic dynamics without signs of stagnation. For the shorter lifespan on the larger grid, 5a and 5c show similar behaviour as in the previously described results. After initial oscillation, the *Cell State* counts stabilizes. The amount of cells with state 1 and 0 converge over time.

As can be seen in 4b and 4d of 4, for the longer lifespan of $a_{max} = 50$ and $a_{dec} = 70$, the *Cell State* counts oscillate quite substantially in the beginning, but the oscillation's amplitude decrease exponentially in the first roughly 3,000 generations. Afterwards, the behaviour is again almost constant, with only little inter and intra run deviation. Alive cells are more numerous than decaying cells, which in turn occur more often than dead cells. For the *Grid State*, the plot for the dead cells is naturally the same. The count of alive cells with state 0 oscillates for the first 2,000 generations, but there is only little variability for the alive cells with state 1. The curves for the alive cells with different grid state seem to converge up until the 9,000 generations, interestingly however this tendency does not seem to continue. For the longer lifespan of $a_{max} = 50$ and $a_{dec} = 70$, results are in 5b and 5d. The oscillation continues over the full $1,000$ generations of the experiment for both the cell and *Grid State* counts. Overall, the plots are again similar to corresponding generations of 4b and 4d. After 100 generations, the amount of alive cells with state 1 and dead cells are still roughly of the same size. In those plots, it becomes quite clear that the oscillations have different amplitudes: the count of alive and decaying cells are oscillating in similar strength, but oppositional. The same is true for the count of alive cells with state 0 and the count of dead cells, but the overall amplitude is smaller. For the alive cells with state 1 there is only small oscillation.

In plot 4e, the fluctuation of *Cell States* are depicted. The two benchmarks of GoL-like implementations have an early spike of *Grid State* movement, but go down to a low, constant value. Our implementation with shorter lifespan has a noisy curve that, from about the 1,500 generation, seems to slowly fluctuate around a value of 400. The *Grid State* movement corresponding to the longer lifespan is unsurprisingly unstable in the first 2000 generations, before the oscillations fades. However, surprisingly the longer lifespan curve shows more phenotypic fluctuations than the shorter lifespan. Both show rather high dynamic behaviors of around 400 to 600 grid changes between consecutive generations. For the larger grid and shorter duration, the fluctuation of *Cell States* in 5e increases over time for both our configurations. After 200 generations, the value is almost always higher for the configuration with a small lifespan than for those of the other configurations. In case of the longer lifespan, the fluctuations oscillate for the full 1000 generations as in the experiment before.

As can be seen in plot 4f and 5f, the genotypic diversity increases over time for both of our implementations in both experimental setups, and when extrapolated one would expect more unique genomes to occur. The model with a shorter lifespan on the larger grid discovers the most unique genomes with almost 175,000 of the possible 262,144. On the smaller grid, 70,000 genomes are discovered, but in a time span ten times longer. Also, it seems that the curve here is slowly saturating. The implementation with the longer lifespan discovers just over 20,000 and 35,000 genomes in the two experiments respectively.

The tiles in 6, 7, 8, and 9 qualitatively illustrate the behaviour of the CA over the generations in a single run. The first row of the plots shows the almost constant amount of alive, decaying and dead cells in later generations, while the actual positions of those states change over time. In the second row, the genome of the cells is color-coded. Of course, the combinational possibilities make it impossible to have a humanly recognizable unique color for every genome, but it can be seen how there are small populations of similar genomes spread over the grid with no single population dominating it and with dead cells that allow for (re)growth in between. Especially for the longer lifespan, the convergence of cell with state 0 and 1 and the increasing amount of cell with state 1 can be observed in the lower row.

6 Conclusion and Future Work

We have demonstrated an heterogeneous life-like CA, where long-term phenotypic dynamics are possible thanks to an evolutionary inner loop. Such long-term behaviors are an important ingredient for open-endedness, which is typically lacking in other homogeneous life-like CA such as the Game of Life. Additionally, in this work we incorporate the concept of cell ageing to ensure that cell aliveness and actual cell computation are kept conceptually separated. We plan to conduct further studies where cell ageing can be perturbed by an environment or a task to be solved, and the substrate should show further adaptation.

Acknowledgments. This work was partially financed by the Research Council of Norway's DeepCA project, grant agreement 286558.

A Appendix Figures

Fig. 6. Example execution of 50 × 50 grid, $a_{max} = 10$, $a_{dec} = 15$. First Row: *Cell States*, Second Row: Genomes, and Third Row: *Grid States* - Columns: snapshots at generation 1, 2,500, 5,000, 7,500, 10,000.

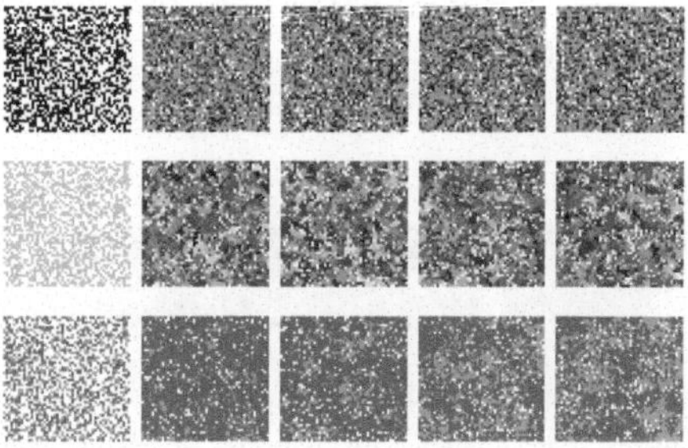

Fig. 7. Example execution of 50 × 50 grid, $a_{max} = 50$, $a_{dec} = 70$. First Row: *Cell States*, Second Row: Genomes, and Third Row: *Grid States* - Columns: snapshots at generation 1, 2,500, 5,000, 7,500, 10,000.

Fig. 8. Example execution of 500 × 500 grid, $a_{max} = 10$, $a_{dec} = 15$. First Row: *Cell States*, Second Row: Genomes, and Third Row: *Grid States* - Columns: Snapshots at generation 1, 250, 500, 750, 1,000.

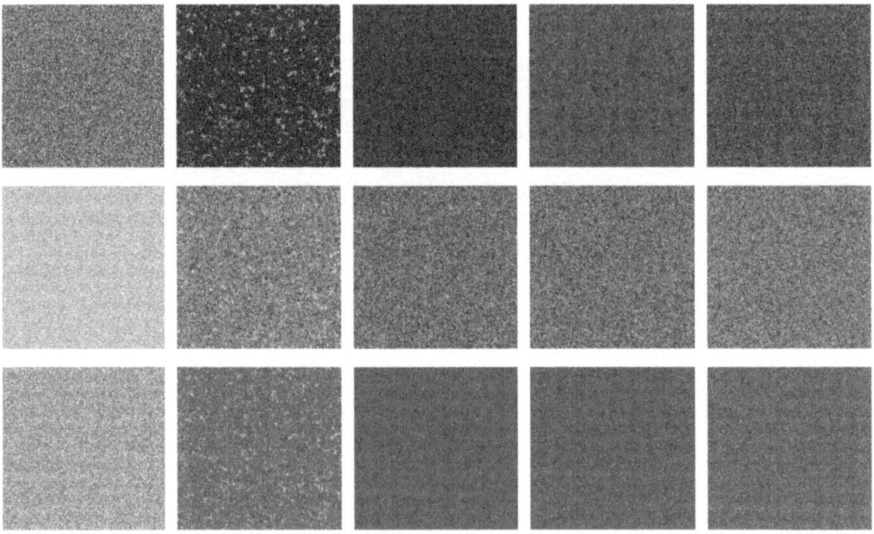

Fig. 9. Example execution of 500 × 500 grid, $a_{max} = 50$, $a_{dec} = 70$. First Row: *Cell States*, Second Row: Genomes, and Third Row: *Grid States* - Columns: Snapshots at generation 1, 250, 500, 750, 1,000.

References

1. Bak, P., Chen, K., Creutz, M.: Self-organized criticality in the' game of life. Nature **342**(6251), 780–782 (1989)
2. Bays, C.: Candidates for the game of life in three dimensions. Complex Syst. **1**(3), 373–400 (1987)
3. Berlekamp, E.R., Conway, J.H., Guy, R.K.: Winning Ways for Your Mathematical Plays, vol. 4. AK Peters/CRC Press, Boca Raton (2004)
4. Biehl, M., Virgo, N.: The game of life in a glider's frame of reference. In: Workshop The Distributed Ghost, ALIFE 2023 (2023)
5. Eppstein, D.: Searching for spaceships. arXiv preprint cs/0004003 (2000)
6. Eppstein, D.: Growth and decay in life-like cellular automata. In: Game of Life Cellular Automata, pp. 71–97. Springer, Heidelberg (2010). https://doi.org/10.1007/978-1-84996-217-9_6
7. Gardner, M.: Mathematical games. Sci. Am. **222**(6), 132–140 (1970)
8. Gregor, K., Besse, F.: Self-organizing intelligent matter: a blueprint for an ai generating algorithm. arXiv preprint arXiv:2101.07627 (2021)
9. McCaskill, J.S., Packard, N.H.: Analysing emergent dynamics of evolving computation in 2D cellular automata. In: Martín-Vide, C., Pond, G., Vega-Rodríguez, M.A. (eds.) TPNC 2019. LNCS, vol. 11934, pp. 3–40. Springer, Cham (2019). https://doi.org/10.1007/978-3-030-34500-6_1
10. Medernach, D., Carrignon, S., Doursat, R., Kowaliw, T., Fitzgerald, J., Ryan, C.: Evolution of heterogeneous cellular automata in fluctuating environments. In: Artificial Life Conference Proceedings, pp. 216–223. MIT Press, Cambridge (2016)
11. Medernach, D., Kowaliw, T., Ryan, C., Doursat, R.: Long-term evolutionary dynamics in heterogeneous cellular automata. In: Proceedings of the 15th Annual Conference on Genetic and Evolutionary Computation, pp. 231–238 (2013)
12. Packard, N., et al.: An overview of open-ended evolution: editorial introduction to the open-ended evolution ii special issue. Artif. Life **25**(2), 93–103 (2019)
13. Packard, N.H., McCaskill, J.S.: Open-endedness in genelife. Artif. Life, 1–34 (2024)
14. Peña, E., Sayama, H.: Life worth mentioning: complexity in life-like cellular automata. Artif. Life **27**(2), 105–112 (2021)
15. Randazzo, E., Mordvintsev, A.: Biomaker ca: a biome maker project using cellular automata. arXiv preprint arXiv:2307.09320 (2023)
16. Rendell, P.: Turing universality of the game of life. In: Collision-Based Computing, pp. 513–539. Springer, Heidelberg (2002)
17. Shapiro, A.: Sproutlife (2022). https://github.com/ShprAlex/SproutLife. Accessed 2 June 2024
18. Sipper, M.: Co-evolving non-uniform cellular automata to perform computations. Physica D **92**(3–4), 193–208 (1996)
19. Sipper, M.: Computing with cellular automata: three cases for nonuniformity. Phys. Rev. E **57**(3), 3589 (1998)
20. Sipper, M., Tomassini, M.: Computation in artificially evolved, non-uniform cellular automata. Theor. Comput. Sci. **217**(1), 81–98 (1999)
21. Suzuki, K., Ikegami, T.: Homeodynamics in the game of life. In: ALIFE, pp. 600–607 (2008)
22. Vichniac, G.Y., Tamayo, P., Hartman, H.: Annealed and quenched inhomogeneous cellular automata (INCA). J. Stat. Phys. **45**, 875–883 (1986)
23. Wuensche, A.: Exploring Discrete Dynamics. Luniver Press (2011)
24. Zamaraev, A.: Evolife (2016). https://github.com/a5kin/evolife. Accessed 2 June 2024

Autonomous Traffic Management: Integrating Vissim Traffic Model with a Swarm of Drones

Davoud Alahvirdi[1(✉)], Julien Pietquin[2], Alexandre Mauroy[2], and Elio Tuci[1]

[1] Robotics Lab, Computer Faculty of Namur University, 5000 Namur, Belgium
{davoud.alahvirdi,elio.tuci}@unamur.be
[2] NaXys Institute, Department of Mathematics, Namur University, 5000 Namur, Belgium
{julien.pietquin,alexandre.mauroy}@unamur.be
https://www.naxys.be, https://www.naxys.be/robotics

Abstract. In this research, we propose a system for monitoring and efficiently managing traffic congestion at the intersection of Park Avenue and E 72nd Street, New York, USA. This methodology is applied to a realistic traffic scenario where all intersections are controlled by fixed traffic signals. The method is based on a data-driven approach deploying a swarm of drones to measure the number of vehicles on roads. The collected information by the drones is sent to traffic lights and simultaneous perturbation stochastic approximation (SPSA) method is used to minimize traffic congestion by adapting the green traffic light durations. More precisely, we simulate the scenario thanks to the Vissim traffic software and Python. The simulation results highlight the effects of traffic light optimization to reduce traffic jams in contrast to the baseline case, where the duration of green lights is fixed.

Keywords: Traffic Monitoring and Management · Swarm of Drones · Vissim Software · Simultaneous perturbation Stochastic Approximation

1 Introduction

The rapid increase in the number of vehicles has created some challenges such as traffic congestion, higher fuel consumption, and gas emissions in large urban areas [6]. Traditional traffic management systems based on fixed traffic light controllers and stationary sensors have proven inadequate efficiency in addressing the challenges [8]. Unmanned aerial vehicles (UAVs) known as drones play a crucial role in surveillance and data collection by covering larger areas more effectively because of their features such as their mobility, adaptability, or coverage efficiency, compared to fixed camera. In [7], the authors explore data collected by small drones to design a highway management system. Advanced computer vision algorithms are used to monitor the traffic conditions from videos and

T. Carletti et al. (Eds.): WIVACE 2024, CCIS 2532, pp. 16–27, 2025.
https://doi.org/10.1007/978-3-031-93631-9_2

images acquired by a UAV. This analysis is used to evaluate the actions to be taken to reduce traffic flow, assess the risk, and plan assistance during road accidents. In [5], a cloud-based approach is designed in a ground station unit for managing drones in a smart city for traffic monitoring. The proposed algorithm helps to cover a larger area while conserving battery energy. Continuously monitoring of a simulated smart city as Unity game by using an aerial swarm is presented in [3]. In another work [9], a swarm of drones is deployed to gather data to create a general map containing data on traffic, pedestrians, climate, and pollution.

In a smart city, optimal traffic management is critical in optimizing traffic flow, improving overall transportation efficiency, and reducing travel times and congestion. In [16], the authors design an intelligent traffic control system based on a fuzzy logic controller to reduce average delay time. In [15], an adaptive real-time traffic light control algorithm employs the YOLO image processing technique to optimize traffic signal phases, using data captured by a fixed camera. In contrast to the traffic management design based on fixed sensors data with machine learning [4] or reinforcement learning [13], in this paper, we aim to design and analyse an autonomous optimized traffic light control system based on SPSA algorithm by using a swarm of drones.

Due to the challenges in acquiring real traffic data for the selected scenario in simulation, it is necessary to use micro traffic simulators such as Vissim or Sumo software. In this regard, in [2,12], the authors deploy these softwares to simulate the traffic in urban areas to demonstrate the effectiveness of their described traffic control method. A detailed example can be found in [10] where traffic lights are added to a non-signalized intersection in Vissim to show the significant reduction of traffic conflicts in the intersection. In [18], a microsimulation model is developed in the Vissim software to analyze the effect of different pedestrian densities on the traffic flow in an intersection. In [11], the authors simulate an intelligent traffic management system (ITMS) based on a deep neural fuzzy model in Sumo. The simulation software allow to explore of traffic-related issues such as route choice, simulating traffic lights, or vehicular communication in the ITMS.

In this work, we transfer the number of vehicles on the roads network from the Vissim software to a cellular model in Python and use a data-driven approach with the swarm of drones to collect the data. To solve the communication problem between drones and traffic lights such as limited bandwidth and data security, some specific meeting points [17] are considered to share the traffic information. The traffic lights receive this data from the UAVs to adapt the green traffic lights duration to minimize congestion by using the SPSA-optimization method. Finally, the optimal green time durations are applied to the Vissim software. The integration of the Vissim traffic model with a swarm of drones provides an efficient autonomous traffic management system for smart cities that not only alleviates congestion but also contributes to safer and more sustainable urban transportation ecosystems.

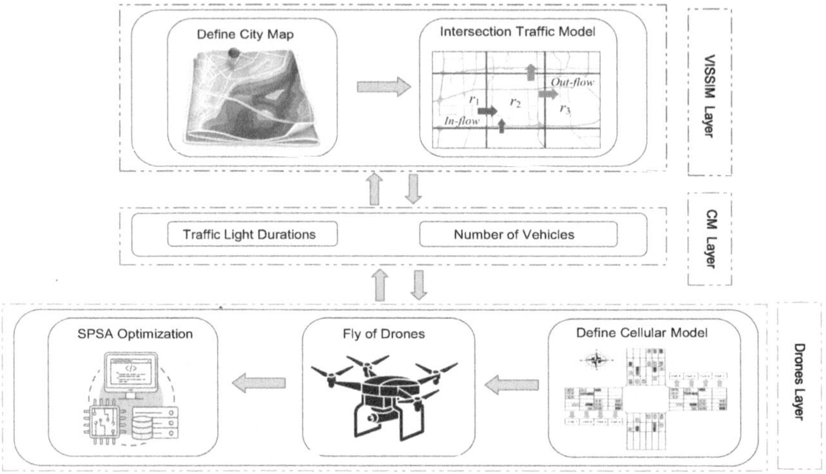

Fig. 1. Total layers of the simulation with Vissim layer, communication layer, and drones layer.

This paper is organized as follows. The main methodology of the simulation is discussed in Sect. 2. Subsections 2.1 and 2.2 introduce the urban road network and possible ways for movements of vehicles in the intersections, respectively. In Subsect. 2.3, the traffic model created by the Vissim software and the cellular map in Python is investigated. Then, Sect. 3 defines the route path planning for drones navigation. The SPSA method for traffic signals optimization is presented in Sect. 4 and Simulation results are provided in Sect. 5. Finally, a brief conclusion is given in Sect. 6.

2 Methodology

This section describes the whole scenario for simulating an optimal signalized intersection network where the Vissim software is used to simulate the real traffic model and where a swarm of drones is deployed to monitor and manage the traffic. As depicted in Fig. 1, the simulation model consists of three layers: i) the Vissim layer, ii) the communication (CM) layer, and iii) the drones layer. The Vissim layer consists of a geographically accurate representation of a real signalized intersection in New York, USA, that is, the junction of Park Avenue and E 72nd Street as shown in Fig. 2. The traffic flow is executed by the Vissim application, which enhances accuracy by considering some realistic features such as vehicle acceleration and deceleration, lane changing, and the presence of various types of vehicles. The CM layer is considered as a communication layer to transfer the generated number of vehicles on each road from the Vissim layer to the drones layer, and simultaneously to transfer the optimized duration of green traffic lights using the SPSA optimization model from the drones layer to the Vissim layer. In the third layer, the cellular map is created by using the

Fig. 2. Satellite map by Google.

dynamic propagation data of traffic in the Vissim layer. We employ four drones to continuously monitor the cellular map at different speeds. Next, the swarm of drones fly individually above the road cells to collect the number of vehicles based on the predefined trajectory. Designated meeting points have been established to address the communication issues between the drones and the traffic lights. Whenever a drone reaches a meeting point, it transfers collected data on traffic to the traffic lights. The traffic unit runs the SPSA optimization algorithm based on the information collected by the drones to minimize traffic congestion by changing the green duration of traffic lights.

2.1 Urban Road Network

The urban road network considered in this paper is the real-world intersection of Park Avenue and E 72nd Street in New York, USA as shown in Fig. 2. The intersection has four legs: West, East, North, and South directions. Each leg includes two entry points and two exit points. The entry point is designated for vehicles moving forward and turning right, while the other entry point is for vehicles turning left. The first exit point is considered for receiving vehicles traveling forward and for those leaving the intersection with a right turn, and the other exit point is for vehicles leaving the intersection with a left turn. In the intersection network, we refer to $R_{\{j,k\}}$; $j = \{IN_{fr}, IN_l, OUT_{fr}, OUT_l\}$ as the lanes of each leg, using IN_{fr} to indicate the lane for vehicles entering the intersection to move forward and turn right, IN_l is assigned to the lane for vehicles entering the intersection to turn left, OUT_{fr} denotes the lane for vehicles leaving the intersection with forward movement and right turn, and OUT_l represents the lane used for the vehicles leaving the intersection with left turn; $k = \{W, E, N, S\}$ refers to the four legs direction of the intersection

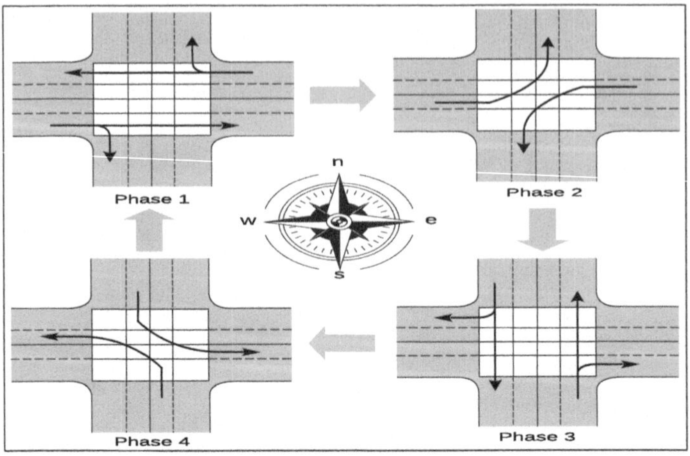

Fig. 3. Traffic light phases.

using the cardinal points. The length of Park Avenue in the north and south is approximately 60 m, while E 72nd Street extends around 120 m in the west and east directions.

2.2 Traffic Lights Phases

In the intersection network, the traffic lights operate through four phases, as shown in Fig. 3. All the traffic lights in the designed phases are referred to as a single traffic unit. Each phase consists of two sets of three signal lights (i.e., red, yellow, and green), with one set controlling the fr-lane and the other controlling the l-lane. We denote signal controls as $S_{\{i,k,c\}}$, where $i = \{fr, l\}$ represents the lane, $k = \{W, E, N, S\}$ corresponds to the cardinal directions at the intersection, and $c = \{red, yellow, green\}$ indicates the traffic light colors. A detailed description of the phases is provided as follows.

- Phase 1: In this phase, vehicles movie forward from the W to the E leg and vice versa. Since the traffic light set is the same for both forward-moving and right-turning, vehicles move from the W to the S leg and from the E to the N leg. Hence, the signal controls for legs W and E are $S_{\{fr,W,green\}}$ and $S_{\{fr,E,green\}}$ in fr-lanes, $S_{\{l,W,red\}}$ and $S_{\{l,E,red\}}$ in l-lanes. All the other phases for the N and the S legs are in red light situation: $S_{\{fr,N,red\}}$ and $S_{\{l,N,red\}}$ for the N leg, $S_{\{fr,S,red\}}$ and $S_{\{l,S,red\}}$ for the S leg.
- Phase 2: In this phase, vehicles turn left from the W to the N leg and from the E to the S leg. Hence, the signal controls for legs W and E are $S_{\{l,W,green\}}$ and $S_{\{l,E,green\}}$ in l-lanes, $S_{\{fr,W,red\}}$ and $S_{\{fr,E,red\}}$ in fr lanes. All the other phases in the N and the S legs are in red light situation: $S_{\{fr,N,red\}}$ and $S_{\{l,N,red\}}$ for the N leg, $S_{\{fr,S,red\}}$ and $S_{\{l,S,red\}}$ for the S leg.

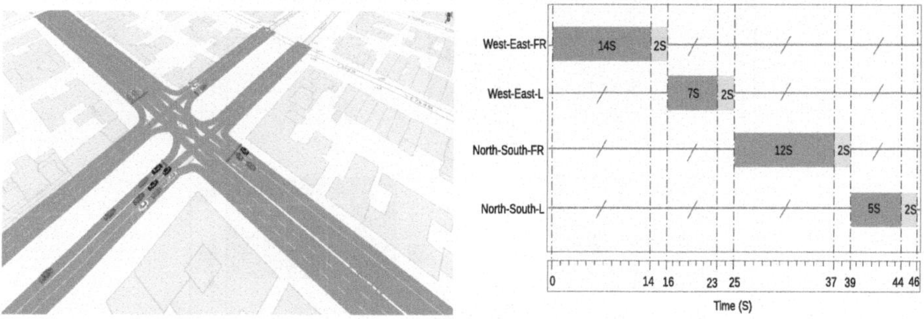

(a) Simulated intersection (b) Fixed control timing of the traffic phases

Fig. 4. Implementation in the Vissim software.

- Phase 3: In this phase, vehicles move forward from the N to the S leg and vice versa. Since the traffic light set is the same for both forward-moving and right-turning, vehicles move from the N to the W leg and from the S to the E leg. Hence, the signal controls for legs N and S are $S_{\{fr,N,green\}}$ and $S_{\{fr,S,green\}}$ in fr-lanes, $S_{\{l,N,red\}}$ and $S_{\{l,S,red\}}$ in l-lanes. All the other phases for the W and the E legs are in red light situation: $S_{\{fr,W,red\}}$ and $S_{\{l,W,red\}}$ for the W leg, $S_{\{fr,E,red\}}$ and $S_{\{l,E,red\}}$ for the E leg.
- Phase 4: In this phase, vehicles turn left from the N to the E leg and from the S to the W leg. Hence, the signal controls for legs N and S are $S_{\{l,N,green\}}$ and $S_{\{l,S,green\}}$ in l-legs, $S_{\{fr,N,red\}}$ and $S_{\{fr,S,red\}}$ in fr-legs. All the other phases in the W and the E legs are in red light situation: $S_{\{fr,W,red\}}$ and $S_{\{l,W,red\}}$ for the W leg, $S_{\{fr,E,red\}}$ and $S_{\{l,E,red\}}$ for the E leg.

2.3 Traffic Model

In this paper, we will use the Vissim software [1] to accurately simulate the traffic dynamics (see Fig. 4(a)). The Vissim software provides a microscopic, discrete, and stochastic representation of vehicles as separate entities, modeling each vehicle by specific characteristics such as speed, acceleration, deceleration, and driver behavior. Moreover, it considers vehicle interactions, including car-following, lane-changing, and gap acceptance in simulation. Fixed input sources are defined for each lane to run vehicle flow in the simulated network. It should be noted that we use a cellular map in Python to transfer vehicle counts from the Vissim software for drone monitoring purposes. In the cellular map, Park Avenue is divided into 2 subcells, each 30 m long, while E 72nd Street is divided into 4 subcells of the same length.

In this paper, two separate scenarios such as fixed and optimized traffic control will be simulated by the Vissim software. Figure 4(b) illustrates the fix timing schedule for the defined traffic phases in the Vissim software. The green light durations for the forward/right and left lanes in the west/east directions are

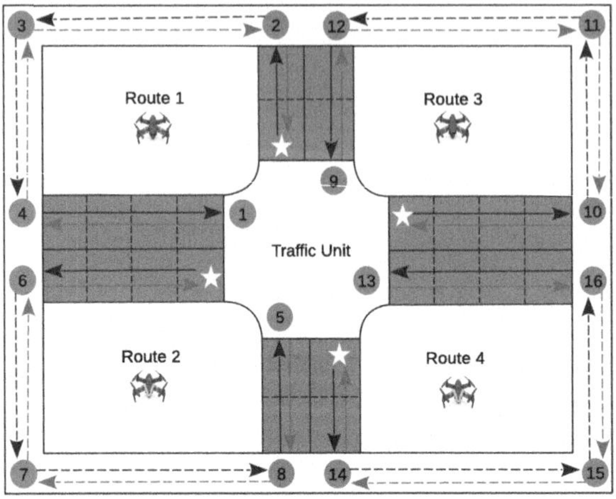

Fig. 5. Path planning of the drones (black arrows), start points of the drones (white stars), meeting points of the drones (grey circles), and traffic flows (blue arrows). (Color figure online)

set to 14 s and 7 s, respectively. Similarly, the green light durations for the forward/right and left lanes in the north/south directions are set to 12 s and 5 s, respectively. The sum of the green light duration for all traffic phases in the intersection $S_{\{i,k,green\}}$ is defined as the cycle time C:

$$C = \sum_{k=\{W,N\}} S_{\{fr,k,green\}} + S_{\{l,k,green\}} \tag{1}$$

3 Path Planning for Drones

We assign four drones to monitor the intersection network by traveling along designated flight paths. Four distinctive rectangle routes are defined for each drone as shown in Fig. 5. The black and blue arrows depict the directions of drone movement within each route and the flow of traffic by the Vissim, respectively. Continuous black arrows indicate the portions of each route that correspond to roads in the monitored network, while parts outside the monitored network are shown with dashed arrows. Moreover, the white stars indicate the flight starting point of each route, the grey circles mark a road change within the route, and the circles numbered 1, 5, 9, and 13 are the meeting points placed at the intersections. Whenever each drone arrives at a meeting point, it informs the traffic unit with the number of vehicles in each assigned route during its most recent loop on the route. Drones travel along the routes cell by cell at varying speeds, collecting data from four cells on the west-east roads and two cells on the north-south roads as defined in Subsect. 2.3.

4 Traffic Control Optimization

To reduce traffic congestion on the intersection legs, the durations of green traffic lights are optimized using the simultaneous perturbation stochastic approximation (SPSA) algorithm, which uses information (i.e., the number of cars in the monitored cells) gathered from drones. The SPSA algorithm is a powerful heuristic method that gradually approximates the optimal solution by iteratively estimating the gradient information of the objective function. The algorithm employs two estimated values of the objective function, which is independent of the dimension of the optimization problem. The feature of perturbing each variable in both positive and negative directions allows the algorithm to estimate the gradient of the objective function more robustly. After perturbing the parameters, SPSA monitors the resulting variations in the objective function. By averaging these perturbations, SPSA calculates an estimated gradient and adjusts its steps to iteratively improve the solution. The SPSA equation is defined as

$$\hat{x}_{k+1} = \hat{x}_k - a_k \cdot \hat{g}(x_k) \tag{2}$$

where $\hat{x}_k$ is the estimated value at the k-th iteration, a_k is a scalar gain coefficient, and $\hat{g}(x_k)$ is the estimated value of the gradient $g(x_k)$. The algorithm is summarized by the following steps:

1. Parameters a_k and C_k are defined as:

$$a_k = a/(A + k + 1)^\alpha \tag{3}$$

$$C_k = m/(k + 1)^\gamma \tag{4}$$

   where a, α, A, m and γ are non-negative coefficients as the inputs of the algorithm.
2. Based on Eqs. (3) and (4), two estimated values of the objective function are calculated as $L(\hat{x}_k + C_k \cdot \delta_k)$ and $L(\hat{x}_k - C_k \cdot \delta_k)$. δ is non-negative coefficients.
3. The approximation of the gradient is given by

$$\hat{g}(x_k) = (L(\hat{x}_k + C_k \cdot \delta_k) - L(\hat{x}_k - C_k \cdot \delta_k))/(2C_k)\delta_k \tag{5}$$

4. Equation (2) is updated. The algorithm continues to run until it solves the traffic congestion or reaches the maximum number of iterations to achieve the optimal result. Otherwise, $k = k+1$ and returns to step 1 until the conditions are fulfilled.

More details on the SPSA algorithm can be found in [14].

4.1 Objective Function

The objective function is formulated to minimize the total number of vehicles in the intersection legs as follows:

$$\min D(t) = \min \sum_{k=1}^{Y} \Delta t(n_k(t)) \tag{6}$$

where Δt is the step time and $n_r(t)$ is the number of vehicles in cell r at time t. Note that $Y = 4$ and $Y = 2$ are the numbers of the cellular sub-cells for the west/east and the north/south legs, respectively. S_{Drone} denotes the drone's speed, allowing it to gather information from each subcell in 2S. The objective function aims to reduce the overall street congestion by optimizing the timing of green traffic lights in the four directions.

5 Simulation Results

In this section, we evaluate the performance of the system by analyzing the intersection congestion under (i) fixed control timing, as illustrated in Fig. 4(b) and (ii) the optimization timing model described in Sect. 4. In this simulation, the length of each cell in the cellular map is set to 30 m, and the maximum capacity of each cell is 10 vehicles for the $R_{\{IN_{fr},k\}}$ lane and 5 vehicles for the $R_{\{IN_l,k\}}$ lane, with $k = \{W, E, N, S\}$ referring to the cardinal points. The optimization parameters are given in Table 1.

Table 1. Values of the optimization parameters for the simulation used during the evaluation tests.

parameters	values
step time Δt	2 s
simulation run T	552 s
cycle time C	46 s
drone speed S_{Drone}	16 m/s

In both scenarios, the network input flows are set at a fixed rate of 1800 vehicles per hour entering $R_{\{IN_{fr},W\}}$ and $R_{\{IN_{fr},E\}}$ lanes, and 900 vehicles per hour entering $R_{\{IN_{fr},N\}}$ and $R_{\{IN_{fr},S\}}$ lanes. Notably, the inflow for the west/east roads is twice larger than in the north/south roads. It is assumed that, at the beginning of each simulation run, a total number of 16 and 8 vehicles are distributed randomly among $R_{\{IN_{fr},W/E\}}$ and $R_{\{IN_l,N/S\}}$, respectively. We have executed 10 seeded simulation runs, with different initial distributions of the vehicles in the roads. Figure 6 and Fig. 7 illustrate the mean value and standard deviation of the remaining total number of vehicles in $R_{\{IN_{fr},W/E\}}$ at the

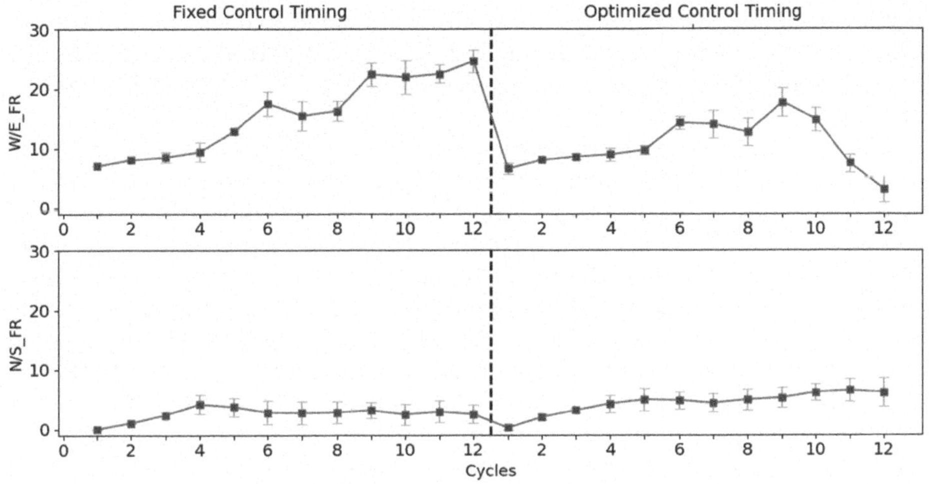

Fig. 6. Number of remaining vehicles at the end of the assigned green traffic light duration in fr lane for the west/east and the north/south directions, respectively. In the first 12 cycles, fixed control timing is used, and the SPSA optimization algorithm starts operating during the second 12 cycles. (Color figure online)

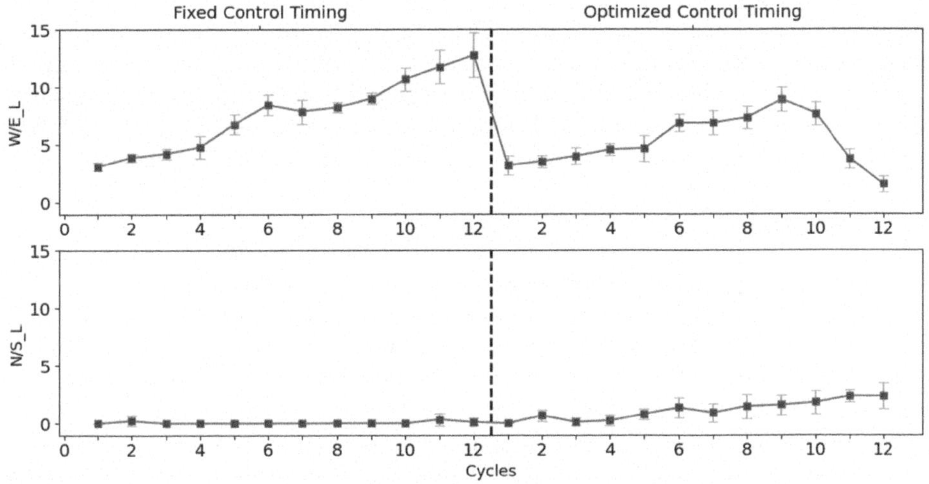

Fig. 7. Number of remaining vehicles at the end of the assigned green traffic light duration in l lane for the west/east and the north/south directions, respectively. In the first 12 cycles, fixed control timing is used, and the SPSA optimization algorithm starts operating during the second 12 cycles. (Color figure online)

beginning of traffic phase 2, $R_{\{IN_l,W/E\}}$ at the beginning of traffic phase 3, $R_{\{IN_{fr},N/S\}}$ at the beginning of traffic phase 4, and $R_{\{IN_l,N/S\}}$ at the beginning of traffic phase 1. In each simulation run, the traffic lights complete 24 full cycles. During the first 12 cycles, the traffic is not monitored and the duration of each green light phase is fixed. In the last 12 cycles, the traffic is monitored by the drones and the green light duration is set by the SPSA-based optimization control.

The results from our 10 evaluation tests for fr lanes are shown in Fig. 6, where the total remaining number of cars in the west/east directions increases up to 30 cars during the fixed traffic light control. This means that the fixed green duration in traffic phase 1 is insufficient to empty the lanes. In contrast, the mean number of cars in the north/south direction remains within the normal range, averaging about 4 cars per cycle. During the last 12 cycles, traffic congestion in both directions is influenced by the SPSA optimization method. The drones send data to the traffic unit with the higher number of cars in the west/east compared to opposite directions. To tackle the traffic jam in the west/east direction, SPSA optimization extends the green time for phase 1 and reduces it for phase 3, ensuring that congestion decreases to a total number of 4 cars at the 12th cycle in the west/east direction and slightly increases in the north/south direction.

Similar results for l lanes are shown in Fig. 7. Examining the traffic evolution during the first 12 cycles, with unchanged traffic light durations, reveals that the average total number of vehicles in the west/east directions reaches up to 15 cars by the end of the simulation, while traffic density in the other direction remains at 0 cars. Through the SPSA optimization method, the green time in phase 2 is extended by decreasing the duration of phase 4. Figure 7 show a similar trend of congestion reduction in l lanes for the west/east direction, along with a slight increase in the north/south direction. Figure 6 and Fig. 7 illustrate the efficiency of the autonomous optimized traffic monitoring and management in the network to reduce the traffic jams in all roads by considering fixed cycle time.

6 Conclusion

This paper demonstrated the effectiveness of an optimized traffic control system by deploying a swarm of drones to reduce traffic congestion at the intersection of a realistic signalized road intersection. The Vissim software was used to simulate a real traffic flow within the four legs of the network. The numbers of vehicles generated by the Vissim software were transferred to a cellular map in Python, where four drones were deployed to monitor traffic cell by cell and send the collected information to the traffic unit at designated meeting points. The traffic unit then used the real-time data captured by the drones with an SPSA-based optimized traffic light controller. The simulation results clearly showed that optimized traffic light durations using drone-collected data can substantially reduce traffic congestion in the west/east directions while maintaining acceptable conditions for the north/south directions in comparison with fixed signal control timing. This approach not only enhances traffic efficiency but also highlights the potential of drone technology in autonomous urban traffic management systems. In future work, we plan to employ the autonomous system in more complex urban networks by adding more connected intersections and more decentralized traffic units.

References

1. Al-Msari, H., Koting, S., Ahmed, A., El-Shafie, A.: Review of driving-behaviour simulation: vissim and artificial intelligence approach. Heliyon **10** (2024)
2. Fabianova, J., Michalik, P., Janekova, J., Fabian, M.: Design and evaluation of a new intersection model to minimize congestions using vissim software. Open Eng. **10**(1), 48–56 (2020)
3. Garcia-Aunon, P., Roldán, J.J., Barrientos, A.: Monitoring traffic in future cities with aerial swarms: developing and optimizing a behavior-based surveillance algorithm. Cogn. Syst. Res. **54**, 273–286 (2019)
4. Liang, X., Du, X., Wang, G., Han, Z.: Deep reinforcement learning network for traffic light cycle control. IEEE Trans. Veh. Technol. **68**(2), 1243–1253 (2019)
5. Nguyen, D.: Cloud-based drone management system in smart cities. In: Development and Future of Internet of Drones (IoD), pp. 211–230 (2021)
6. Anjum, M.S., et al.: An emerged challenge of air pollution and ever-increasing particulate matter in Pakistan. J. Hazardous Mater. **402**(123943) (2021)
7. Outay, F., Mengash, H., Adnan, M.: Applications of unmanned aerial vehicle (uav) in road safety, traffic and highway infrastructure management: Recent advances and challenges. Transport. Res. Part A: Policy Pract. **141**, 116–129 (2020)
8. Ravish, R., Swamy, S.R.: Intelligent traffic management: a review of challenges, solutions, and future perspectives. Transp. Telecommun. J. **22**(2), 163–182 (2021)
9. Roldán-Gómez, J.J., Garcia-Aunon, P., Mazariegos, P., Barrientos, A.: Swarmcity project: monitoring traffic, pedestrians, climate, and pollution with an aerial robotic swarm: Data collection and fusion in a smart city, and its representation using virtual reality. Pers. Ubiquit. Comput. **26**(4), 1151–1167 (2022)
10. Septyaningrum, I., Anindita, R.: Traffic signalizing application at unsignalized intersection applying vissim software microsimulation. Eng. Technol. **2**(2), 294–306 (2022)
11. Shamim Akhter, M., Quaderi, S.J.S., Al Forhad, M.A., Sumit, S.H., Rahman, M.R.: A sumo based simulation framework for intelligent traffic management system. J. Traffic Logist. Eng. **8**(1) (2020)
12. Vuong, X., Mou, Rand Vu, T., Van Nguyen, H.: An adaptive method for an isolated intersection under mixed traffic conditions in hanoi based on anfis using vissim-matlab. IEEE Access **9**, 166328–166338 (2021)
13. Yang, M., Wang, Y., Yu, Y., Zhou, M.: Mixlight: mixed-agent cooperative reinforcement learning for traffic light control. IEEE Trans. Ind. Inf. (2023)
14. Yao, Z., Jin, Y., Jiang, H., Hu, L., Jiang, Y.: Ctm-based traffic signal optimization of mixed traffic flow with connected automated vehicles and human-driven vehicles. Physica A: Stat. Mech. Appl. **603**(127708) (2022)
15. Zaatouri, K., Ezzedine, T.: Self-adaptive traffic light control system based on yolo. In: IEEE 91st Vehicular Technology Conference, pp. 16–19 (2018)
16. Zhao, Y.H., Li, R.X., Li, J.: Intelligent optimal control of urban traffic lights based on fuzzy control. Telecommun. Radio Eng. **79**(7) (2020)
17. Zhu, P., et al.: Detection and tracking meet drones challenge. IEEE Trans. Pattern Anal. Mach. Intell. **44**, 7380–7399 (2021)
18. Ziemska-Osuch, M., Osuch, D.: Modeling the assessment of intersections with traffic lights and the significance level of the number of pedestrians in microsimulation models based on the ptv vissim tool. Sustainability **14**, 8945 (2022)

Analysing Opinion Dynamics via a Cognitive Model of Structured Beliefs

Eleni Michaelidou[1,2,3](✉), Eirini Ioannou[1,2,3], Keren Tapper[1,2,3], Benjamin D. Goddard[1,2], and Guillermo Romero Moreno[4]

[1] School of Mathematics, University of Edinburgh, Edinburgh EH9 3FD, UK
e.michaelidou-1@sms.ed.ac.uk

[2] The Maxwell Institute for Mathematical Sciences, Edinburgh, UK

[3] Mathematics Department, Heriot–Watt University, Edinburgh EH14 4AS, UK

[4] School of Informatics, University of Edinburgh, Edinburgh EH8 9AB, UK

Abstract. Opinion dynamics models can help explain observed social phenomena that emerge from social interaction, such as echo chambers or polarisation, by using computational and numerical methods. However, to make them mathematically tractable, many such models are based on simplistic and unrealistic assumptions that do not align with the knowledge from the empirical cognitive science field. A significant exception is a recent multidimensional opinion model proposed by Mueller and Tan, which was able to reproduce opinion divergence and group polarisation by establishing a need for consistency between related beliefs that restricts the space of allowed combinations of opinions. Although this contribution is promising in closing the gap between the cognitive sciences and opinion dynamics models, the authors only provided a limited set of experiments to show the behaviour of the model and a more in-depth understanding is lacking. Here, we provide a concise mathematical analysis and a more thorough numerical analysis of the cognitive model of opinion dynamics proposed by Mueller and Tan and show that the outcomes of opinion divergence and group polarisation only emerge under some regimes of the parameter space: medium–term dynamics, a moderate number of topics discussed upon interaction, with opinion divergence emerging under sparse belief spaces and group polarisation under large, dense belief spaces. These scenarios should be contrasted with real–life cases to confirm whether the assumptions of the constrained belief space are enough to explain the observed phenomena or whether additional ingredients should be added for a more realistic representation of social behaviour.

Keywords: Group polarisation · Opinion dynamics · Consensus formation · Agent–based modelling · Collective behaviour

1 Introduction

The field of opinion dynamics initially emerged from the transfer of approaches used in statistical physics—which study systems of particles with simple inter-

T. Carletti et al. (Eds.): WIVACE 2024, CCIS 2532, pp. 28–41, 2025.
https://doi.org/10.1007/978-3-031-93631-9_3

action rules—to the study of social dynamics [2]. For instance, the paradigmatic Ising model, which describes the disposition of spins in a ferromagnetic material, has been shown to be able to describe social phenomena well in various contexts [6]. Although these approaches have been useful in providing a qualitative understanding of the social systems that they model [2], it is common that the interaction rules and representation of beliefs are founded on unrealistic assumptions that do not align with the empirical knowledge of human behaviour from the cognitive sciences [9,11,19]. For example, it is typically assumed that opinions lie on a uni–dimensional continuum [2,5], while cognitive science results suggest that humans do not consistently differentiate between levels of confidence regarding an issue [12]. To explain group opinion divergence, opinion dynamics models typically include *assimilative behaviour* (i.e., individuals' opinions coming closer upon interaction) and *bounded confidence*, by which individuals do not change opinions when interacting with remarkably dissenting others. However, empirical results have shown that large differences in opinion tend to lead to larger changes [20], while narrow differences tend to lead to no change [10], challenging the bounded–confidence assumption. Additionally, as models with assimilative forces always arrive at opinions more moderate than the initial ones—a behaviour which is at odds with many real–life cases that show group polarisation—models typically include either *negative reactions* (individuals' opinions moving away from each other if they are too different) or *stubborn extremists* who do not change their opinion [7,16]. However, there is no evidence that those with extreme views are more reluctant to opinion change than anyone else [1] or that negative reactions are common [20].

Recent attempts have tried to bring knowledge from empirical findings in the cognitive sciences to opinion dynamics models [11,17,18]. Among these cognitively grounded models, the one proposed by Mueller and Tan [11] stands out in that it is able to produce behaviour of opinion divergence and group polarisation without the need for mechanisms such as bounded confidence, stubborn agents, or negative reactions. Instead, their model includes a multidimensional opinion space with binary beliefs and with restrictions on the combinations of beliefs that are allowed—reflecting a need for consistency among related opinions—and this ingredient alone is enough to show behaviours of opinion divergence and group polarisation [11]. Even though this model provides a promising link between cognitive sciences and opinion dynamics models, the authors do not provide a full account of the model behaviour and only show the desired outcomes under a set of experiments with a very limited set of model parameters. Although these are acknowledged to be "somewhat arbitrary" and provided only as a "proof–by–example", it is not clear if these outcomes are robust to the decisions of the chosen model parameters.

Here, we provide a concise mathematical analysis and a more thorough numerical analysis of the cognitive model of structured beliefs proposed in [11] and arrive at a series of observations. First, we provide a mathematical argument about the long–term behaviour of the model, proving that if given enough time, the model dynamics would certainly arrive at the collapse of the opinions of the population to a single viewpoint. Second, we study the robustness of the model outcomes to changes in the number of topics discussed upon interaction and find a phase transition where the group moves from consensus to opinion divergence, while consensus tends to happen around the extremes of the opinion space, showing group polarisation. However, when fewer topics are discussed upon interaction, there is a higher tendency for opinion divergence and groups around more moderate stances. Last, we analyse the effect of variations in the size and density of the belief space and find that the consensus opinion of the group is moderate only when there are few topics or few valid combinations of them, while it tends to gather near the extremes otherwise. Therefore, the model has a variety of outcomes depending on the time scales and parameter choices, and a reference of real cases should be sought to calibrate them and evaluate the suitability of the model in these scenarios.

The remainder of the paper is organised as follows: Sect. 2 describes the belief-space model introduced in [11] and the experimental settings used in our simulations. Section 3 provides a deeper understanding of the behaviour of the model by analysing its long-term behaviour and exploring the parameter space through simulations. Section 4 concludes the paper with a discussion of the results in context and suggestions for future research.

2 Model and Experimental Settings

2.1 Description of the Model

The model we consider in this paper was introduced by Mueller and Tan [11]. In this multi-agent interactive model, opinion-based beliefs are represented as binary vectors in a multidimensional space, where each dimension or *belief feature* represents a topic, agents—without loss of generality—can hold either a positive (1) or negative (0) opinion or *belief* in each topic, and a combination of beliefs over all features is termed as a *belief state*. Due to, e.g., logical or traditional associations between beliefs, only a subset of combinations of beliefs is possible, which constitute *valid* belief states. The set of B valid belief states that agents are allowed to hold over F features is termed as a *knowledge* or *belief space*, $\Lambda_{F,B} = \{1, \ldots, F\} \times \{1, \ldots, B\}$.

To describe the complexity of the knowledge space, we use the concepts of *sparsity*, *density* and *connectivity*. Although for a given number of features, F, it is possible to create 2^F belief states, the set of valid combinations of beliefs may typically be much smaller than 2^F, i.e., $B/2^F \ll 1$, corresponding to a *sparse* knowledge space [11]. In this case, the probability of finding a valid belief state and updating agents' belief states in each round is low. In contrast, the knowledge space is *dense* when $B/2^F \approx 1$. Lastly, a belief state is classified as *well-connected* if it shares many opinions with many other states, allowing easier movement between them [11].

As in the original paper [11], when creating the knowledge space, we always include the two most extreme belief states, i.e., all 0 s and all 1 s. The remaining belief states are generated using the standard uniform distribution. That is, for each belief state, a single value p is generated from the standard uniform distribution, $p \sim U(0,1)$, and then, each feature is set to 0 with a probability of p and 1 with a probability of $1-p$. As in [11], the uniqueness of each belief state is ensured by generating a new state if the same belief state is generated twice. Lastly, once the belief states to be included in the knowledge space are generated, they are ordered based on the number of positive features, with belief state 1 being all 0s and belief state B being all 1s.

Simulations start with a group of N agents randomly assigned belief states from the knowledge space. The opinion formation process is then implemented with no restrictions on interactions. Hence, in each round, a pair of agents is randomly selected to interact, where changes in the belief state of both agents may occur. Considering agent i, each feature is selected for discussion with a probability of $1-\mu$, forming the set of features to be discussed. Here, μ is the probability of each feature not being selected for inclusion in the "discussion", i.e., $1-\mu$ is the probability of inclusion.[1] Then, a new candidate belief state is generated for agent i, $\boldsymbol{s}^*$, where the values of agent j on those features (features included in the "discussion") are copied. Agent i will only adopt the candidate belief state if it exists in the knowledge space, i.e., $\boldsymbol{s}^* \in \Lambda_{F,B}$. This process is then repeated for agent j, considering the initial state of agent i before any potential change. The probability of inclusion $1-\mu$ is fixed throughout each simulation.

2.2 Experimental Settings

Mueller and Tan [11] presented the desired outcomes—the behaviour of opinion divergence and group polarisation—under a set of experiments with a limited set of model parameters: $N = 100$ agents, $B = 15$ belief states, $F = 20$ features, and $\mu = 0.3$. To explore the effects of this parametrical decision, we perform experiments where we vary one or more parameters while fixing the remaining parameters to the original values.

We define a simulation to last 100,000 rounds or until consensus formation—when all agents adopt the same belief state—is produced. This choice was made considering that, in real life, we expect individuals to interact a finite number of times before encountering external factors, making 100,000 interactions a reasonable upper bound. To address the stochastic nature of the model, we run each experiment either 50 or 100 times. This choice was made to keep a balance; while we wish to conduct enough experiments to accommodate the impact of randomness in outcomes, we are also limited by computational time.

2.3 Evaluation Metrics

For us to conclude that a polarisation effect is produced as an output of the model, all groups of agents with size larger than 5% of the total number of

[1] In the opinion formation process followed in the original paper [11], μ was the probability of selecting each feature for inclusion in the "discussion".

agents, N, must hold an extreme belief state. Here, an extreme belief state is defined as any state in the knowledge space that falls within the first or last 20% of the belief states. The results presented in Sect. 3 are robust in the group size threshold (larger than 5%), with the same results achieved with other reasonably small values. They are also robust to different thresholds of extremity.

We are also interested in examining consensus formation—the result where the opinions within an interacting group collapse to a single viewpoint—as an outcome of the model. One way to measure consensus formation is to look at the size of the largest group of agents in the same belief state after 100,000 interactions. In addition, we have an indication of consensus formation by examining whether $G\%$ group agreement is produced. Here, $G\%$ group agreement is achieved when $G\%$ of the population holds the same belief.

3 Results

In this section, we first discuss the long-term behaviour of the model. Then, we conduct numerical experiments to investigate the outcomes of the system in a bigger set of model parameters, namely, (a) how the variety of topics discussed upon interaction affects the probability of polarisation and consensus formation and (b) how the complexity of the knowledge space affects the probability of polarisation and the time required to reach group agreement.

3.1 Temporal Dynamics of the Model

In [11], it is argued that an interacting group with initially distinct opinions may "converge" to distinct belief groups. In this section, however, this claim is contradicted since by modelling the dynamics as a Markov chain, it is proven that the population will converge to consensus with probability 1.

We consider a discrete–time, finite–state Markov chain with state space $S = \{A_B, A_{B-1}, ..., A_2, A_1\}$, where A_k represents the state for which all agents of the population have k distinct belief states. Suppose we have a model with knowledge space $\Lambda_{F,B}$. Observe that, in this case, there exist B different belief states in the knowledge space; hence, A_B is the state of the system where at least one agent holds each possible belief state. Consensus formation corresponds to the state A_1, where the population is in a single belief state.

Let α_k be the transition probability from A_k to itself, β_k be the transition probability from A_k to A_{k-1}, and γ_k be the transition probability from A_k to A_{k+1}. Note that there is always a positive probability of reducing or remaining at the same number of beliefs that exist among the population, i.e., $\alpha_k > 0$ and $\beta_k > 0$. In addition, if everyone in the population adopts the same belief, there is no way for an agent to change belief via interacting with other agents; thus, $\gamma_1 = 0$. Lastly, since γ_k is a probability, it follows that $\gamma_k \geq 0$ for each k.

It is clear that the model introduced in [11] simulates a Markov chain on the state space S, as the probability of the process moving from one state to another depends only on the current state. Moreover, we observe that S has

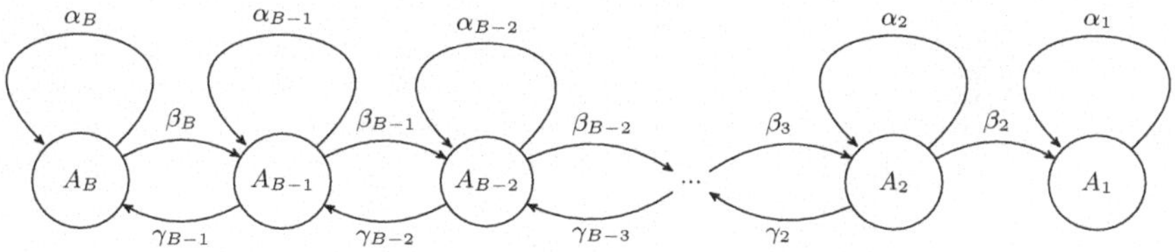

Fig. 1. Markov chain with one–step transition probabilities for opinion dynamics among an interacting group with initially distinct opinions. Each state A_k represents the number of distinct belief states among all agents of the population.

only one absorbing state (A_1)—once the system reaches this state, it stays in this state—and there is a positive probability of reaching this state in a finite number of steps (see Fig. 1). This means that all other states $A_B, A_{B-1}, ..., A_2$ are transient, and the system will reach state A_1 with probability 1.

In conclusion, if given enough time, the model will reach global consensus of beliefs with probability 1. Therefore, at most, *quasi–stable* groups of different opinions can be formed in the model. Quasi–stability in this context means that for a large number of iterations, the number of agents in each state remains approximately the same. It is important to note that the number of steps taken to reach A_1 could be very large. Hence, quasi–stability, as in [11], is worth investigating with numerical simulations.

3.2 The Variety of Topics Discussed

As aforementioned, $1 - \mu$ is the average proportion of topics discussed in each interaction between two agents. The topics to discuss are chosen with uniform probability. When $\mu = 0$, all topics will be discussed in every interaction. This causes the agents to just swap beliefs. Therefore, the beliefs held by the group do not change. Similarly, when $\mu = 1$, no topics are discussed, so interactions do not affect an agent's belief.

We look at the effect of choosing different values of μ on the size of the largest group and whether the group of agents has polarised. We consider 21 different values of $\mu \in [0, 1]$ and perform 100 simulation runs for each value.

Conditions for Group Agreement. The largest group size for each simulation run with different values of μ is shown in Fig. 2a. Larger values of μ are less likely to have large groups of agents holding the same belief, corresponding to fragmented opinions, while lower values of μ have a high chance of consensus, with all 100 agents holding the same belief state. The transition from consensus at $100,000$ interactions to fragmentation occurs at around $\mu = 0.25$. The simulations in [11] considered $\mu = 0.3$, which is slightly above the transition and hence has a high chance to obtain only two groups in the final state.

Increasing μ decreases the number of topics discussed and suggests that agents are less likely to make big changes in their belief state. This also means

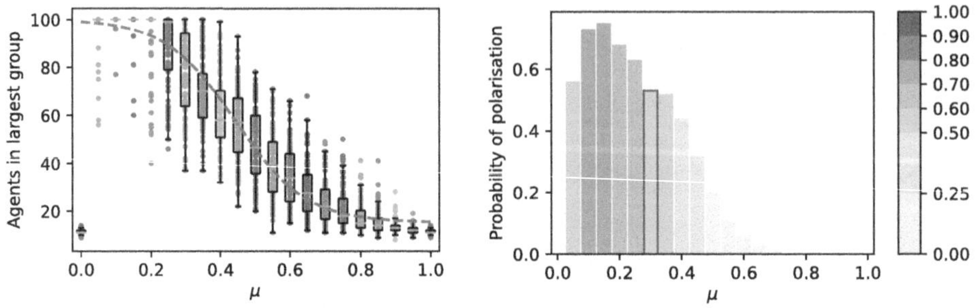

(a) Size of the largest group

(b) Probability of polarisation

Fig. 2. (a) Box plots representing the range of the size of the largest group for each μ value from 100 simulation runs. The dashed curve is $L(\mu)$ (as defined in Eq. 1) with parameters $A = {}^{280}/_{3}$, $\mu_{1/2} = {}^{1}/_{k} \ln({}^{277}/_{3} - \omega)$, $k = 9.985$, and $\omega = 8.554$. The box and whisker plots show the mean, upper, and lower quartiles. (b) Probability of polarisation over the different values of μ from 100 simulation runs. The value $\mu = 0.3$ (highlighted) was used in [11], and is used in the experiments of Sect. 3.3. The colour bar shows the probability of polarisation using the same scale as Fig. 3 for comparison. All other parameters are as in [11]: $N = 100$, $F = 20$, and $B = 15$.

that if a new belief is adopted by an agent, it will not be very different from their original belief. This can cause the population to arrive at a quasi-stable state with fragmented opinions.

The mean values of the size of the largest group for each $\mu \in (0, 1]$, $L(\mu)$, can be well fitted by a logistic curve, given in Eq. 1. The parameter A is dependent on the number of belief states. These are given by

$$L(\mu) = N - \frac{A}{1 + e^{-k(\mu - \mu_{1/2})}}, \tag{1}$$

$$A = N\,(1 - {}^{1}/_{B}) - \omega, \tag{2}$$

where N is the number of agents, B is the number of valid belief states, and ω is a fitting parameter. Initially, the agents are distributed uniformly across the belief states, so we expect all groups to be, on average, of size ${}^{N}/_{B}$. However, there will be some deviation in the distribution, causing the largest group to be above this number, which is incorporated by ω.

The value $\mu_{1/2}$ represents the μ value such that the mean largest group size is halfway between the maximum and minimum possible largest group sizes; $L(\mu_{1/2}) = \frac{1}{2}(N - {}^{N}/_{B})$. Allowing for one agent to not be in the final consensus group, taking $\lim_{\mu \to 0} L(\mu) = N - 1$, we see that $\mu_{1/2} = k^{-1} \ln(A - 1)$. The coefficient k is dependent on the number of interactions and the ratio between the number of features and beliefs, which affects the probability of taking a new belief during an interaction. As aforementioned, when $\mu = 0$, all topics are discussed and so the distribution of agents among the belief states does not change, so $L(0) := L(1) = {}^{N}/_{B} + \omega$, corresponding to the initial conditions.

With the parameters in Fig. 2a, $A = {}^{280}/_{3} - \omega$ and $\mu_{1/2} = {}^{1}/_{k} \ln({}^{277}/_{3} - \omega)$, the remaining parameters, $k = 9.985$ and $\omega = 8.554$, were fitted by minimising residuals. Therefore, average initial maximum group size is ${}^{N}/_{B} + \omega = 15.22$.

Conditions for Polarisation. Having a high chance of consensus also suggests a high chance of having polarisation, as shown in Fig. 2b, as it is likely that the largest group is in an extreme belief state. Having an intermediate low value of μ (above 0.15 and below 0.5) is more likely to cause the population to polarise within 100,000 interactions because it takes fewer interactions for an agent to change to an extreme belief state. However, when the value of μ is too low, i.e., $\mu < 0.15$, many topics are discussed, and the beliefs of the agents converge very quickly to a random belief state, highly linked to the beliefs of the first few agents to interact before the knowledge space has been well sampled. This reduces the chance of polarisation compared with the previously discussed range of μ. High μ values (above 0.5) have a much lower chance of polarisation because the largest group is still quite small, which means that the rest of the population will be well spread across the other belief states.

In addition, for fixed $\mu = 0.3$, reducing the number of agents N results in consensus being reached faster. This is because, with the number of interactions being fixed to 100,000 and a smaller population, a larger number of individual interactions will correspond to each agent. As with very low values of μ, when N is small, polarisation is less likely to occur because consensus can be reached before the whole knowledge space has been explored. Increasing the number of agents (or studying the thermodynamic limit of an infinite number) reduces the stochasticity in the outputs and results in a "cleaner" behaviour. However, humans are limited in the number of people they can interact with on a regular basis, with 100 already being around the upper limits [21]. Therefore, a realistic study of larger populations would also require constraints in agent interactions based on social or geographical structures [14].

3.3 The Complexity of the Belief Space and Its Restrictions

The complexity of the knowledge space is determined by the number of belief states and the number of features. For each scenario $s_{F,B}$ representing a different choice of the knowledge space $\Lambda_{F,B}$, we perform 50 simulation runs. In particular, the number of features considered for the different scenarios is $F \in \{4, 5, 6, ..., 20\}$. The number of belief states for $F \geq 7$ is $B \in \{5, 10, 15, \cdots, 60, 65\}$, while for $F \leq 6$, the number of belief states is $B \in \{5, 10, 15, ..., s_F\}$, where s_F is the largest multiple of 5, smaller or equal to $2^F - 1$. All other parameters remain consistent with the default settings in [11].

Conditions for Polarisation and for Group Agreement. For each scenario, the probability of polarisation and consensus is calculated by finding the

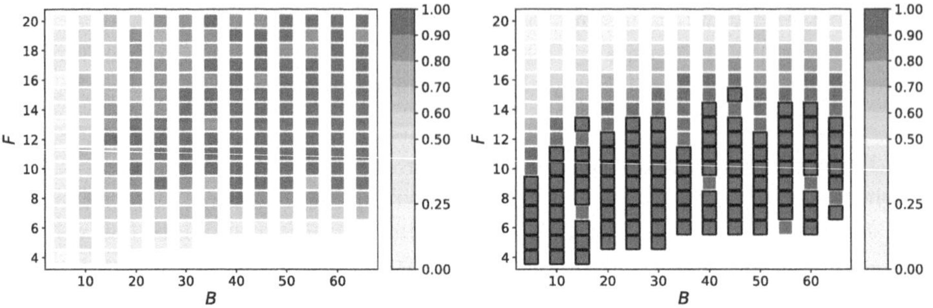

(a) Probability of polarisation (b) Probability of consensus

Fig. 3. Heat map of (a) the probability of polarisation and (b) the probability of consensus (as defined in Sect. 2.3) for different choices of the knowledge space $\Lambda_{F,B}$ over 50 simulation runs and for 100,000 interactions each. All other parameters are consistent with the default settings in [11]: $N = 100$ and $\mu = 0.3$. To the right of each heat map is a gradient bar indicating the probability of polarisation and consensus, respectively, ranging from light yellow (probability $< 25\%$) to bright red (probability $\geq 90\%$). The black outline in (b) indicates cases where consensus was reached in all simulation runs.

proportion (out of 50) of simulation experiments that produce polarisation and consensus formation (as defined in Sect. 2.3), respectively.

As shown in Fig. 3a, the probability of polarisation is low for either a small number of belief states or of features, i.e., $B \in \{5, 10\}$ or $F \in \{4, 5, 6, 7\}$. In particular, a small number of belief states indicates limited available movement within the knowledge space, while a small number of features implies that all valid belief states are approximately equally well–connected. Consequently, under these scenarios, reaching extremes is not favoured. For the remaining scenarios, if we consider intermediate values of F, i.e., $10 \leq F \leq 16$, and large values of B, i.e., $B \geq 40$, the knowledge space is dense. This implies a high frequency for agents updating their belief state and greater exploration of the knowledge space. Hence, with the only well–connected beliefs being at the extremes, a polarisation effect is produced. For even larger values of F, even though the well–connected belief states are at the extremes, the knowledge space is more sparse than before, which results in less exploration and, hence, a lower probability of polarisation. Note that the behaviour of the model for different values of μ (see Sect. 3.2) is generalised across the various scenarios $s_{F,B}$, which was verified experimentally.

We also observe that the probability of consensus formation is near 1 for small values of F, as shown in Fig. 3b, and decreases as we increase the value of F. Keeping B fixed and increasing F increases the sparsity of the knowledge space, which implies fewer interactions that result in valid belief states, agents not updating their belief states frequently, and, consequently, a lower probability of reaching consensus before completing 100,000 simulated rounds.

To further explore the impact of the complexity of the knowledge space on consensus formation, we study the dependence between the knowledge space,

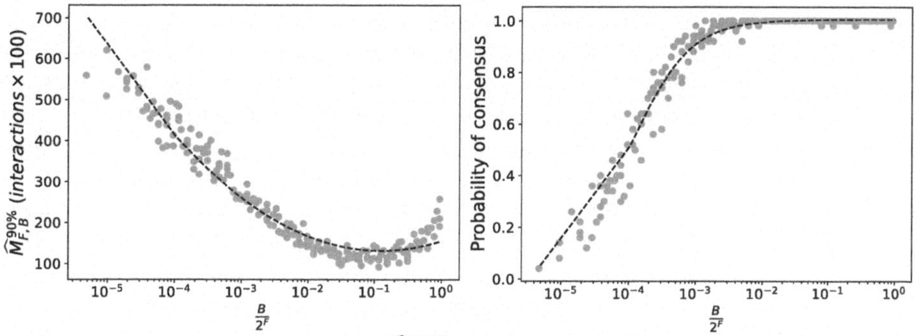

(a) Average number of interactions $\widehat{M}_{F,B}^{90\%}$ (b) Probability of consensus

Fig. 4. (a) Average number of interactions ($\widehat{M}_{F,B}^{90\%}$) required to reach 90% group agreement, considering only the simulation runs where at least 90% of the population converges to the same belief state within 100,000 interactions, out of 50 simulation runs for different scenarios $s_{F,B}$. (b) Probability of consensus formation before 100,000 interactions out of 50 simulation runs for different scenarios $s_{F,B}$. All other parameters are consistent with the default settings in [11].

$\Lambda_{F,B}$, and the number of interactions needed to reach 90% group agreement (as defined in Sect. 2.3). For each scenario, $s_{F,B}$, we calculate $\widehat{M}_{F,B}^{90\%}$, the average number of interactions needed to reach 90% group agreement. This average is computed over the simulation runs (50 in total) where at least 90% of the population holds the same belief state by the end of the run. Similar results are achieved when considering 85% and 95% group agreement.

The average number of interactions $\widehat{M}_{F,B}^{90\%}$ as a function of $B/2^F$ (density of the knowledge space) is shown in Fig. 4a. The fitted line in Fig. 4a (black dotted line) is given as

$$\widehat{M}_{F,B}^{90\%} = 5.54 \log \left(B/2^F\right)^2 + 23.01 \log \left(B/2^F\right) + 155.52. \tag{3}$$

This implies that the average number of interactions has a quadratic relationship with the logarithm of $B/2^F$. In Fig. 4b, we plot the probability of reaching consensus formation as a function of the density of the knowledge space.

As shown in Fig. 4a, in the most dense knowledge space scenario ($B/2^F \approx 1$), the average number of interactions needed for group agreement is approximately 25,000, with the minimum average number of interactions achieved when $B/2^F \approx 0.1$. As the knowledge space becomes more sparse ($B/2^F \to 0$), the average number of interactions required for group agreement increases. However, for scenarios where $B/2^F < 10^{-3}$, the results become less reliable since the proportion of simulation runs that reached consensus within 100,000 interactions decreases (see Fig. 4b). Therefore, the most reliable estimations are the ones with an average estimation of 30,000 interactions or less ($B/2^F > 10^{-3}$).

The explanation for the above observations is the following: A very sparse knowledge space, i.e., $B/2^F \ll 1$, results in a slow update of agents' belief states because many non–valid candidate belief states are formulated during the opinion formation process. This slow update implies that many more interactions will be needed for the system to reach a group agreement. As explained in Sect. 3.1, the system reaches consensus (state A_1) with a probability of 1. However, the number of interactions required to reach state A_1 could be very large. Therefore, not producing consensus formation as the output of the model does not imply that the system will not reach consensus after 100,000 interactions; it just implies that more interactions are required, although if the system arrives at quasi–stable states, the distribution of interactions needed for consensus may have very long tails. As the knowledge space becomes less sparse, i.e., $B/2^F$ approaches 0.1 (where the minimum point occurs), the transition between belief states becomes easier, so agents reach 90% group agreement faster. However, for a very dense knowledge space, the average number of interactions needed to reach group agreement increases again. This is because the knowledge space is too well–connected, and agents jump between belief states too quickly (constant update of agents' beliefs), making the time to reach group agreement slightly larger than when the knowledge space is less well–connected.

4 Discussion

Overall, simulations of the model of Mueller and Tan [11] illustrate an alternative approach to accounting for opinion dynamics processes. The model relies on the notion of a belief or knowledge space, which provides a rich and realistic way to simulate belief transmission. They argue that social behaviours such as consensus formation, opinion divergence, and group polarisation can be reproduced as consequences of a complex belief space without the need for any additional assumptions. An interesting insight that this model provides is that the largest obstacle to attitude change is conceptual coherence [13]. Group polarisation, the phenomenon at which people reach agreement at more extreme positions, in the model of Mueller and Tan [11], can be explained from the "connectivity" of the knowledge space. More specifically, the generation mechanism of the knowledge space, which provides it with structure by oversampling the extremes, makes it harder for intermediate beliefs to be well–connected, enabling the extreme beliefs to be the most "central" within each pole.

We have performed an extensive parameter space analysis, which was lacking in [11], and have illustrated and specified how their observations can be extended to more general cases. We have shown that the model is able to quickly converge to group agreement when many topics are included in the "discussions" or when belief spaces are dense (many legal combinations). This behaviour is akin to opinion dynamics models with a purely assimilative force [5], which reflect occurrences of the real–world such as fads, fashions and herd mentality [22]. However, there is an important difference in that in those models, the consensus opinion occurs at the average point of the initial opinion distribution, while in the cognitive model with structured beliefs, consensus tends to occur at the extremes, showing a stark group polarisation.

We have also seen that group divergence is present in the short term when not many topics are included in the "discussions" or the knowledge space becomes more sparse. This is seen as few polarised groups (in scenarios with a moderate number of topics discussed or with moderately sparse knowledge spaces), or as many fragmented groups over the whole opinion space (when only a few topics discussed upon interaction or with very sparse knowledge spaces). These transitions in the number of groups are similar to the effects of increasing the bounded confidence parameter in bounded confidence models [4] or the number of possible values that each cultural trait has in Axelrod's model of cultural dynamics [3], although without the outcome of group polarisation that is shown by the cognitive model of structured beliefs. Furthermore, those models arrive at "frozen" states of group divergence from which no subsequent change in opinion is possible. In contrast, through our mathematical analysis, it is easy to conclude that consensus formation will occur with certainty in our model of interest given enough time. Therefore, even though two or more stable groups may form at the beginning, creating an effect of group divergence, these actually are quasi–stable, as they will eventually merge given long enough time. This behaviour is also seen in bounded–confidence and social impact models when noise—i.e., random moves in individuals' opinions—is included, showing "staircase dynamics", where the number of opinion groups remains the same for long periods, interleaved by sharp descends until a single group is reached [8,15]. It would be of interest to explore the transient dynamics of model from [11], although some differences to the other models could be expected as the cognitive model encourages "wholesale adoptions" instead of "breadcrumb" approximations [11]. Other interesting model alterations are heterogeneity in the number of topics discussed or the presence of stubborn agents [7,16].

We have numerically and mathematically investigated the novel model proposed by Mueller and Tan [11]. Their innovation relies on the notion of a belief or knowledge space, which provides a richer representation through which to simulate belief transmission and is consistent with sociological and psychological theories. We have performed extensive parameter space analyses, which were lacking in [11], and have shown that one of their main characteristics is the arrival at group polarisation whenever discussions are not limited to a few topics under specific conditions in the knowledge space. However, we have also shown that the arrival at consensus or group divergence also depends on the variety of topics discussed at each interaction, the size and density of the knowledge space, and the time scales of the dynamics. We have provided guidelines on the expected behaviours of the model in different parameter regimes, which will be useful when adapting the model to specific empirical scenarios, completing the three main ingredients—theoretical foundations, numerical understanding, and empirical validation—required for a sound approximation to complex phenomena via modelling.

Acknowledgments. EM, EI, and KT were funded by the Engineering and Physical Sciences Research Council Grant EP/S023291/1 (MAC–MIGS Centre for Doctoral Training).

Disclosure of Interests. The authors have no competing interests to declare that are relevant to the content of this article.

References

1. Carney, D.R., Jost, J.T., Gosling, S.D., Potter, J.: The secret lives of liberals and conservatives: personality profiles, interaction styles, and the things they leave behind. Polit. Psychol. **29**(6), 807–840 (2008). https://doi.org/10.1111/j.1467-9221.2008.00668.x
2. Castellano, C., Fortunato, S., Loreto, V.: Statistical physics of social dynamics. Rev. Mod. Phys. **81**(2), 591–646 (2009). https://doi.org/10.1103/revmodphys.81.591
3. Castellano, C., Marsili, M., Vespignani, A.: Nonequilibrium phase transition in a model for social influence. Phys. Rev. Lett. **85**(16), 3536–3539 (2000). https://doi.org/10.1103/physrevlett.85.3536
4. Deffuant, G., Neau, D., Amblard, F., Weisbuch, G.: Mixing beliefs among interacting agents. Adv. Complex Syst. **03**(01n04), 87–98 (2000). https://doi.org/10.1142/s0219525900000078
5. Flache, A., et al.: Models of social influence: towards the next frontiers. J. Artif. Soc. Soc. Simul. **20**(4) (2017). https://doi.org/10.18564/jasss.3521
6. Galesic, M., Stein, D.: Statistical physics models of belief dynamics: theory and empirical tests. Physica A: Stat. Mech. Appl. **519**, 275–294 (2019). https://doi.org/10.1016/j.physa.2018.12.011
7. Hegselmann, R., Krause, U.: Opinion dynamics under the influence of radical groups, charismatic leaders, and other constant signals: a simple unifying model. Netw. Heterogeneous Media **10**(3), 477–509 (2015). https://doi.org/10.3934/nhm.2015.10.477
8. Hołyst, J.A., Kacperski, K., Schweitzer, F.: Social Impact Models of Opinion Dynamics, pp. 253–273. World Scientific (2001). https://doi.org/10.1142/9789812811578_0005
9. Jędrzejewski, A., Sznajd-Weron, K.: Statistical physics of opinion formation: is it a SPOOF? C R Phys. **20**(4), 244–261 (2019). https://doi.org/10.1016/j.crhy.2019.05.002
10. Moussaïd, M., Kämmer, J.E., Analytis, P.P., Neth, H.: Social influence and the collective dynamics of opinion formation. PLoS ONE **8**(11), e78433 (2013). https://doi.org/10.1371/journal.pone.0078433
11. Mueller, S.T., Tan, Y.-Y.S.: Cognitive perspectives on opinion dynamics: the role of knowledge in consensus formation, opinion divergence, and group polarization. J. Comput. Soc. Sci. **1**(1), 15–48 (2017). https://doi.org/10.1007/s42001-017-0004-7
12. Mueller, S.T., Weidemann, C.T.: Decision noise: an explanation for observed violations of signal detection theory. Psychonomic Bull. Rev. **15**(3), 465–494 (2008). https://doi.org/10.3758/pbr.15.3.465
13. Murphy, G.L., Medin, D.L.: The role of theories in conceptual coherence. Psychol. Rev. **92**(3), 289–316 (1985). https://doi.org/10.1037/0033-295x.92.3.289
14. Newman, M.: Networks. Oxford University Press, Cambridge (2018). https://doi.org/10.1093/oso/9780198805090.001.0001
15. Pineda, M., Toral, R., Hernández-García, E.: Noisy continuous-opinion dynamics. J. Stat. Mech: Theory Exp. **2009**(08), P08001 (2009). https://doi.org/10.1088/1742-5468/2009/08/p08001

16. Schweighofer, S., Garcia, D., Schweitzer, F.: An agent-based model of multi-dimensional opinion dynamics and opinion alignment. Chaos: Interdisc. J. Nonlinear Sci. **30**(9) (2020). https://doi.org/10.1063/5.0007523
17. Schweighofer, S., Schweitzer, F., Garcia, D.: A weighted balance model of opinion hyperpolarization. J. Artif. Soc. Soc. Simul. **23**(3) (2020). https://doi.org/10.18564/jasss.4306
18. Sikder, O., Smith, R.E., Vivo, P., Livan, G.: A minimalistic model of bias, polarization and misinformation in social networks. Sci. Rep. **10**(1) (2020). https://doi.org/10.1038/s41598-020-62085-w
19. Sobkowicz, P.: Whither now, opinion modelers? Front. Phys. **8** (2020). https://doi.org/10.3389/fphy.2020.587009
20. Takács, K., Flache, A., Mäs, M.: Discrepancy and disliking do not induce negative opinion shifts. PLoS ONE **11**(6), e0157948 (2016). https://doi.org/10.1371/journal.pone.0157948
21. Tamarit, I., Cuesta, J.A., Dunbar, R., Sánchez, A.: Cognitive resource allocation determines the organization of personal networks. Proc. Natl. Acad. Sci. **115**(33), 8316–8321 (2018). https://doi.org/10.1073/pnas.1719233115
22. Weisbuch, G., Stauffer, D.: Hits and flops dynamics. Physica A: Stat. Mech. Appl. **287**(3–4), 563–576 (2000). https://doi.org/10.1016/s0378-4371(00)00393-9

When is Reservoir Computing with Cellular Automata Beneficial?

Tom Eivind Glover[1(✉)], Evgeny Osipov[3], and Stefano Nichele[1,2]

[1] Department of Computer Science, Oslo Metropolitan University, Oslo, Norway
tomglove@oslomet.no
[2] Department of Computer Science and Communication, Østfold University College, Halden, Norway
stefano.nichele@hiof.no
[3] Department of Computer Science, Electrical and Space Engineering, Luleå University of Technology, Luleå, Sweden
Evgeny.osipov@ltu.se

Abstract. Reservoir Computing with Cellular Automata (ReCA) is a relatively novel and promising approach. It consists of 3 steps: encoding the problem into the CA, the CA iterations step, and a simple classifying step. This paper demonstrates that the ReCA concept is effective even in arguably the simplest implementation of a ReCA system. However, we also report a failed attempt on the UCR Time Series Classification Archive where ReCA seems to work, but only because of the encoding scheme, not the CA. This highlights the need for ablation testing, i.e., comparing internally without sub-parts of one model, but also raises an open question on what kind of tasks ReCA is best suited for.

Keywords: Cellular Automata · Reservoir Computing · ReCA · MNIST · UCRArchive

1 Introduction

Reservoir Computing (RC) is a method that relies on an untrained reservoir and a linear and trained simple classifier, making it a very energy-efficient method to train. CA is a regular vector of discrete values that interact with Boolean operations. Therefore, CA is relatively simple to implement in hardware, e.g., FPGA, making it energy-efficient in execution. When RC and CA are combined into Reservoir Computing with Cellular Automata (ReCA), it becomes, in theory, an energy-efficient method to both train and run. Such features are important for EdgeAI [31]. The property of energy efficiency can enable capabilities closer to or on the edge devices. This, in turn, can improve the system's privacy, cost, capabilities, and availability.

Most ReCA works explore ReCA on time-dependent benchmarks, such as chaotic time series prediction or the 5-bit memory benchmark [13,16]. These tasks require time to be taken into account, and the ReCA solutions are designed

T. Carletti et al. (Eds.): WIVACE 2024, CCIS 2532, pp. 42–55, 2025.
https://doi.org/10.1007/978-3-031-93631-9_4

such that the CA is the only place it can be contained. This makes it infeasible to do a common analysis called ablation testing, where a compontent (CA) is removed to assess its contribution. One can still demonstrate value by comparing to other known methods, but as different encoding schemes can react differently to different substrate and classifiers, it is less clear how much value the CA reservoir is actually contributing. Furthermore, the ReCA method has in a few cases also been explored on benchmarks such as MNIST [10,18,25] that do not depend on time. Therefore, there is a need to clarify what role the CA serves. As far as we can tell, only [18] did an ablation test, where accuracy was 89.58% without and 91.86% with CA. While these results are informative, it is worth mentioning that our trivial solution achieves higher accuracy on the MNIST benchmark.

This work's main contribution is exploring ReCA on MNIST, using a simple design and demonstrating that it works for many simple CA rules. The simple design is a form of Minimum Viable Product (MVP), a design principle where only the vital features are implemented. This provides value as a feasibility study and clarifies that the CA can be useful. The simple design also serves as an example of ReCA that can easily be understood and replicated. Additionally, we show that MNIST can be solved relatively easily even if all the grayscale information is removed. We also explored the method on the UCR dataset, a benchmark of several time series classification tasks. As RC works well with time series, it seemed a natural benchmark to test on. We observed that it seems to work with a similarity-preserving encoding scheme. However, we show that the encoding scheme itself does all the work, making this a great example of the importance of ablation testing.

2 Background

2.1 Cellular Automata

Cellular Automata (CA) are a simple model consisting of a grid of cells in a limited set of k discrete states. The grid is uniformly connected, typically in 1 or 2 dimensions. The cell state changes iteratively, depending on the state of the neighbouring cells. The combination of the neighbour states deterministically defines the next state via a lookup table, typically called the Transition Table (TT). CA was first used to study self-replication by John von Neumann in 1940 but published in 1966 [26]. It can be considered an idealised system for parallel and decentralised computation [23].

Elementary Cellular Automata (ECA). Elementary Cellular automata (ECA) is a subset of CA that has 1-dimension, binary states ($S = 2$) and 3 neighbours ($K = 3$) (left, right and centre). Therefore, ECA only has $S^{S^K} = 2^{2^3} = 256$ possible rules (unique TT), and the whole set of these rules is often named the rule space. It is a convention to name individual rules in a rule-space after the output states of the TT, in Fig. 1 the transition table is represented by the binary

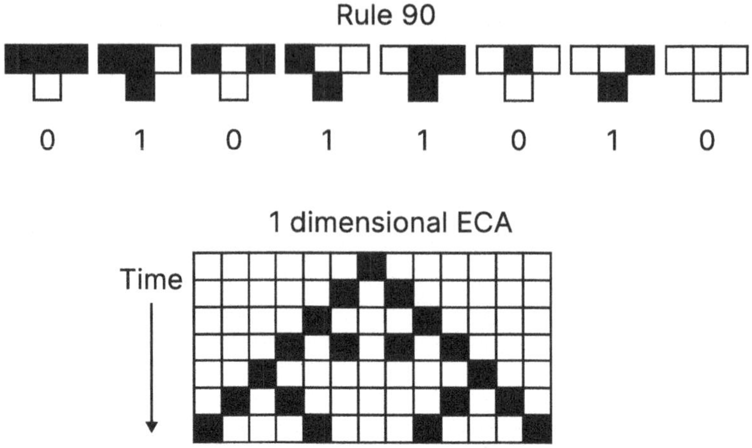

Fig. 1. Example of ECA with rule 90 with TT, starting from a central cell on, executing 7 time-steps.

sequence 01011010. When this sequence is translated from binary to decimal, it becomes 90, hence the name Rule 90. CA is deterministic, and the rule, together with the initial condition, leads the CA into a set of subsequent states called the trajectory. Rule 110 has even been shown to be computationally universal [5], but one can question whether that is a useful definition of computation for a parallel and distributed computational substrate [14].

Due to symmetries in the rule-space, one can use trivial transformations to reduce the 256 ECA rules to just 88 that represent the entire rule-space, called the minimum equivalent rules. The trivial transformations, often called reflection and complement, were first pointed out in [34, p. 51, p. 176], but was popularised and demonstrated for CA later in [17,35–37]

2.2 Reservoir Computing (RC)

Reservoir Computing (RC) is a substrate-independent framework for computing. RC is independent because it works on many different substrates, but to be clear different substrates would of course have different capabilities. The RC framework consists of 3 parts, the encoding, the untrained reservoir and the output.

The encoding part inputs information into the untrained reservoir and typically into higher dimensions. The untrained reservoir typically expands, modifies or changes the information, but could, in the context of the framework, be considered a black box as seen in Fig. 2. The output part is typically linear, does dimensional reduction and extracts useful features.

The RC concept originated in echo state networks (ESN) (Fig. 3) using recurrent neural networks as a substrate [15] and in liquid state machines (LSM) using a spiking neural network for a substrate [19]. Since then, both ESM and LSM

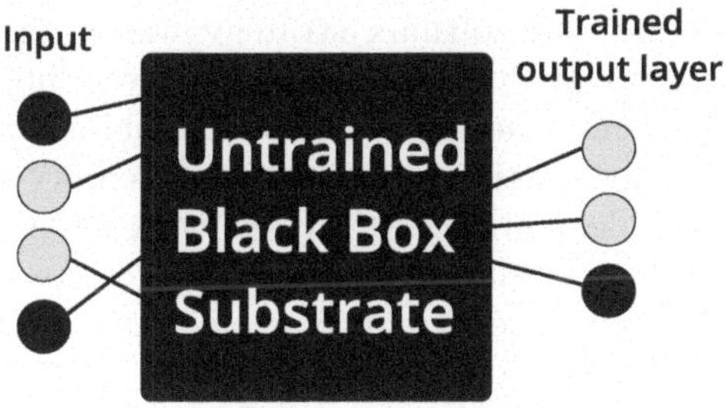

Fig. 2. RC as a substrate-independent framework

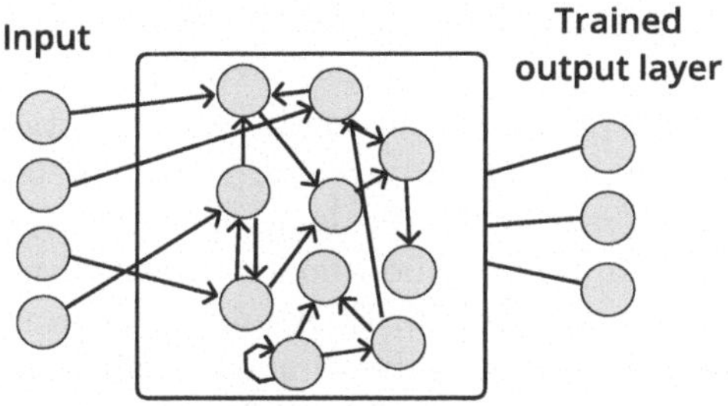

Fig. 3. Basic network Architecture of an ESN.

and a host of other substrates have been categorised under the umbrella term of RC. Due to RC substrate-independent nature, many different substrates have been explored and compared [32]. Some explore different topology configurations as in [9], where deep layered sub-reservoirs were analysed instead of the typical one large reservoir. RC is also a prevalent method with physical reservoirs [32], as an extreme example in [8], it was demonstrated that RC could use the surface waves on a bucket of water as a reservoir, and they successfully solved speech recognition and xor tasks using this substrate. One interesting substrate is real biological neural networks (BNN), specifically disassociated neurons that self-organise over a microelectronic array [1]. In different direction is ReCA, where instead of using a more complicated Reservoir (BNN), a very simple reservoir is used instead (CA).

2.3 Reservoir Computing with CA (ReCA)

The first study that introduced CA as a substrate in reservoir computing is [39]. This study investigated Game of Life and several ECA rules as reservoir substrates and tested them on a 5-bit and 20-bit memory benchmark. In addition, it presents a theoretical comparison of CA vs ESN, using the metric of the number of operations needed to solve the benchmark, which documents a clear advantage of using CA.

As an ECA reservoir only relies on simple discrete binary interactions between cells, it affords a hardware-friendly implementation. In [24], ReCA using

ECA with a max-pooling and softmax strategy was implemented on a Field Programmable Gate Array (FPGA). In [29], a CA was implemented on Complementary metal-oxide-semiconductor (CMOS) combined with a custom hardware SVM implemented in resistive random-access memory (ReRAM). In [18], a synthesised hardware implementation of ReCA using ECA with a max-pooling and ensemble bloom filter classifier. It showed impressive results compared to "state-of-the-art" in terms of energy efficiency, memory usage and area(number of gates) usage, but with some loss in accuracy [24].

Many works have studied ReCA using the 5-bit memory benchmark. In [28], the structure of the CA was changed to a deep layered architecture and compared to a single layer, which resulted in noticeable performance improvements. In [27], the CA substrate was organised as consisting of two regions of different ECA rules. Different combinations of rules were explored, and some combinations showed great promise. In [20], different methods of selection of cell history are used for the classification model and are tested on the 5-bit memory task, a temporal order task and arithmetic and logic operation tasks. In [2], CA rules with multiple states and larger neighbourhoods were evolved and then tested on the 5-bit memory benchmark. In [33], ECA and asynchronous ECA are tested and compared on the 5-bit memory benchmark, mainly in the context of the distractor period. In [21], it was pointed out that the benchmark has no train test split. They modified the benchmark by training on just a few (2 or 3) of the 32 possible input streams, and some of the rules with more ordered behaviour could still solve this version of the benchmark. In [12], the full ECA set was tested using key parameters of number of bits (N_b), redundancies (R) and Grid size. This work was also extended in [13] to include more parameters such as Iterations (I) and Distractor Period (D_p). This paper also explained many of the unexpected results in the previous study, but perhaps as important, it similarly to [21] pointed out some weaknesses in the 5-bit memory benchmark.

ReCA is explored on benchmarks other than the 5-bit memory benchmark. In [10,18,24], ReCA is implemented (sometimes in hardware) and tested using MNIST. In [22], solved tasks of sine and square wave classification non-linear channel equalization, Santa Fe Laser Data and iris classification.

In [16], a method for Rule Selection for ReCA was presented. Limiting the search space to only linear rules that obey a list of specific mathematical properties (see paper for details), the paper demonstrates that the method selects for rules in the high performance (95-80 percentile) bracket on several time-series prediction benchmarks compared to the entire Linear CA space of same neighbourhood and number of states.

3 Methodology

3.1 MNIST and bMNIST Datasets

Perhaps the most well-known benchmark in AI is MNIST [3]. It originally contained binary handwritten digits. However, it was grey-scaled to reduce the effect of aliasing.

Here, as shown in Fig. 4, we simplify MNIST further by making the values of each image binary by simply rounding to 0 or 1 for each pixel, creating a binary MNIST (bMNIST). We adopt this approach not because we believe it's the most effective way to preserve information in terms of encoding. Rather, we choose it for its simplicity, as it allows for the most straightforward encoding strategy we could devise.

It's important to note that as we remove some of the information in the MNIST to make the bMNIST, any solution in terms of performance on the bMNIST is also applicable to the MNIST. However, it's crucial to understand that any solution developed for the MNIST may not necessarily be valid for the bMNIST (Fig. 4).

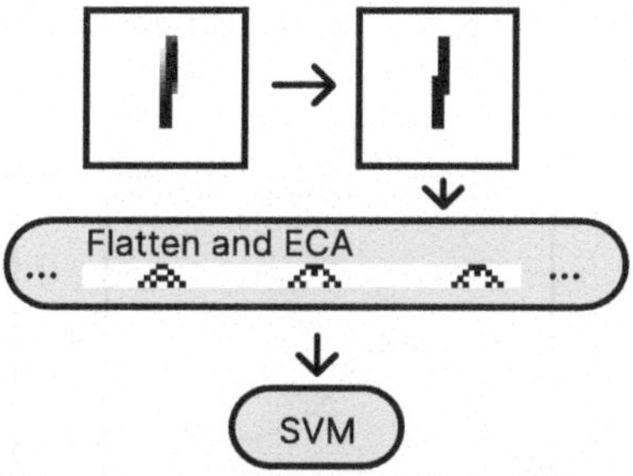

Fig. 4. Example and architecture of MNIST ReCA solution

To encode the bMNIST into the CA we simply flatten it and set it as the initial condition of the CA. We acknowledge that flatting only retains relational information in the x plane and not in the y plane, but it is done for the sake of simplicity.

The MNIST dataset consists of 70000 labelled images of 28×28 pixels. We keep 33% in a holdout set, intended to be used later. The remaining 46900 are split into a 90%/10% train test split.

3.2 UCR Dataset

The UCR Time Series Classification Archive [7] is a set of several smaller datasets that are time-series classification tasks. It is an improved version that considers criticisms of the previous UCR archive [6], such as a lack of benchmarks with varied length time-series. The authors point out some salient criticisms on the field of AI itself and give recommendations on best using the UCR datasets and other datasets, e.g., the importance of ablation testing when introducing multiple working parts.

3.3 Similarity Preserving Expanded Encoding (SimExp)

Binarising the MNIST dataset via rounding retains much of the useful information, but the same is not true for the UCR datasets. We used a similarity-preserved encoding that converts the data into a higher-dimensional binary vector, see Fig. 5. First, we make a random vector n times larger than the largest vector in the dataset, representing a unique fixed value for every index position in the dataset's vectors. Then, we normalise the dataset between 0 and 1. Each value in a dataset vector was transformed into an n-sized vector, with 0 s and 1 s representing the float value (e.g., $n = 4, 0.75 = 1110$). Both binary vectors are then XORed together. These values were concatenated into a vector, resulting in a vector n times larger than the original vector, which is binary. In our case, we used expansions of $n = \{16, 32, 64\}$, labelled as SimExp{n}, e.g., SimExp16. For this example, a vector of 10 float values would be transformed into 160 binary values.

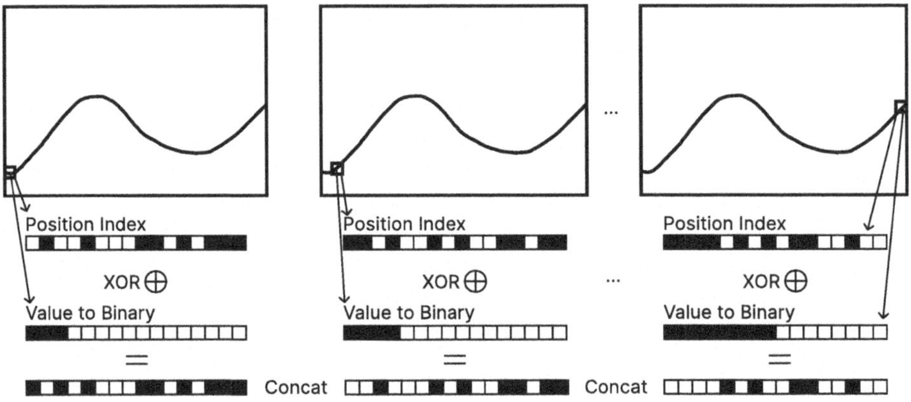

Fig. 5. Example of time series being expanded into binary, without losing similarity.

Note that typically, this method does not concatenate every n-sized vector but rather embeds all the n-sized vectors into a binary cube of the same size as n for the sake of dimensionality reduction. This type of method has many names, but it is sometimes called Binary Embedding [38]. We call it something else to highlight the difference in terms of dimensionality reduction vs expansion. In our case, we started with concatenation as embedding into the cube might create some loss of information if the randomised index vectors were unlucky or the cube size was too small to contain the information. This way, one could have a baseline and, in a sense, another form of ablation test. As the method did not work well with our intended reservoir, the binary embedding method was not performed as intended. We still report it for the sake of scientific integrity but also because the method worked surprisingly well on its own.

3.4 CA Reservoir and Classifier

In all experiments reported, the encoded vector is set as the initial condition of an ECA of the same size as the vector. 3 CA steps are computed and concatenated to the initial condition. The 3 CA steps and the initial condition are then used to train a classifier, typically SVM with a linear kernel unless specified. Source code and results can be found at [11].

We tested briefly with other alternative number of CA steps to see if it would greatly impact performance. We kept the original parameters of 3 as the variation yielded no significant differences.

4 Results

4.1 bMNIST

As mentioned, ReCA has been tested on MNIST before. Both [10,25] explored it with much more sophisticated methods and got better accuracy than our experiment (although they are not too far away). It is important to note that we will likely remove some useful information (though less than expected) from the benchmark by turning it into bMNIST. This makes it a different benchmark. Furthermore, we aim to make a minimal and simple solution, not to achieve the best performance. As shown in Fig. 6, most ECA improve the performance significantly compared to no ECA Reservoir. The baseline (ablation) solution of no ECA had a 8.2% error rate. The best rule (94) nearly halves the error rate from 8.2% to 4.6%. We also see that a wide range of rules improves the results. This improvement indicates that a parallel non-uniform design [27] might be worth investigating for the bMNIST benchmark.

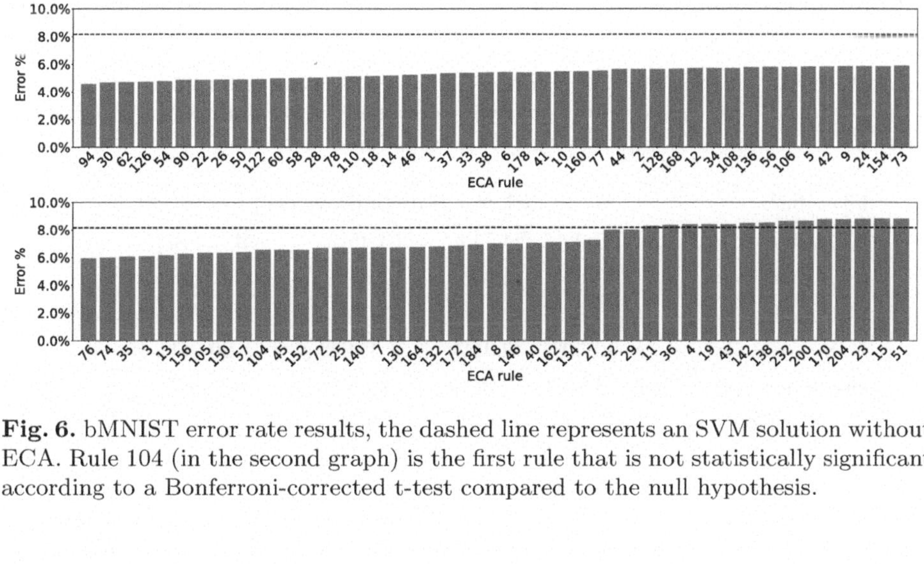

Fig. 6. bMNIST error rate results, the dashed line represents an SVM solution without ECA. Rule 104 (in the second graph) is the first rule that is not statistically significant according to a Bonferroni-corrected t-test compared to the null hypothesis.

We also briefly tested on the digits benchmark from SciKit-learn [30], a smaller version of MNIST where the images are 8×8 pixels. This Digits benchmark was also binarised the same way as the bMNIST benchmark. ReCA worked well for this benchmark, but fewer rules performed well, and overall, the performance impact of the ECA was smaller.

Additionally, we briefly tried other classifiers on the bMNIST benchmark; we used Rule 126 because it worked well on the binary Digits and the bMNIST benchmarks. The results can be seen in Table 1. The results show that CA also worked with an SVM using a Radial basis function (RBF) kernel. RBF kernel is not linear. Therefore, it is not typically used together with RC. The RBF kernel default parameters were used, and no hyperparameter tuning was performed. Therefore, we can not rule out that, for some parameters, the ECA could be redundant. We hypothesised that the ECA would only improve performance if the classifier had some form of selection of individual cells/pixels, allowing it to filter out the noise. Therefore, it was surprising that the method worked with a simple centroid classifier. In contrast KNN ($n = 5$) the ECA Reservoir made the performance significantly worse, which was more in line with our hypothesis. It worked with a single perceptron layer but did not improve performance when paired with larger multiple-layered perceptrons. This was expected as RC is a more energy-efficient replacement for a trained model. The multi-layered perceptron model received the highest performance of any classifier, which is unsurprising considering it had to be, in contrast, trained longer to settle the NN. We do not expect this to be the highest possible score, as we did close to no hyperparameter tuning. Therefore, it should not be hard to achieve a higher score. One can view this result as being closer to the theoretical limit of what can be done with the bMNIST benchmark, as we expect some information to have been lost in the binarisation, getting scores such as 99.83 [4], might be impossible. Even so, this is surprisingly high, considering the binarisation reduced the precision in the data from 256 to 2.

Table 1. Results of other classifying models for bMNIST using rule 126. "Improvement" is the improved performance contrasted with not using ECA. "Improvement %" is the improved performance as a percentage of the remaining error rate.

Classifier	Accuracy	Error Rate	Improvement	Improvement %
SVM linear	95.29%	4.71%	3.45%	42.28%
SVM RBF	95.63%	4.37%	2.37%	35.16%
Centroid	75.62%	24.38%	5.70%	18.95%
KNN ($n = 5$)	90.59%	9.41%	-1.65%	-20.26%
single perceptron layer	93.55%	6.45%	2.85%	30.65%
4 perceptron layers	97.31%	2.69%	-0.05%	-1.82%

4.2 UCR

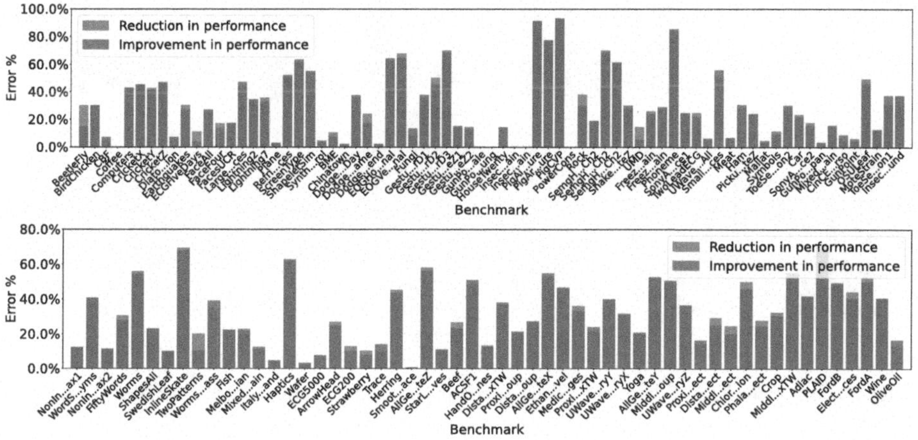

Fig. 7. UCR dataset results with Rule 90 ECA Reservoir contrasted with its ablation. The first graph features all the benchmarks where the training set is perfectly solvable with an SVM with a linear kernel (linearly solvable). The second graph features the benchmarks that were not linearly solvable and are sorted from best to worst based on how well an SVM with a linear kernel solved them.

In Fig. 7, Rule 90 is shown. While it is performing relatively well considering the simplicity of the solution, when compared directly to the same solution without the ECA steps, it is, in fact, performing better on average. This result means that the ECA part of the ReCA solution is, in general, functionally useless in this implementation, though not necessarily for individual datasets. None of the ECA rules manged to overall improve results, 90 was picked as an example due to its popularity in ReCA. In contrast, the SimExp16 method works better than just a Linear SVM, as seen in Fig. 8. This difference becomes more significant in SimExp32 and furthermore in SimExp64, where it got 4.6% lower error on average than SVM with a linear kernel; this is shown in Table 2. Additionally, as we trained the SVM ablation test on the original benchmark where the numbers are floating point values, we essentially would have some loss of precision even in the SimExp64 as it only has 64 precision values in the encoding, meaning it is on par for precision with a 6-bit integer for precision. We split the benchmarks into linear solvable and not linearly solvable because we hypothesised that in benchmarks where the training data was already linearly separable, the SVM would find no benefit from the ECA-developed features. We, therefore, predicted no performance gain from the first graph in Fig. 7. This prediction was technically confirmed, but as we saw no benefit from the second graph, the prediction should be viewed in spirit as false for this experiment. In [6], it was recommended to use a Texas sharpshooter plot, as a model is only good if one can predict what benchmarks it will perform well on or if it improves overall performance. We

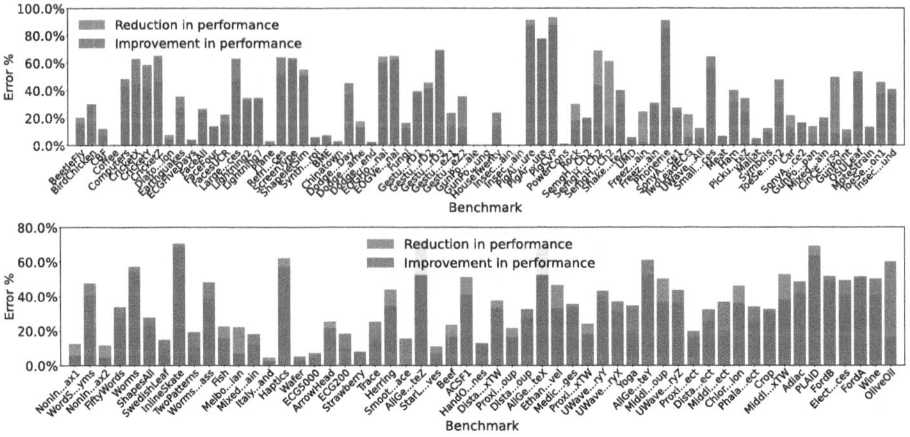

Fig. 8. Same as Fig. 7, but the SimExp16 experiment contrasted with its ablation.

instead used a more straightforward bar graph, and as the hypothesis failed, there was no need to prove the hypothesis in the graph. A bar graph allows for a better overview.

Interestingly, we also observed that the SimExp64 performed marginally better than dynamic time warping (DTW) (W = 100), with a mean error rate reduction of 0.54%; this might be low, but it warrants further study.

Table 2. Average UCR archive results using only SimExp and SVM

Expansion size	Accuracy	error rate	Improvement
SimExp16	70.79%	29.21%	2.15%
SimExp32	72.37%	27.63%	3.73%
SimExp64	73.32%	26.68%	4.68%

5 Discussion

5.1 Locality vs Globality

Considering that in the UCR dataset, the DTW method does well, and the DTW essentially allows global properties of a time series to be non-linearly mapped and compared [6], we would assume that global properties of the datasets are essential for performance. If we contrast this to our 3 CA step solution for bMNIST, global features are less crucial for bMNIST, as CA's speed of light means that no information can move beyond the 3 nearest cells in the CA. Considering these two observations, we argue that ReCA works best for local features and poorly for global features. As most ReCA explore temporal problems or problems that can be solved with local features, an open question is how to configure ReCA to handle global features better.

5.2 MNIST vs bMNIST

Our results indicate that turning MNIST into bMNIST can be done trivially, reducing the precision of the data from 256 (uint8) to 2 (binary) while losing accuracy capability of at most 2.69% (probably much less) as MNIST can be reduced to $\frac{1}{8}$ of it memory size and $\frac{1}{128}$ of its precision, without losing much accuracy. This can have implications for any work that binarises the MNIST dataset. Essentially, there is not much necessary or non-redundant information encoded into the grayscale dimension, so any work that tries to preserve it might naturally assume the methods used are more successful than they are.

Quantization is a common practice to reduce the precision of values in datasets, e.g., float64 to float16. Perhaps many other benchmarks can also be reduced to binary. This would be great news for binary models such as CA.

5.3 Deception of Good Encoding

SimExp works very well for preserving useful information in the UCR datasets, while combining it with ECA did not improve the results. This showcases the value of ablation tests. Otherwise, we might have mistakenly claimed that ReCA is effective for UCR classification when, in fact, it is the SimExp encoding that does all the work. Why it is so effective is a question that remains open for further research.

6 Conclusion and Future Work

We have demonstrated a simple ReCA system in action. We have also shown a deceptive example where ReCA might seem to work but the actual work is done by the encoding scheme, highlighting the importance of ablation testing.

A possibility for future work would be a comparison with a ReCA system using linear regression or ridge regression readouts instead of SVM. A more difficult task would allow rigorous identification of suitable methods for encoding global features into the CA. We also have left 33% of MNIST in a holdout set for further use. Considering the relatively good results, a more sophisticated encoding method, e.g., including the y-dimension of MNIST would be prudent. Additionally, the promising results of the SimExp encoding warrant further study.

Acknowledgements. This work was partially financed by the Research Council of Norway's DeepCA project, grant agreement 286558.

Disclosure of Interest. The authors have no competing interests.

References

1. Aaser, P., et al.: Towards making a cyborg: a closed-loop reservoir-neuro system. In: Artificial Life Conference Proceedings, vol. 14, pp. 430–437. MIT Press (2017)
2. Babson, N., Teuscher, C.: Reservoir computing with complex cellular automata. Complex Syst. **28**(4), 433–455 (2019)
3. Bottou, L., et al.: Comparison of classifier methods: a case study in handwritten digit recognition. In: Proceedings of the 12th IAPR International Conference on Pattern Recognition, vol. 3-Conference C: Signal Processing (Cat. No. 94CH3440-5), vol. 2, pp. 77–82. IEEE (1994)
4. Byerly, A., Kalganova, T., Dear, I.: No routing needed between capsules. Neurocomputing **463**, 545–553 (2021)
5. Cook, M., et al.: Universality in elementary cellular automata. Complex Syst. **15**(1), 1–40 (2004)
6. Dau, H.A., et al.: The UCR time series archive. IEEE/CAA J. Automatica Sinica **6**(6), 1293–1305 (2019)
7. Dau, H.A., et al.: The ucr time series classification archive (2018). https://www.cs.ucr.edu/~eamonn/time_series_data_2018/
8. Fernando, C., Sojakka, S.: Pattern recognition in a bucket. In: European Conference on Artificial Life, pp. 588–597. Springer, Heidelberg (2003)
9. Gallicchio, C., Micheli, A.: Deep reservoir computing: a critical analysis. In: ESANN (2016)
10. García-Arias, Á.L., Yu, J., Hashimoto, M.: Low-cost reservoir computing using cellular automata and random forests. In: 2020 IEEE International Symposium on Circuits and Systems (ISCAS), pp. 1–5. IEEE (2020)
11. Glover, T.E.: Osf link for source code and full results (2024). https://osf.io/8ayc9/?view_only=f9c2bdd1a6c3493db66f1e466b5c2593
12. Glover, T.E., Lind, P., Yazidi, A., Osipov, E., Nichele, S.: The dynamical landscape of reservoir computing with elementary cellular automata. In: ALIFE 2021: The 2021 Conference on Artificial Life. MIT Press (2021)
13. Glover, T.E., Lind, P., Yazidi, A., Osipov, E., Nichele, S.: Investigating rules and parameters of reservoir computing with elementary cellular automata, with a criticism of rule 90 and the five-bit memory benchmark. Complex Syst. (2023)
14. Hudcová, B., Mikolov, T.: Computational hierarchy of elementary cellular automata. In: ALIFE 2021: The 2021 Conference on Artificial Life. MIT Press (2021). https://doi.org/10.1162/isal_a_00447
15. Jaeger, H.: The "echo state" approach to analysing and training recurrent neural networks-with an erratum note. Bonn, Germany: German National Research Center for Information Technology GMD Technical Report **148**(34), 13 (2001)
16. Kantic, J., Legl, F.C., Stechele, W., Hermann, J.: Relicada: reservoir computing using linear cellular automata design algorithm. Complex Intell. Syst., 1–24 (2024)
17. Li, W., Packard, N.: The structure of the elementary cellular automata rule space. Complex Syst. (1990)
18. Liang, D., Hashimoto, M., Awano, H.: Bloomca: a memory efficient reservoir computing hardware implementation using cellular automata and ensemble bloom filter. In: 2021 Design, Automation & Test in Europe Conference & Exhibition (DATE), pp. 587–590. IEEE (2021)
19. Maass, W., Natschläger, T., Markram, H.: Real-time computing without stable states: a new framework for neural computation based on perturbations. Neural Comput. **14**(11), 2531–2560 (2002)

20. Margem, M., Gedik, O.S.: Reservoir computing based on cellular automata (reca) in sequence learning. J. Cellular Automata **14** (2019)
21. Margem, M., Gedik, O.S.: Feed-forward versus recurrent architecture and local versus cellular automata distributed representation in reservoir computing for sequence memory learning. Artif. Intell. Rev. **53**(7), 5083–5112 (2020). https://doi.org/10.1007/s10462-020-09815-8
22. McDonald, N.: Reservoir computing & extreme learning machines using pairs of cellular automata rules. In: 2017 International Joint Conference on Neural Networks (IJCNN), pp. 2429–2436. IEEE (2017)
23. Mitchell, M.: Life and evolution in computers. In: History and Philosophy of the Life Sciences, pp. 361–383. JSTOR (2001)
24. Morán, A., Frasser, C.F., Roca, M., Rosselló, J.L.: Energy-efficient pattern recognition hardware with elementary cellular automata. IEEE Trans. Comput. **69**(3), 392–401 (2019)
25. Morán, A., Frasser, C.F., Rosselló, J.L.: Reservoir computing hardware with cellular automata. arXiv preprint arXiv:1806.04932 (2018)
26. von Neumann, J.: Theory of self-reproducing automata. Math. Comput. **21**, 745 (1966)
27. Nichele, S., Gundersen, M.S.: Reservoir computing using non-uniform binary cellular automata. arXiv preprint arXiv:1702.03812 (2017)
28. Nichele, S., Molund, A.: Deep learning with cellular automaton-based reservoir computing. Complex Syst. **26**(4), 319–339 (2017)
29. Olin-Ammentorp, W., Beckmann, K., Cady, N.C.: Cellular memristive-output reservoir (cmor). arXiv preprint arXiv:1906.06414 (2019)
30. Pedregosa, F., et al.: Scikit-learn: machine learning in python. J. Mach. Learn. Res. **12**, 2825–2830 (2011)
31. Singh, R., Gill, S.S.: Edge AI: a survey. Internet Things Cyber-Phys. Syst. **3**, 71–92 (2023)
32. Tanaka, G., et al.: Recent advances in physical reservoir computing: a review. Neural Netw. **115**, 100–123 (2019)
33. Uragami, D., Gunji, Y.P.: Universal criticality in reservoir computing using asynchronous cellular automata. Complex Syst. **31**(1) (2022)
34. Walker, C.C.: A Study of a Family of Complex Systems–an Approach to the Investigation of Organisms' Behavior. Ph.D., University of Illinois at Urbana-Champaign, United States – Illinois (1965). https://www.proquest.com/docview/302133130/citation/38D4271784FF4B1EPQ/1. iSBN: 9781084587113
35. Wolfram, S.: Theory and Applications of Cellular Automata. World Scientific (1986)
36. Wolfram, S.: Tables of cellular automaton properties: 1986. In: Cellular Automata and Complexity, pp. 513–584. CRC Press (2018)
37. Wuensche, A., Lesser, M., Lesser, M.J.: Global Dynamics of Cellular Automata: An Atlas of Basin of Attraction Fields of One-Dimensional Cellular Automata, vol. 1. Andrew Wuensche (1992)
38. Yi, X., Caramanis, C., Price, E.: Binary embedding: fundamental limits and fast algorithm. In: International Conference on Machine Learning, pp. 2162–2170. PMLR (2015)
39. Yilmaz, O.: Reservoir computing using cellular automata. arXiv preprint arXiv:1410.0162 (2014)

Chemotactic Swarming of *C. elegans* as a Collective Decision-Making Process

Nemanja Antonic(✉), Aymeric Vellinger, and Elio Tuci

University of Namur, Rue de Bruxelles 61, 5000 Namur, Belgium
nemanja.antonic@unamur.be

Abstract. Collective decision making is a process observed both in natural and artificial swarms, in which agents converge and select a single option without the influence of leading individuals. In this study, we look at a collective decision making process in a group of *C. elegans*, a model organism in biology, characterised by a fully mapped neural system of 302 neurons, and a pheromone-based system that regulates the interactions among the worms. In particular, we model and optimise swarms of *C. elegans* through sets of differential equations. The experimental setup compares the performances of two different types of swarms: one in which the agent interactions are modulated by self-released pheromone production, and the other in which this means of interaction is absent. The worms' task is to choose to selectively swarm towards one of the two target locations containing an attractive odour. We analyse and compare the swarming behaviour and the collective decision making process of the agents in the two models. Our findings suggest that the presence of pheromones is of paramount importance for the capabilities of the swarm to reliably choose one of the two attractive targets.

Keywords: Decision-Making · Evolutionary Strategy · Partial Differential Equations · Caenorhabditis Elegans

1 Introduction

Swarms of natural organisms display collective behaviour as a result of the interactions of the individuals inside the swarm to some external or social stimulus. A plethora of collective behaviours has been observed and analysed both in very small organisms such as colonies of bacteria [14,19] and in bigger animals such as fishes [2,7,16], birds [6], and troops of mammals [10,17]. In this study, we focus on the collective behaviour of *C. elegans*, a 1mm long roundworm, thus characterised by a size in between the microscopic scale of bacteria and the macroscopic scale of mammals, referred to as mesoscale [5]. Despite the simplicity of its fully mapped synaptic connections, featuring only 302 neurons in the hermaphrodite animals [22], the natural *C. elegans* worm displays social behaviour when certain conditions, such as sub-optimal ambient oxygen levels and food scarcity, are met.

T. Carletti et al. (Eds.): WIVACE 2024, CCIS 2532, pp. 56–69, 2025.
https://doi.org/10.1007/978-3-031-93631-9_5

Through genetic modifications (i.e., genes editing) which aim at reconfiguring specific neural circuits of the worm, it is possible to alter the natural behaviour of *C. elegans*.

Within the EU funded project BABOTS[1], an interdisciplinary team of scientists exploit the possibility to reprogram the behavioural repertoire of *C. elegans*, in order to induce large groups of worms to display complex collective behaviours. This study contributes to the scientific challenge of the BABOTS project by describing a model that allows identifying mechanisms that underpin complex collective decision making processes in genetically modified *C. elegans* worms.

We are interested in the development of a model for the collective swarming of *C. elegans* towards the chemotactic gradient of an attractive chemical odour placed in a target area distant from where the worms are initially placed. Examples of chemotactic assays can be found in [15,21], where experiments are consistently performed with a relatively low number of worms to shed light on the mechanisms underpinning the individual response to chemical stimuli. Contrary to this type of studies, we are interested in exploring the collective response of a large number of worms to a chemotactic gradient. Authors in [12] performed chemotactic assays with different strains of *C. elegans* obtaining useful insights on the correlation between the proportion of worms that reach the odour source and the total number of worms. Nevertheless, they did not attempt to develop models of this phenomenon.

Models of collective behaviour of *C. elegans* have been developed to look at aggregation [3,18,20] and collective motion [23]. Aggregation guided by local levels of oxygen has been studied in [4] through a Keller-Segel [9] macroscopic model. Authors in [5] model both swarming and aggregation in *C. elegans* in presence of food through a microscopic, agent-based model. In this study, we leverage a macroscopic model from the literature [1] for the aggregation of the worms to self-produced attractive and repulsive pheromones. We adapt it so that only an attractive, non self-produced odour is present at a fixed location at a certain distance from where the worms are initially placed. Moreover, we extend the model so that worms produce attractive and repulsive pheromones in order to assess how this might influence the collective swarming. Within this type of models, where partial differential equations (PDEs) govern the evolution of worm and odour densities, parameters representing the interaction of worms to the chemical and the chemical properties of the gradient play a crucial role in the accuracy and time needed for the worms to succeed in reaching the target area. We are interested in analysing the decision-making capabilities of a swarm of *C. elegans* modeled through the aforementioned macroscopic model.

Motivated by previous work [11,12] where the authors study the chemotaxis of individual *C. elegans* when two odour sources are present, we are interested in the behaviour of a swarm of worms when two odour sources, with distinct chemical properties, are present. Such an experiment would represent a first attempt at modeling the decision-making capabilities of a swarm of *C. elegans*

[1] www.babots.eu.

solving a best-of-n problem, where a set of n options, each with an associated quality, is present and the swarm needs to reach a consensus on the best option in terms of quality. In particular, the difficulty of the problem would be given by two main factors, namely the chemical properties of the gradient and the distance at which the gradient is placed with respect to the initial position of the worms. Analysing the influence of these two factors would help in building a decision-making model for the chemotactic swarming of *C. elegans*, which, to the best of our knowledge, has not been yet developed.

2 Model

Chemotaxis, i.e. movement towards a chemical source, of a swarm of *C. elegans* can be described through a reaction-diffusion model, where the evolution of the worm density ρ and of the concentration of attractant u is described by two PDEs, similarly to [1]:

$$\frac{d\rho}{dt} = \nabla \cdot [\sigma \nabla \cdot \rho + \rho \nabla \cdot V] \tag{1}$$

$$\frac{du}{dt} = D\nabla^2 u - \gamma u \tag{2}$$

where

$$V = V(u) = -\beta \log(\alpha + u) \tag{3}$$

is the function describing the potential of the attractant source; σ is the random movement of the worms, D and γ are, respectively, the diffusion coefficient and the decay rate of the attractant. Notably, this model is adapted in such a way that the odour source is not produced nor consumed by the worms and worms respond only to the odour gradient, since we ignore density-induced responses. Thus, this model approximates the ideal behaviour of genetically modified worms such that their response level to an attractant odour is pronounced so strongly, that other factors can be ignored. We believe that accounting for a single stimulus can be biologically justifiable, hence giving a reasonable model for the chemotactic swarming of *C. elegans*. The terms in Eq. 1 take into account the diffusion of the worms due to their random movement and their taxis towards the odour source potential, while the terms in Eq. 2 model the variation in odour concentration by its diffusion and its evaporation.

Moreover, we are interested in expanding Eqs. 1, 2 so to also include self-produced attractive and repulsive pheromones. Let u_a, u_r be the concentration of attractive and repulsive pheromone. Then the evolution of pheromone i can be described as in [1]:

$$\frac{du_i}{dt} = D_i \nabla^2 u_i - \gamma_i u_i + s_i \rho \tag{4}$$

where $i \in \{a, r\}$ represents the attractive or repulsive pheromone, γ_i is the evaporation rate, D_i the diffusion constant and s_i the secretion rate. In order to account for the reaction of the worms to the pheromones, we modify the potential V of Eq. 1:

$$V = V(u) + V(u_a) + V(u_r) + V_\rho \tag{5}$$

where $V(u)$ is the same as in Eq. 3. The last term of Eq. 5 is speculated by the authors in [1] to be a potential created by the worms trying to squeeze themselves as compactly as possible:

$$V_\rho = \sigma \text{scale} \left(1 + \tanh\left(\frac{\rho - \rho_{max}}{\text{cushion}}\right)\right) \tag{6}$$

We solve these systems of PDEs by means of a finite difference scheme, where the environment is a square with length $l = 0.02$ m discretised in $N = 512$ cells for each of the two spatial coordinates x, y and with periodic boundary conditions. In particular, we resort to a Euler integration scheme with a time step $dt = 0.01$ and a stopping criterion based on the maximum residual $d\rho_{max}$ per integration step, with $\epsilon = 10^3$ worm/m^2, where the unit of measure is expressed in terms of the number of worms per m^2, together with a maximum allowed time $t_{max} = 5000$.

In order to evaluate the performance of a given model, we define a chemotactic index c which is simply given by the sum of worm densities inside the target area divided by the total sum of worm densities:

$$c = \frac{\sum_{i=T_1}^{T_2} \sum_{j=T_1}^{T_2} \rho(i,j)}{\sum_{i=0}^{N-1} \sum_{j=0}^{N-1} \rho(i,j)} \tag{7}$$

where $T_1, T_2 \in [0, N-1]$ such that $T_1 < T_2$ are the indices of the target area, assumed to be a square.

3 Methods

We begin by defining the initial area where worms are placed as a square of 40×40 cells centered at $(380, 300)$ and the target area, filled with the attractive odour, as a square of 40×40 cells centered at $(256, 256)$. Then, we seek to find the best parameters that optimise the swarming towards the target area, thus we employ an evolutionary strategy (**ES**) which is able to find optimal values for this task. An ES is a type of evolutionary algorithm where the parameter values of a particular solution are evaluated in terms of how well they perform with respect to some metric. In our case, we select the metric defined in Eq. 7 as the fitness of a solution. The set of solutions, after evaluation, is sorted, then cross-over and mutation are applied and finally the top-λ solutions are kept for the next generation, while the top-m solutions, called the elites, are always kept. In particular, we leverage the self-adaptative ES (**SA-ES**) as in [8]. All meta-parameters are listed in Table 1. One of the most important aspects of such a strategy is the selection of reasonable bounds for the parameter values: the broader, the less likely the numerical simulation is able to converge; the narrower, the harder it is to find an optimum. Based on these numerical limitations, we choose to iteratively apply SA-ES, broadening the range of the parameters and augmenting the initial number of solutions until the improvement of the best solution is negligible. We begin by using the values in Table 2 and define the lower (l_i) and upper (u_i) bounds of each parameter i as:

$$l_i = \frac{t_i}{\mu} \tag{8}$$

$$u_i = t_i \mu \tag{9}$$

where t_i is the value of the parameter in Table 2 and μ is an integer. The strategy employs iteratively higher values of μ following an exponential trend where the power is 2 and with an initial value of 2.

Table 1. Meta-parameters for the SA-ES leveraged for the optimisation process as in [8].

Meta-parameter	Value
maximum number of generations	10
population size	10
mutation rate	0.1
elitism	3
minimum number of offspring λ_{min}	2
maximum number of offspring λ_{max}	100
mutation boost factor c	2
memory of previous epochs m	10

4 Results

We first establish a baseline behaviour, where chemotaxis is driven solely by the diffusion of the attractive odour placed in the center of the environment, thus solving Eqs. 1, 2 with the initial parameters listed in Table 2, with $\rho_0 = 1.2 \times 10^8$. Numerical solutions of ρ without pheromone production are shown in the top row of Fig. 1: at $t = 0$ (a), the worms are placed on the initial area; at $t = 2500$ (b) the worms have sensed the diffused odour and begun to aggregate in the target area; finally, at $t = 5000$ (c), the chemotactic index of Eq. 7 gives $c \approx 0.52$. This indicates that the worms first diffuse around their initial area, until the odour has diffused enough for them to sense it; then they initiate movement to this source, while still partially diffusing. However, the value reached by the chemotactic index with these parameters is well below any optimum, as they are taken from [1] and represent basic $L1$ worms. Since it is possible to genetically alter the synaptic connections of *C. elegans* and thus modify their natural behaviour, we propose to optimise the parameters of

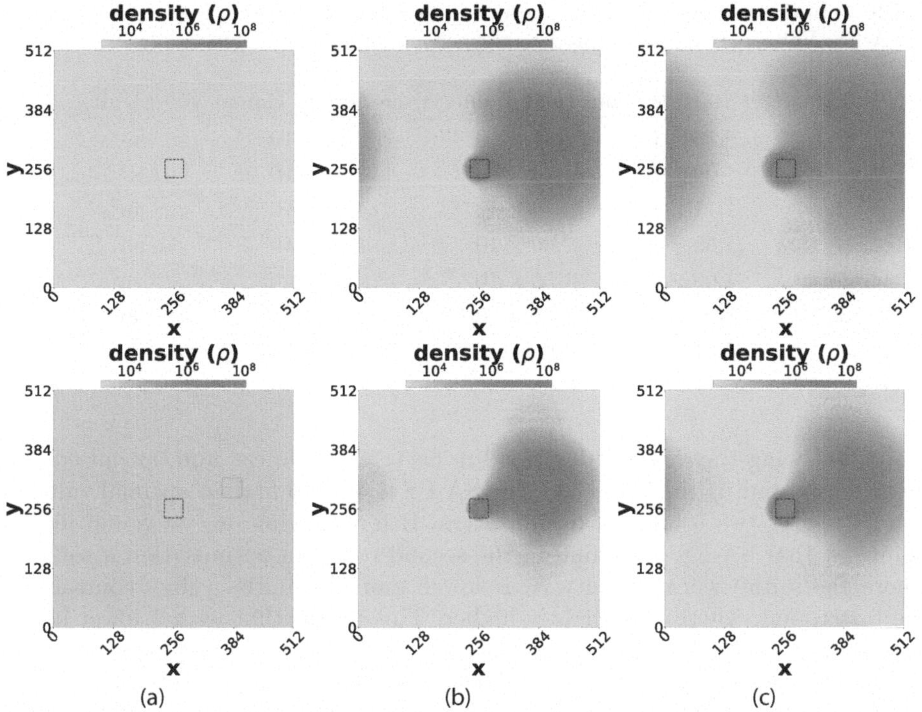

Fig. 1. Heatmap representing the worm density ρ at $t = 0$ (a), $t = 2500$ (b) and $t = 5000$ (c) when an attractive odour is placed in the center of the environment without pheromone release (top) and with pheromones (bottom). Initial worm density: $\rho = 120 \times 10^6 \text{worm/m}^2$. The heatmap is normalised with a logarithmic scale with a cut-off for values below 10^3.

the models in order to achieve ideal swarming, such that all worms reach the target area. Hence, we can design worms by employing the SA-ES as described in Sect. 3, which aims at maximising the density in the target area by evolving a population of individuals each possessing a genome composed of the model parameters. The evolved parameters are then fed to the numerical solver and the final relative density in the target area is evaluated. The evolutionary strategy leverages the hyperparameters in Table 1. After two rounds of optimisation, we find a value of $c \approx 0.996$ with the parameters as in Table 2. The latter shows as well the intermediate optimisation step parameter values and its fitness.

The same method is applied to the model featuring the self-produced pheromones composed of Eqs. 1, 2 and 4, thus with the potential V as in Eq. 6. The bottom row of Fig. 1 shows the numerical solution of ρ: at $t = 0$ (a) the worms are placed in the initial square; at $t = 2500$ (b) they begin to diffuse and at $t = 5000$ (c) the chemotactic index of Eq. 7 has a value of $c \approx 0.63$. When pheromones production is enabled, the worm density visually appears less sparse and results in a higher chemotactic index. We then optimise the pheromone

Table 2. Initial and optimised parameters in the odour model after one and two rounds of SA-ES. Values are approximated to the second significant digit.

Parameter	Starting Value	First Round Value	Second Round Value	Unit
σ	5.56×10^{-10}	3.35×10^{-10}	2.52×10^{-10}	m^2s^{-1}
γ	1.00×10^{-4}	1.74×10^{-4}	1.40×10^{-4}	s^{-1}
β	1.11×10^{-8}	2.19×10^{-8}	4.39×10^{-8}	m^2s^{-1}
α	1.50×10^{6}	7.60×10^{5}	1.25×10^{6}	m^{-2}
D	1.00×10^{-9}	1.56×10^{-9}	3.16×10^{-9}	m^2s^{-1}
ρ_0	1.2×10^{8}	1.08×10^{8}	4.33×10^{8}	m^{-2}
c	0.52	0.9	0.996	

model following the method described in Sect. 3. The initial and optimised values are reported in Table 3, where the SA-ES is able to find an optimal value of $c = 0.98$ after two rounds of optimisation. Before moving on, we would like to point out that SA-ES has found in the second round of optimisation a solution where the initial worm density ρ_0 is lower than the initial value, contrary to the first round, where it is slightly higher. This means that with a lower initial density, we are able to achieve a higher degree of swarming, which does not seem to be the case when pheromone production is disabled, as per Table 2.

With these parameters, we now seek to assess the performance of the model with and without self-produced pheromones in different environmental setups. First, we begin by fixing two areas of 40×40 cells containing the attractive odour, namely A and B, with the same chemical properties. Target A is centered in the cell $(380, 300)$, while target B is centered at $(b_0, 300)$, with $b_0 \in [20, 256]$. In our experiments, we moved b_0 20 cells at a time. The worms are placed in a 40×40 cells area in the center of the environment $O = (256, 256)$. In particular, we find that the distance of O from the two targets influences the target to where the worms aggregate: if b_0 is placed in such a way that B and A are equidistant from O, then we expect that an equal number of worms will aggregate on each target. In Fig. 2 we show the final density distribution when A and B are equidistant from the center, thus with $b_0 = 132$, both with and without pheromones using the starting parameter values. The chemotactic indices of A and B are roughly equal, suggesting that when the distance of the two odour sources is equal, the attraction is equal, thus divides the population in two, as can be seen by the value of c inside the two target areas: $c_A \approx c_B$. Distance is an important factor as well as the diffusion of the chemical odour: this has been studied *in-vitro* in [13], where the authors find that a stronger diffusion constant of the attractive odour is correlated to a stronger chemotactic response. To test this idea, we decided to run numerical simulations across a range of diffusion constants D from 1.12 to $4.47 \times 10^{-9} m^2s^{-1}$ while varying b_0 as before. This range has been chosen as to accommodate the optima of both the pheromone and the pheromone-deprived models (blue cell in Tables 2 and 3). The key observation is that worms are performing a collective decision-making task: their collective swarming towards

Table 3. Initial and optimised parameters in the pheromone model after one and two rounds of SA-ES. Values are approximated to the second significant digit.

Parameter	Starting Value	First Round Value	Second Round Value	Unit
σ	5.56×10^{-10}	1.04×10^{-9}	2.74×10^{-10}	m^2s^{-1}
γ	1.00×10^{-4}	1.38×10^{-4}	8.28×10^{-5}	s^{-1}
β	1.11×10^{-8}	1.83×10^{-8}	2.82×10^{-8}	m^2s^{-1}
α	1.50×10^{6}	7.60×10^{5}	1.32×10^{6}	m^{-2}
D	1.00×10^{-9}	1.83×10^{-9}	2.23×10^{-9}	m^2s^{-1}
ρ_0	1.2×10^{8}	1.58×10^{8}	5.54×10^{7}	m^{-2}
β_a	4.111×10^{-10}	6.59×10^{-10}	2.82×10^{-10}	m^2s^{-1}
α_a	1.5×10^{6}	1.8×10^{6}	1.6×10^{6}	m^{-2}
γ_a	10^{-3}	1.38×10^{-3}	2.7×10^{-3}	s^{-1}
s_a	10^{3}	5.05×10^{2}	3.5×10^{3}	$m^{-2}s^{-1}$
D_a	10^{-10}	1.56×10^{-10}	2.7×10^{-10}	m^2s^{-1}
β_r	-1.111×10^{-13}	-9×10^{-13}	-3.1×10^{-12}	m^2s^{-1}
α_r	1.5×10^{6}	2.87×10^{6}	3.6×10^{6}	m^{-2}
γ_r	10^{-3}	1.67×10^{-3}	1.1×10^{-3}	s^{-1}
s_r	10	18.99	22	$m^{-2}s^{-1}$
D_r	10^{-10}	1.04×10^{-10}	3.8×10^{-10}	m^2s^{-1}
scale	2	3.21	5.56	
ρ_{max}	2.8×10^{8}	4.58×10^{8}	10^{9}	m^{-2}
cushion	2×10^{8}	2.95×10^{8}	6×10^{7}	m^{-2}
c	0.63	0.85	0.98	

A or towards B corresponds to them making a decision, and if a sufficient number of them reaches one of the two targets, we can say that they have decided that to be the best target. In particular, this resembles a best-of-2 problem, where the swarm is presented with two options and they need to agree on the best option in terms of its quality. Imposing the distances from O and the diffusion constants of A and B to be equal, we incur in a symmetry breaking problem, since these two properties can represent the quality of the two options. Hence, we arbitrarily impose a quorum $\theta = 0.7$, which represents the minimum relative density ρ in either of the two spots, which is equivalent to the metric in Eq. 7 applied to target A (c_A) and B (c_B), such that a decision has been made. In other words, when $c_A > \theta$, a decision has been made for option A, while when $c_B > \theta$, a decision has been made for option B. When neither of these conditions is met, we define the outcome of the decision-making process as a decision deadlock.

Figure 3 shows the phase separation diagram for the decision-making process initiated by the swarming of the simulated worm densities when pheromone production is disabled (Fig 3a) and when it is enabled (Fig 3b). The blue region corresponds to an agreement on option A ($c_A > \theta$), while the green region cor-

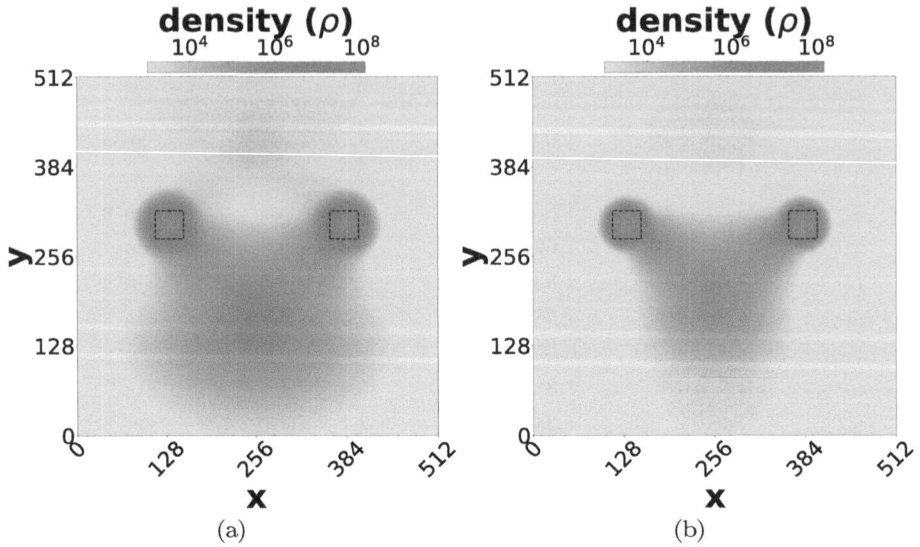

Fig. 2. Heatmap representing the worm density ρ at $t = 5000$ when an attractive odour is placed in two square spots equidistant from the center, where the worms are initially placed, without pheromone release (a) and with pheromones (b). In (a) $c_A = 0.40$ and $c_B = 0.37$, while in (b) $c_A = 0.45$ and $c_B = 0.41$. Initial worm density: $\rho = 120 \times 10^6 \text{worm/m}^2$. The heatmap is normalised with a logarithmic scale with a cut-off for values below 10^3.

responds to an agreement on option B ($c_B > \theta$). The yellow region corresponds to a decision deadlock. We can observe that when B is equidistant from A, since the two spots have the same chemical properties, neither the pheromone-enabled nor the pheromone-deprived system is able to break the symmetry. This is in line with *in-vitro* results from [12], where the authors find that in the presence of two attractive sources placed at the same distance from the initial position of the worms, the chemotactic index is fixed at around 0.5. In the case of B being closer to the origin than A (lower half of Fig. 3), there tends to be an agreement on B, while the opposite is true for the case of A being closer to the origin than B (upper half of Fig. 3). Indeed, distance is a key factor influencing the chemotactic preference of optimised *C. elegans* since the odour that is able to reach them faster will be targeted by a greater proportion of worms. This concept can also be seen by the tendency to incur in a decision deadlock when the diffusion constant of both odours increases, leading more frequently to a deadlock. In particular, the proportion of points resulting in a deadlock for Fig. 3a is 45.83%, while for Fig. 3b it is 28.33%. The overall lower deadlock proportion suggests that the presence of pheromones contributes greatly to the symmetry-breaking capabilities of the swarm.

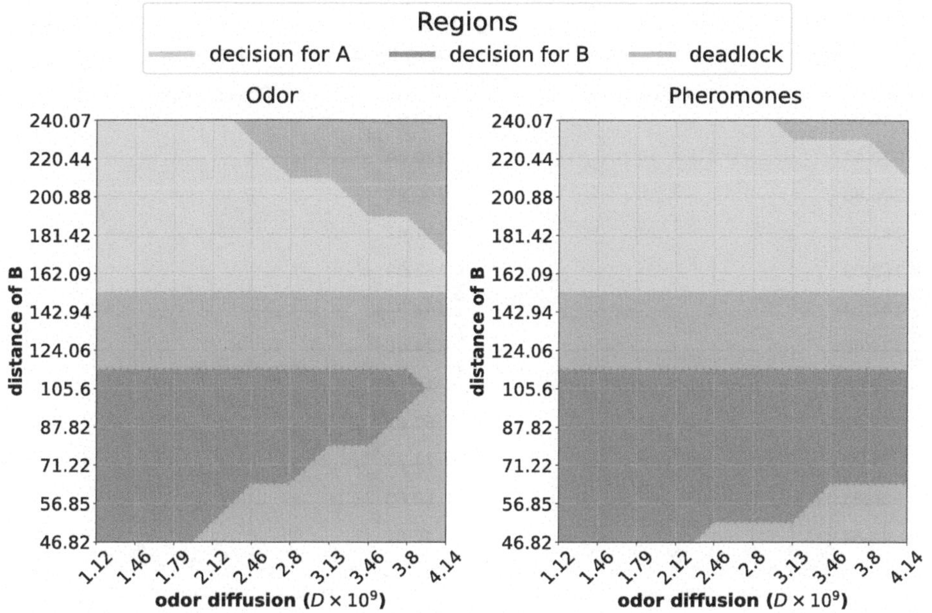

(a) (b)

Fig. 3. Phase separation diagrams for the decision-making process when two equivalent odour sources are present, namely A, whose position is fixed, and B, whose position is varied, without (a) and with pheromones (b). Each cell in the grid represents the numerical integration of the corresponding system (with/without pheromones), with a specific diffusion constant of the odour and a distance of B from the center in units each of length $\frac{l}{N}$. Regions in blue correspond to $c_A > \theta$, while regions in green correspond to $c_B > \theta$ and the regions in yellow to a decision deadlock.

Figure 4 shows the phase separation diagram for the decision-making process when target A has a fixed position and a fixed diffusion constant, while the position and diffusion constant of B are varied. In particular, Fig. 4a corresponds to the case when no pheromones are produced and Fig. 4b to the case when pheromones are produced. For both cases, a vertical line shows the corresponding value for the diffusion constant of A, as per Tables 2 and 3. We can observe that when the diffusion constant of B is close to the diffusion constant of A (D_A), both phase separation diagrams around the vertical line of Fig. 4 resemble the corresponding phase diagram of Fig. 3. For example, the diagram in Fig. 4a around D_A shows a decision has been made for the target that is closer to the center, with some deadlocks due to the relatively high diffusion constant, while it incurs in a deadlock when the two targets are equidistant from the center. These observations are consistent with those for the diagram in Fig. 3a. Analogous observations can be made for the case when pheromones are present (Figs. 4b, 4b). Moreover, we notice that pheromone production is disabled, the decision for A is made mostly when A is closer to the origin than B, while

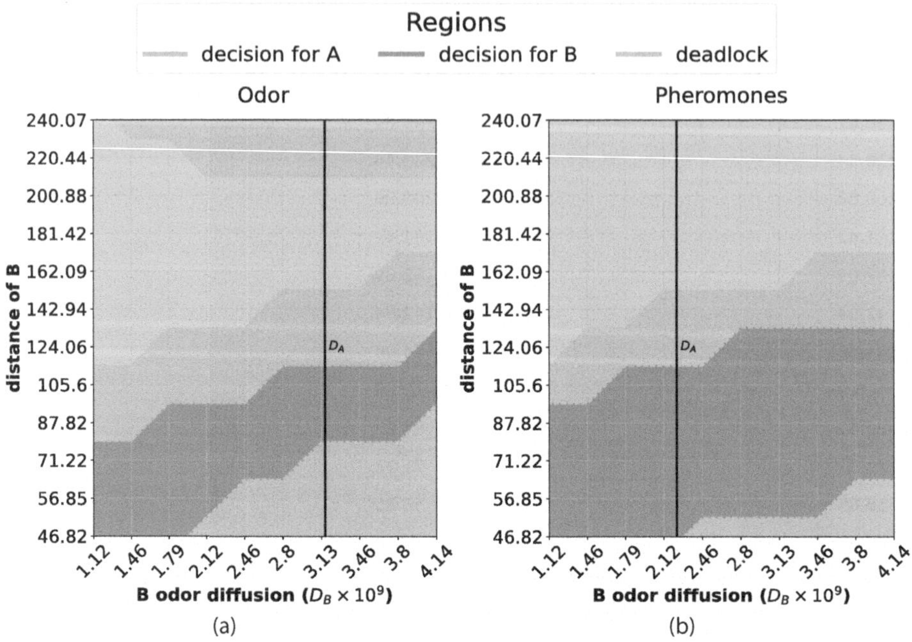

Fig. 4. Phase separation diagrams for the decision-making process when two odour sources are present, namely A, whose position and diffusion constant (D_A) are fixed, and B, whose position and diffusion constant are varied, without (a) and with pheromones (b). Each cell in the grid represents the numerical integration of the corresponding system, with a specific diffusion constant of the odour and a distance of B from the center in units each of length $\frac{l}{N}$. Regions in blue correspond to $c_A > \theta$, while regions in red correspond to $c_B > \theta$ and the regions in yellow to a decision deadlock.

the decision for B is located in a region following a linear trend between the distance of B from the origin and its diffusion constant. The proportion of points resulting in a decision deadlock is 45.83% when no pheromones are present and 20.0% when pheromones are present. The case with pheromones has the same trends, however it features a lower deadlock area, reinforcing the suggestion that pheromones play a crucial role in the symmetry-breaking properties of the system.

5 Conclusion

In this paper, we demonstrate how the swarming guided by chemotaxis of *C. elegans* resembles a collective decision-making process in the best-of-n problem. We consider a macroscopic Keller-Segel model from the literature [1] which we modify to account for the presence of one or more attractive odours and which allows the worms to produce attractive and repulsive pheromones. We thus consider two cases: when the only source of attraction is given by the attractive source(s) and when pheromone production is enabled as well. The parameters governing the models with and without pheromones are optimised through an evolutionary strategy with respect to the same swarming task, where our desired behaviour is for all the worms to reach the designated target area, containing the attractive odour. Then, those parameters are used to test the ability of the swarm in making a decision within a best-of-2 problem. In particular, the two options are given by two target areas each containing an odour source, namely A and B. In a first experiment we vary the distance of target B from the origin, while keeping the distance of A fixed. Furthermore, we designed an experiment where in addition to the position, we also fix the diffusion constant of A and vary both the distance and the diffusion constant of B. The resulting worm densities were then categorised into outcomes favoring target A or B, where a given quorum is able to reach one of the two targets, and the inability to favour either, which we define as a decision deadlock. Our findings imply that the presence of self-produced pheromones plays a pivotal role in the capabilities of the swarm to break decision deadlocks. This difference can be explained by the biological phenomenon that made the worms evolve to display pheromones: as in any organism, communication between organisms is an important factor regulating the way they behave. Through the optimisation of the parameters representing the properties of these chemicals, we show that this type of communication, a type of stigmergy, allows worms to choose an odour source over another one more robustly. Future work should be devoted to a more accurate analysis of the two different types of behaviours: from a preliminary analysis we find that in the model without pheromones, worms are able to reach the desired quorum faster than the model with pheromones, although the latter is able to perform better in terms of symmetry-breaking. This leads to the conclusion that a trade-off between speed and accuracy exists even in swarms of *C. elegans*. One possible direction for future work could be the analysis of heterogeneous swarms, composed of both worms that produce pheromones and of worms that do not. However, the current model may be inadequate for such a task given the ever-increasing amount of equations that are needed to be solved. Hence, it would be more adequate to move to an agent-based model, leveraging parallelisation and allowing for different parameter values within the same simulation.

Acknowledgments. The BABots project has received funding from the Horizon Europe, PathFinder European Innovation Council Work Programme under grant agreement No 101098722. Views and opinions expressed are however those of the authors only and do not necessarily reflect those of the European Union or European Innovation Council and SMEs Executive Agency (EISMEA). Neither the European Union nor the granting authority can be held responsible for them.

Disclosure of Interests. The authors have no competing interests to declare that are relevant to the content of this article.

References

1. Avery, L., Ingalls, B., Dumur, C., Artyukhin, A.: A keller-segel model for c elegans l1 aggregation. PLOS Comput. Bio. **17**(7), 1–25 (2021). https://doi.org/10.1371/journal.pcbi.1009231
2. Berdahl, A., Torney, C.J., Ioannou, C.C., Faria, J.J., Couzin, I.D.: Emergent sensing of complex environments by mobile animal groups. Science **339**(6119), 574–576 (2013). https://doi.org/10.1126/science.1225883. https://www.science.org/doi/abs/10.1126/science.1225883
3. Chen, Y., Ferrell, J.E.: C. elegans colony formation as a condensation phenomenon. Nat. Commun. **12**(1), 4947 (2021). https://doi.org/10.1038/s41467-021-25244-9
4. Demir, E., Yaman, Y.I., Basaran, M., Kocabas, A.: Dynamics of pattern formation and emergence of swarming in caenorhabditis elegans. Elife **9**, e52781 (2020)
5. Ding, S.S., Schumacher, L.J., Javer, A.E., Endres, R.G., Brown, A.E.X.: Shared behavioral mechanisms underlie c. elegans aggregation and swarming. eLife **8** (2019 2019). https://www.proquest.com/scholarly-journals/shared-behavioral-mechanisms-underlie-i-c-elegans/docview/2233053764/se-2
6. Emlen, J.T.: Flocking behavior in birds. Auk **69**(2), 160–170 (1952). http://www.jstor.org/stable/4081266
7. Georgopoulou, D.G., King, A.J., Brown, R.M., Fürtbauer, I.: Emergence and repeatability of leadership and coordinated motion in fish shoals. Behav. Ecol. **33**(1), 47–54 (2021). https://doi.org/10.1093/beheco/arab108
8. Hubermont, A., Vellinger, A., Antonic, N., Tuci, E.: Feature selection for multi-label classification in predictive maintenance (2024)
9. Keller, E.F., Segel, L.A.: Initiation of slime mold aggregation viewed as an instability. J. Theor. Biol. **26**(3), 399–415 (1970)
10. King, A.J., Sueur, C.: Where next? Group coordination and collective decision making by primates. Int. J. Primatol. **32**(6), 1245–1267 (2011). https://doi.org/10.1007/s10764-011-9526-7
11. Matsuura, T., Endo, S., Iwamoto, R., Takahashi, H., Ichinose, M.: Developmental changes in chemotactic response and choice of two attractants, sodium acetate and diacetyl, in the nematode caenorhabditis elegans. Comparat. Biochem. Physiol. Part A: Molec. Integrat. Physiol. **147**(4), 920–927 (2007). https://doi.org/10.1016/j.cbpa.2007.02.023. https://www.sciencedirect.com/science/article/pii/S1095643307008732
12. Matsuura, T., Oikawa, T., Wakabayashi, T., Shingai, R.: Effect of simultaneous presentation of multiple attractants on chemotactic response of the nematode caenorhabditis elegans. Neurosci. Res. **48**(4), 419–429 (2004)
13. Pandya, D., Blanar, C., Smith, R., Haskell, E.: Modeling and simulation of caenorhabditis elegans chemotaxis in response to a dynamic engineered bacteria. In: Proceedings - 29th European Conference on Modelling and Simulation, ECMS 2015 pp. 100–106 (2015). https://doi.org/10.7148/2015-0100
14. Phillips, B.R., Quinn, J.A., Goldfine, H.: Random motility of swimming bacteria: Single cells compared to cell populations. Aiche J. **40**, 334–348 (1994). https://api.semanticscholar.org/CorpusID:95593501
15. Queirós, L., Marques, C., Pereira, J.L., Gonçalves, F., Aschner, M., Pereira, P.: Overview of chemotaxis behavior assays in caenorhabditis elegans. Curr. Protoc. **1**(5), e120 (2021)
16. Reebs, S.: Time-place learning in golden shiners (pisces: Cyprinidae). Behav. Process. **36**(3), 253–262 (1996). https://doi.org/10.1016/0376-6357(96)88023-5. https://www.sciencedirect.com/science/article/pii/0376635796880235

17. Smith, J.E., Estrada, J.R., Richards, H.R., Dawes, S.E., Mitsos, K., Holekamp, K.E.: Collective movements, leadership and consensus costs at reunions in spotted hyaenas. Animal Behav. **105**, 187–200 (2015). https://doi.org/10.1016/j.anbehav.2015.04.023. https://www.sciencedirect.com/science/article/pii/S0003347215001633
18. Sugi, T., Ito, H., Nishimura, M., Nagai, K.H.: C. elegans collectively forms dynamical networks. Nat. Commun. **10**(1), 683 (2019)
19. Tyson, R., Lubkin, S.R., Murray, J.D.: Model and analysis of chemotactic bacterial patterns in a liquid medium. J. Math. Biol. **38**(4), 359–375 (1999)
20. Vellinger, A., Antonic, N., Tuci, E.: Hybrid modelling of stimuli-based and density-induced aggregation in c. elegans. In: Artificial Life Conference Proceedings, vol. 36, p. 41 (2024)
21. Ward, S.: Chemotaxis by the nematode <i>caenorhabditis elegans: </i> identification of attractants and analysis of the response by use of mutants. Proc. Natl. Acad. Sci. **70**(3), 817–821 (1973). https://doi.org/10.1073/pnas.70.3.817. https://www.pnas.org/doi/abs/10.1073/pnas.70.3.817
22. White, J.G., Southgate, E., Thomson, J.N., Brenner, S., et al.: The structure of the nervous system of the nematode caenorhabditis elegans. Phil. Trans. R. Soc. Lond. B Biol. Sci. **314**(1165), 1–340 (1986)
23. Yuan, J., Raizen, D.M., Bau, H.H.: Gait synchronization in caenorhabditis elegans. Proc. Natl. Acad. Sci. USA **111**(19), 6865–6870 (2014)

Factors Impacting Landscape Ruggedness in Control Problems: A Case Study

Michel El Saliby[1], Eric Medvet[1(✉)], Giorgia Nadizar[2], Erica Salvato[1], and Sarah L. Thomson[3]

[1] DIA, University of Trieste, Trieste, Italy
michel.elsaliby@studenti.units.it, emedvet@units.it, erica.salvato@dia.units.it

[2] MIGe, University of Trieste, Trieste, Italy
giorgia.nadizar@phd.units.it

[3] SCEBE, Edinburgh Napier University, Edinburgh, UK
s.thomson4@napier.ac.uk

Abstract. Understanding fitness landscapes in evolutionary robotics (ER) can provide valuable insights into the considered robotic problems as well as into the strategies found by evolutionary algorithms (EAs) to address them, ultimately guiding practitioners toward better design choices. However, most studies on fitness landscape analysis have been conducted on theoretical benchmarks, lacking direct relevance to practical robotics applications. This work aims to bridge this gap by (1) gathering a variety of measures to assess the *ruggedness* of a fitness landscape, (2) validating them on a simple benchmark problem with a known and tunable fitness landscape, and (3) applying these measures to a continuous control problem—a robotic navigation task. Using a highly customisable task, we investigate how various factors, including environmental conditions (*i.e.*, the arena configuration), agent perception (*i.e.*, the robot sensors), controller design (*i.e.*, the structure of the artificial neural network controlling the robot), and fitness shaping (*i.e.*, how the robot is rewarded for its behaviour) influence the ruggedness of the fitness landscape. Our findings suggest that simple measures can be sufficiently informative of the ruggedness of a given fitness landscape. Regarding the considered factors, we find that the ruggedness is primarily affected by the fitness shaping, followed by the controller features, while other factors tend to have a minor impact.

1 Introduction and Related Works

The notion of evolution is simple, but evolutionary *fitness landscapes* can be complex; this is particularly true in the context of robotics. A fitness landscape [16] is a mathematical object capturing the interplay between an algorithm and a problem. The act of investigating the anatomy of these objects is called fitness landscape analysis (FLA), a field which has brought insight into search difficulty across several domains in both discrete and continuous optimisation [8].

T. Carletti et al. (Eds.): WIVACE 2024, CCIS 2532, pp. 70–83, 2025.
https://doi.org/10.1007/978-3-031-93631-9_6

Despite consistent and increasing research interest in evolutionary robotics (ER) [1,6,7], there have been comparatively few works analysing fitness landscapes in this domain. This might be because the nature of ER presents complications: system complexity can come from many sources, such as the controller, the body, the environment, or from interactions of multiple agents. It is not clear how these factors affect the landscape topology. Additionally, in ER the search space is typically vast and contains many not feasible robots. Another possible consideration is that—unlike most other evolutionary optimisation contexts—the goal in ER is not necessarily to find a global optimum. It is not clear whether FLA methods from other domains are even useful in ER. It follows that the existing literature for fitness landscapes in ER, particularly with respect to the notion of *ruggedness*, seems scattered. There is a lack of agreement or consistency when looking at the literature as a whole: how to define and measure ruggedness, the nature of the task and environment in the experiments, and the findings surrounding this landscape phenomenon.

Ruggedness is a popular notion in FLA across many optimisation problems; although exact definitions can vary, it generally refers to the manifestation of local optima in the space, namely their cardinality and their distribution [10]. One of the most well-established ways to compute ruggedness on a landscape is with the related notions of *autocorrelation* and *correlation length* [23]. For these, random walks are conducted and the encountered fitness values are tracked. The autocorrelation captures the correlation between fitness values of solutions on the walk at different distances; the correlation length is a single value, and is defined as the largest distance at which fitness values on the walk are still correlated (with respect to a defined minimum threshold). Larger values for both the autocorrelation series and the correlation length scalar indicate a less *rugged* landscape. Another set of measurements for ruggedness are information-theoretic in nature; they are the *information content*, *partial information content*, and *information stability* [21]. For these methods, random walks are conducted and the associated time series of fitness values are logged. After that, a string is constructed which represents the walk as a series of symbols. There are three symbols in the alphabet: one to represent increasing fitness between solutions on the walk, one to represent deteriorating fitness, and the last to capture equal fitness. The three information-theoretic measurements quantify ruggedness from this string of symbols—for example, the *information content* relates to the diversity of patterns in this string.

In the field of ER, a study on joint body-brain evolution in 2001 sought to compare two search spaces with known performance differences through the lens of landscape ruggedness measures [18]. In doing so, the authors computed autocorrelation, correlation length, and the information stability (all of which we have now introduced). They found that these metrics did not distinguish between the two spaces to any significant degree. Although the information stability was not helpful in this particular case, the related metric named *information content* (sometimes referred to as *entropy* in the literature) seems to have gained some popularity. This measure was originally proposed in the context of combinatorial optimisation; Malan and Engelbrecht later adapted the *information content* for continuous spaces, proposing it as a standalone metric for ruggedness [9]. This measurement has recently been used for evolving controllers in evolutionary robotics [15]—in this case, the authors found that the measure contained useful information for configuring the evolutionary strategy which was used to evolve the controller. Another version of the information content was introduced for continuous spaces, where the distance between solutions is considered in the calculation [13]. We do not consider this version as suitable for our use case: we are seeking metrics which can be computed from a fitness series alone (this metric requires each genotype to be tracked alongside fitness); additionally, the notion of distance for our neural network genotype space does not seem naturally suited.

Other approaches have been proposed for quantifying ruggedness in robotics: for example, one study measured the change in (dynamic) fitness when the robot began from different starting angles, and computed the fractal dimension of this relationship when plotted as a line graph [4]. Another work defined ruggedness directly from the coupling characteristics of the robot system itself [19]. We do not take this approach, as we would instead prefer to measure ruggedness empirically through search dynamics. Another method used to estimate ruggedness in a robotics context used *fitness evolvability portraits* [5]. The authors found that this approach showed a huge amount of ruggedness and lack of correlation in the landscape. We therefore think that perhaps this method is too coarse and may not lead to useful information about landscape navigability.

In this work, we take steps towards understanding the phenomenon of landscape *ruggedness* in carefully-constructed ER scenarios. To this end, we systematically vary one factor (for example: the number of robot sensors) at a time and observe the change in ruggedness with several different measurements. In light of the ER literature just discussed we select the *information content* measure to quantify ruggedness; additionally, because it is not well understood which metric to use, we choose four other measurements to explore—including a metric from the field of physical topography. Our ultimate aim is to answer the question: in which manner do factors in simple robotic control problems influence fitness landscape ruggedness?

2 Methods

A *fitness landscape* [16] is a triplet (S, N, f), where S is a *search space*, $N : S \to \mathcal{P}(S)$ is a *neighbourhood function* that assigns a set of adjacent solutions $N(s)$ to every solution $s \in S$, and $f : S \to \mathbb{R}$ is a *fitness function* that provides a mapping from solution to associated fitness. That fitness can be thought of as solution height within the landscape metaphor.

In this work, we consider the case where the search space is $\mathbb{R}^p$. Since for $\mathbb{R}^p$ there is not a natural definition of adjacency, *i.e.*, of N, we propose a method for measuring the ruggedness of a fitness landscape which requires only the search space (here $\mathbb{R}^p$) and f, not N.

In brief, our method consists of two steps. First, we collect a number of segments, *i.e.*, sequences of points in the search space, and map them, using f, to sequences of values in $\mathbb{R}$, which we call fitness profiles. Second, we measure the ruggedness of each fitness profile and associate the mean ruggedness of profiles with the fitness landscape. We detail the two steps below.

2.1 Fitness Profiles Construction

We define a *segment* $[\boldsymbol{x}_i, \boldsymbol{x}_f]_n$ as sequence $(\boldsymbol{x}_j)_{j=1}^{j=n} \in \mathbb{R}^{pn}$ of n points in $\mathbb{R}^p$ such that $\boldsymbol{x}_j = \frac{n-j}{n-1}\boldsymbol{x}_i + \frac{j-1}{n-1}\boldsymbol{x}_f$, *i.e.*, $\boldsymbol{x}_1 = \boldsymbol{x}_i$, $\boldsymbol{x}_n = \boldsymbol{x}_f$, and points are evenly spaced. We define as *fitness profile* of a segment $(\boldsymbol{x}_j)_j$ the sequence $(f_j)_j = (f(\boldsymbol{x}_j))_j \in \mathbb{R}^n$.

Our definitions of segment and fitness profile are a special case of *random walk*. While random walks (such as those conducted in [21]) typically follow a random direction in $\mathbb{R}^p$ at every move, in our case, we instead follow a single specific direction through the space.

We adopt two alternative ways to build segments. In both cases, given an initial point $\boldsymbol{x}_i$, we select randomly a point $\boldsymbol{x}_f$ such that its Euclidean distance $\|\boldsymbol{x}_i - \boldsymbol{x}_f\|$ to $\boldsymbol{x}_i$ is r, which is also the length of the resulting segment $[\boldsymbol{x}_i, \boldsymbol{x}_f]_n$. The two alternatives differ in the way we choose $\boldsymbol{x}_i$, for which we either (a) sample this randomly within a certain hypervolume $[x_{\text{min}}, x_{\text{max}}]^p \subseteq \mathbb{R}^p$, or (b) seed $\boldsymbol{x}_i$ as a solution obtained by running differential evolution (DE). The rationale for the latter is to sample the search space in the regions close to (local) optima, which are interesting from the perspective of optimisation.

For determining r, we consider two cases: with the static sampling rate (SSR) we define $r = w_s n$, n being the number of points in the resulting segment; with the adaptive sampling rate (ASR) we define $r = w_a n \sqrt{p}$. These scenarios differ in that ASR takes into account the dimensionality p of the search space, while SSR does not. For both, the parameters w_s and w_a indicate the resolution of the segment, *i.e.*, the desired distance between samples. The rationale for ASR is to link the resolution to the expected length of a step of a walk performed with Gaussian mutation, which is one of the most commonly used mutation operators when dealing with $\mathbb{R}^p$ as search space: indeed, this expected length depends linearly on σ and $\sqrt{p}$.

For each fitness landscape, we sample a total of n_{segments} segments. When sampling the initial point $\boldsymbol{x}_i$ at random, we consider n_{segments} with different initial points. When the initial point $\boldsymbol{x}_i$ is set as a solution obtained through DE, we take a different approach: rather than performing n_{segments} distinct optimisation runs, we run DE n_{runs} times, thereby finding n_{runs} distinct initial points. Then, for each of them we sample $\frac{n_{\text{segments}}}{n_{\text{runs}}}$ final points to finally obtain a total of n_{segments} segments. This way, we obtain segments and corresponding fitness profiles which are close to (namely, centred on) local optima.

Figure 1 shows an example of fitness profiles for one of the fitness landscapes considered in the experimental analysis (see Sect. 3).

2.2 Profile Ruggedness Metrics

Given a fitness profile $(f_j)_j$, we consider the following five metrics to measure its ruggedness. They convey slightly different information about the ruggedness of the landscape. We remark that applying them to the same samples does not imply a significant growth in the computational effort, which is mostly determined by the fitness evaluation. For all the metrics, the larger the value, the more rugged the profile.

Standard deviation (SD) assesses the ruggedness as the standard deviation $\mathrm{SD}\left((f_j)_{j=1}^{j=n}\right) = \sqrt{\frac{1}{n}\sum_{j=1}^{j=n}(f_j - \bar{f})^2}$ of the fitness profile, where $\bar{f} = \frac{1}{n}\sum_{j=1}^{j=n} f_j$ is the average value of the fitness profile elements.

Information content (IC) considers the amount of information in terms of slope changes in the fitness profile [21]. For computing IC, we first build a string $(s_j)_{j=1}^{j=n-1}$ where $s_j = -1$ if $f_{j+1} - f_j < -\epsilon|\bar{f}|$, $s_j = 1$ if $f_{j+1} - f_j > \epsilon|\bar{f}|$, or $s_j = 0$ otherwise, *i.e.*, if $|f_{j+1} - f_j| \leq \epsilon|\bar{f}|$, with ϵ being a parameter, which we set at 0.05. Then we compute $\mathrm{IC}\left((f_j)_{j=1}^{j=n};\epsilon\right) = H\left((s_j)_{j=1}^{j=n-1}\right) = -\sum_{s'\neq s''} P([s's''])\log_6 P([s's''])$, where $s', s'' \in \{-1, 0, 1\}$ are symbols and $P([s's''])$ is the frequency of the substring $[s's'']$ in $(s_j)_j$. The base 6 for the log is because there are six possible substrings of two differing symbols.

First difference sign changes (FDSC) considers the number of changes of slope in the profile, as $\mathrm{FDSC}\left((f_j)_{j=1}^{j=n}\right) = \frac{1}{n-1}\sum_{j=1}^{j=n-1}\frac{1}{2}|\operatorname{sign} f_j - \operatorname{sign} f_{j+1}|$, where $\operatorname{sign} x$ is 1 if $x \geq 0$ and -1 otherwise.

Energy percentile threshold (EPT) measures the ruggedness of a fitness profile by analysing its energy in the frequency domain. For computing EPT, we first compute the energy spectrum $(e_j)_j$ by applying the fast Fourier transform (FFT) to the profile. Then, we compute the index $j_{75\,\%}$ at which 75 % of the spectrum energy accumulates, *i.e.*, $j_{75\,\%} = \max_j j$ such that $\sum_{k=1}^{k=j} e_k \leq 0.75\sum_{k=1}^{k=n} e_k$. Finally, we compute $\mathrm{EPT}\left((f_j)_{j=1}^{j=n}\right) = \frac{j_{75\,\%}}{n}$.

Topography ruggedness (TR) consists of an adaptation of the terrain ruggedness index proposed in the field of topography [17]. Originally, TR was applied to groups of data points centred on a specific location. The idea was to estimate how rough the terrain was by comparing the elevation values in that group to the central elevation point. We adapt it to our scenario by setting $\mathrm{TR}\left((f_j)_{j=1}^{j=n}\right) = \sqrt{\frac{1}{n}\sum_{j=2}^{j=n}(f_j - f_1)^2}$, *i.e.*, with the first point f_1 of the profile considered as central point. We remark that, when we use the solution obtained through an optimisation run as the initial point $\boldsymbol{x}_i$, all the $\frac{n_{\mathrm{segments}}}{n_{\mathrm{runs}}}$ segments starting from $\boldsymbol{x}_i$ share the same f_1, *i.e.*, our TR resembles very closely the corresponding metric used in topography.

2.3 Optimisation Problems

We consider two parametric optimisation problems, corresponding to many fitness landscapes: a synthetic benchmark problem, which we use to validate the overall methodology and the metrics proposed in Sect. 2.2, and a continuous control problem, often used as a benchmark in ER, in which we want to study the fitness landscape in different settings.

The "Stretchable" Rastrigin Function. We consider the Rastrigin function [12] to validate our methodology. We choose this function as it is highly multimodal with multiple local optima, which are regularly distributed in the space. We slightly alter the original formulation of the function by introducing a stretch factor γ, which controls the ruggedness of the fitness landscape acting as a sort of "zoom"—larger values of γ mean a more rugged fitness landscape. The fitness function thus becomes $f(\boldsymbol{x}) = 10p + \sum_{i=1}^{p} \left((\gamma x_i)^2 - 10\cos(2\pi\gamma x_i)\right)$, with p being the dimensionality of the search space. Figure 1 shows a portion of the fitness landscapes for two instances of the modified Rastrigin function (with $p = 2$ and $\gamma = 1$ or 5) with two segments and the corresponding fitness profiles. It is worth noting that the fitness value of this problem is dependent on p.

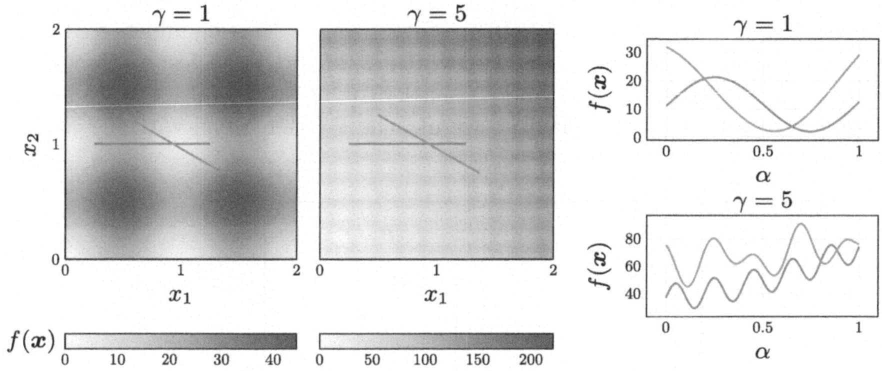

Fig. 1. The "stretchable" Rastrigin function for $\gamma = 1$ and $\gamma = 5$, plotted for $p = 2$ in the domain $[0, 2]^2$ in the form of a heat map. For the two segments (and) of length $r = 1$, the two line plots on the right show the corresponding fitness profiles. Both fitness profiles are more rugged for $\gamma = 5$ than for $\gamma = 1$.

For this problem, we vary γ to control the ruggedness of the landscape, and we vary p to assess the effectiveness of the metrics with increasingly high-dimensional search spaces: we use $\gamma \in \{1, 2, 3, 4, 5\}$ and $p \in \{2, 100, 500, 1000\}$.

Robot Navigation Task. As a control problem, we consider a 2D navigation task that has often been used, in many variants, as a benchmark problem in ER [2,3,14,22]. In this task, the simulated robotic agent has to reach a target position in an arena with obstacles (see Fig. 2): we perform discrete time simulation with time step $\delta t = 0.1$ s and duration of 60 s.

The robot is a differential-drive wheeled robot with a circular shape and two wheels on its sides. Thus it has two control outputs regulating the velocity of the two wheels, which have a maximum of $10\,\mathrm{cm\,s^{-1}}$. Regarding the input, *i.e.*, the agent perception, the robot is equipped with m proximity sensors evenly distributed around half of its circumference (the frontal part, see Fig. 2d), along with two sensors perceiving the distance and the angle to the target position. Hence, the goal of this problem is finding a control function $g : \mathbb{R}^{m+2} \to \mathbb{R}^2$, *i.e.*, a policy, that drives the agent to the goal. We implement g with a feed-forward, fully connected artificial neural network (ANN) where only the weights $\boldsymbol{\theta} \in \mathbb{R}^p$ can be subject to optimisation; that is, the search space contains weight vectors for the ANN.

We assess the impact of four kinds of factors on the ruggedness of the fitness landscape: environmental, perception, controller, and fitness function employed. The *environmental factors* regard the features of the arena. In all cases, we use a square arena of $1\,\mathrm{m} \times 1\,\mathrm{m}$ size (the radius of the robot is $5\,\mathrm{cm}$, for reference), where we either place a *Barrier* between the initial position of the robot and the target (see Fig. 2a) or we place two obstacles constituting a *Maze* that the robot needs to navigate (see Fig. 2b). At the beginning of the simulation, we place the robot at $\boldsymbol{p}^{(0)} = (p_x^{(0)}, p_y^{(0)})$, with $p_x^{(0)} \sim U([0.45, 0.55])$ and $p_y^{(0)} \sim U([0.80, 0.85])$ while the target is at $\boldsymbol{p}^\star = (0.50, 0.15)$. For both the Barrier and the Maze we vary the width l_b of the barrier and the width l_m of the obstacles: we experiment with $l_b \in \{0.2, 0.3, 0.4, 0.5, 0.6\}$ and $l_m \in \{0.45, 0.52, 0.59, 0.66, 0.73\}$. When not assessing the impact of these factors we use the Barrier arena with $l_b = 0.2$.

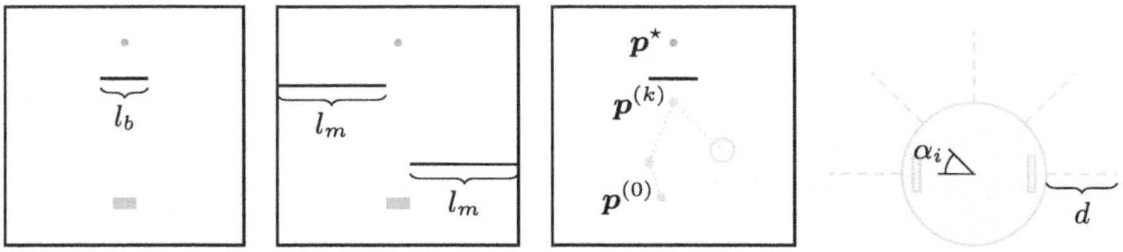

(a) Barrier (b) Maze (c) Trajectory (d) Proximity sensors

Fig. 2. (2a, 2b) Schematic of the Barrier and Maze arenas; in both, the grey area represents the random starting point of the robot, the green circle the position $\boldsymbol{p}^\star$ of the target, the black lines the boundaries and the obstacles of the arena. (2c) An example of trajectory $\left(\boldsymbol{p}^{(k)}\right)_k$. (2d) Schematic of the robot with its m proximity sensors, all with range d, each i-th one placed at angle $\alpha_i = \frac{i-1}{m-1}\pi$.

Concerning the *perception factors* we vary the *number of sensors* m and their *range of perception* d: we experiment with $m \in \{3, 5, 7, 9, 11\}$ and $d \in \{0.1, 0.25, 0.5, 0.75, 1\}$. We default their values to $m = 5$ and a range of $d = 1$. Each proximity sensor returns a normalized value, *i.e.*, it reads 1 when no obstacles are within the range and reads 0 when it is in direct contact with an obstacle.

For the *controller factors* we change the *width* and the *depth* of the ANN controlling the robot. When assessing the width we keep the depth of the ANN fixed to one hidden layer, and we change its size $\rho(m+2)$ according to different values proportional to the input size: we experiment with $\rho \in \{1,3,5,7,9\}$. Conversely, when we study the depth of the ANN we keep the size of the hidden layers fixed to the input size $m+2$ and we increase the number h of layers: we experiment with $h \in \{1,3,5,7,9\}$. In all other scenarios, *i.e.*, when assessing other factors, we keep a single hidden layer ($h=1$) with a fixed size of three times the input size ($\rho=3$). In all cases, we use tanh as the activation function in all neurons.

We consider three *fitness functions* to promote the emergence of navigation behaviour in the robot. They are all based on the robot-target distance measured during the simulation, lasting k_{final} time steps and resulting in the robot trajectory $\left(\boldsymbol{p}^{(k)}\right)_{k=1}^{k=k_{\text{final}}}$: we use the minimum distance $\min_k \left\|\boldsymbol{p}^{(k)}-\boldsymbol{p}^\star\right\|$, the average distance $\frac{1}{k_{\text{final}}}\sum_k \left\|\boldsymbol{p}^{(k)}-\boldsymbol{p}^\star\right\|$, and the final distance $\left\|\boldsymbol{p}^{(k_{\text{final}})}-\boldsymbol{p}^\star\right\|$ as fitness functions $f(\boldsymbol{\theta})$. Since all three fitness functions can be computed from the same simulation, we appraise the ruggedness of the landscape induced by each of them when varying all the previously mentioned factors.

3 Experiments and Results

We applied our method for measuring the ruggedness of a fitness landscape several times, by varying the parameters defining the method itself and those defining the two problems described in the previous sections. For all the cases, we set $w_s=0.01$ (for SSR), $w_a=0.007$ (for ASR), $n_{\text{segments}}=1000$, $n=100$, and $\epsilon=0.05$ (for the IC metric). Moreover, we set $x_{\min}=-3$ and $x_{\max}=-3$ when randomly choosing the segment initial point $\boldsymbol{x}_i$, hence sampling it from $[-3,3]^p$. We made the code for the experiments publicly available at https://github.com/elsalibymichel/2024-WIVACE-ControlFitnessLandscapeFactors/.

3.1 Methodology Validation on "stretched" Rastrigin

For the "stretched" Rastrign problem, we varied the search space dimensionality p and the stretch factor γ. We here used only random sampling for choosing the initial point.

The goals of this experiment were to (a) verify that the method is indeed able to capture the ground-truth changes in ruggedness induced by the γ parameter, (b) compare the SSR and ASR variants concerning the resolution of the segment, (c) compare the different ruggedness metrics; namely, their robustness with respect to p.

Figure 3 summarises the results of this experiment in terms of ruggedness vs. γ. It shows one row of plots for SSR and one for ASR, one column of plots for each ruggedness metric, and one line for each value of p.

The main finding is that in (almost) all the cases, *i.e.*, regardless of the ruggedness metric, p, and with both SSR and ASR, the ruggedness never

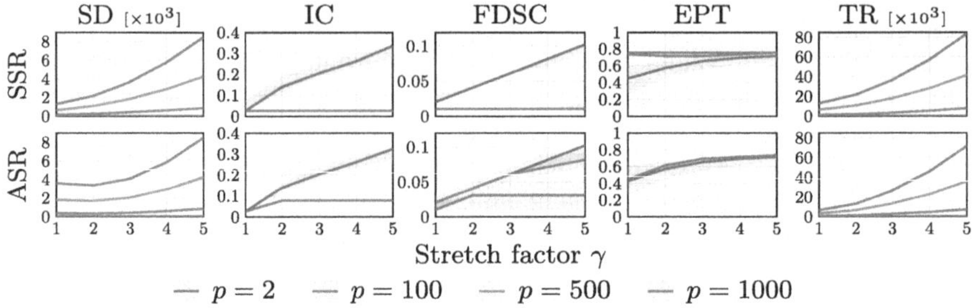

Fig. 3. Ruggedness (mean and interquartile range across $n_{\text{segments}} = 1000$ fitness profiles) on the "stretchable" Rastrigin function, one plot row for each of the variants SSR and ASR, one plot column for each ruggedness metric, one line for each value of p.

decreases with γ increasing. However, the expected effect of γ on the measured ruggedness is more visible for some combinations of the method and problem parameters.

Concerning the impact of p, Fig. 3 shows that impact of γ on the ruggedness is more evident for small values of p with IC, FDSC, and EPT. Conversely, with SD and TR, the impact is more visible for large values of p. This finding can be explained by the way the five metrics work: with SD and TR, the ruggedness value depends on the value of the fitness along the segments, which itself depends on p, for the Rastrigin function: hence, different ruggedness values for different p are not comparable. On the other hand, IC, FDSC, and EPT are scale-invariant with respect to the fitness values and the ruggedness can be compared across different problem dimensionality values. While we remark that we are here comparing the ruggedness for different γ values, taking into account the impact of p on the ruggedness metric is important when the studied factor (e.g., the number of proximity sensors for the robot problem) implicitly impacts on the problem dimensionality p. For reference, in the robot problem p goes from 72 to 632.

Concerning the difference between SSR and ASR, Fig. 3 shows that the adaptive strategy seems to mitigate the detrimental effect of p on the ruggedness metrics—we recall that we expected to see, intuitively, that the larger γ, the greater the ruggedness. This is particularly apparent for IC, FDSC, and EPT, for which ASR makes the lines for large values of p closer to the expectation.

Based on the results of this experiment, we chose to apply our methodology to the robot navigation problem only in the ASR variant.

3.2 Problem Factors and Ruggedness in Robot Navigation

For this problem, we varied (one at a time) the arena parameter (l_b for Barrier and l_m for Maze), the sensor range d, the number m of proximity sensors, and the ANN width ratio ρ and number h of inner layers (corresponding to the depth).

We repeated the experiments twice: with the initial segment point $\boldsymbol{x}_i$ chosen randomly in $[-3, 3]^p$ or with $\boldsymbol{x}_i$ being the outcome of a DE run. In the latter case, we performed $n_{\text{runs}} = 10$ runs of a simple variant of DE (commonly known as DE/rand/1 [20]), replicating the settings of [3]: we set the population size $n_{\text{pop}} = \text{NP} = 15$, the differential weight $w_{\text{diff}} = \text{F} = 0.5$, the crossover probability $p_{\text{xo}} = \text{CR} = 0.8$, and we sampled the initial population in $[-1, 1]^p$ note that along the evolution, solutions can fall outside this region of the search space. We performed the evolutionary optimisation using JGEA [11].

The goals of these experiments were to (a) understand what, if any, is the impact of the considered problem factors on the ruggedness of the fitness landscape and (b) check whether the (impact on) ruggedness is different in the neighbourhood of solutions obtained through optimisation. We remark that, differently from the case of the Rastrigin problem, we here did not have any specific expectation on the trend of ruggedness curves.

Figures 4 and 5 shows the results of these experiments, respectively for the random and optimised $\boldsymbol{x}_i$. Both figures show one plot row for each problem factor (*i.e.*, l_b, l_m, d, m, ρ, and h), one plot column for each ruggedness metric, and one line for each fitness function.

By observing the figures, it can be seen that the main factor impacting the ruggedness is the fitness function itself: the final distance results in a more rugged landscape than those associated with the average distance and the minimum distance. This is apparent in most of the cases, *i.e.*, regardless of other problem parameters and the ruggedness metric. It is slightly less apparent with the SD and TR metrics, which indeed are affected by the scale of the fitness values, which is in general different between the three fitness functions, even though they share the same theoretical domain ($\approx [0, 1]$). We believe this finding is sound: small changes in the ANN parameters $\boldsymbol{\theta}$ result in large changes in the final position (hence its distance to the target) of the robot more likely than in radical changes in the trajectory (hence the average or minimum distance). More broadly, this finding confirms the importance of selecting the proper fitness function for a given robotic task when solving it through evolutionary optimisation [2].

For what concerns the other factors, no clear and consistent impact on the ruggedness is visible: neither when randomly sampling the search space (Fig. 4), nor when sampling it close to optimised solutions (Fig. 5). Only in a few cases, a weak trend may be spotted. With the sensor range d and all the metrics and fitness functions, the shortest range results in a smooth (*i.e.*, not rugged) fitness landscape: we looked at the behaviours of some of the solutions and saw that the robots barely move, being unable to perceive the environment, hence resulting in a flat fitness landscape. With the ANN depth h and the minimum distance, some ruggedness metrics appear to show a decreasing trend: the deeper the ANN, the smoother the landscape. In general, however, the distribution of ruggedness across the 1000 fitness profiles is always large, suggesting that the fitness landscape is diverse, with some portions of the search space being rather flat and others being irregular.

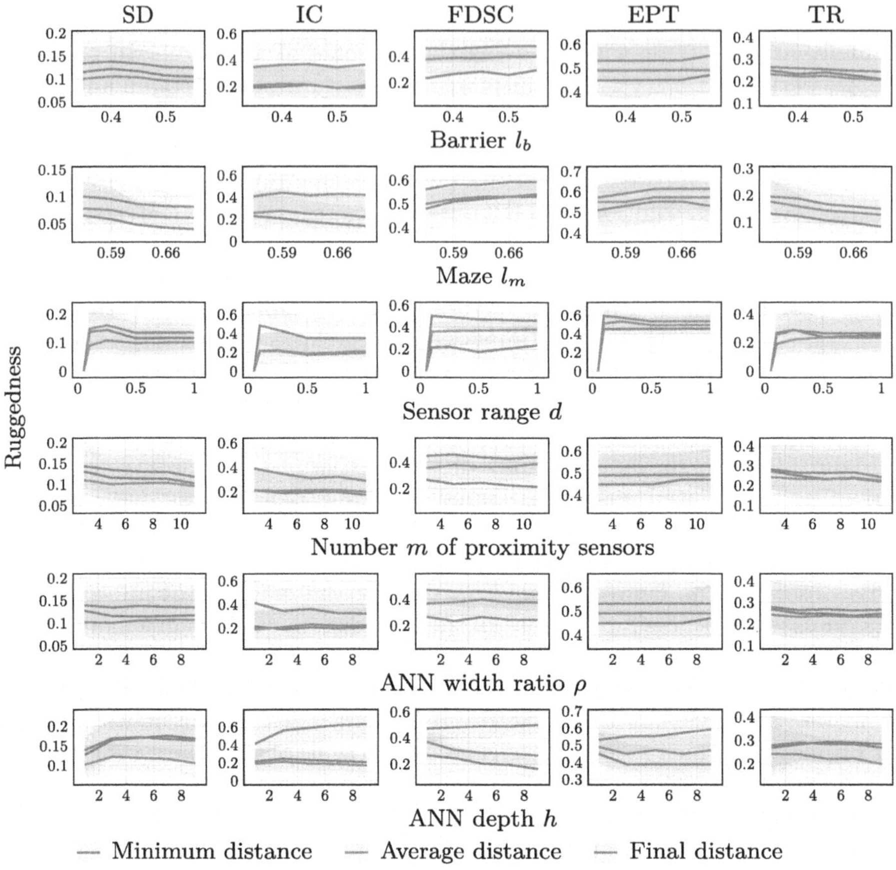

Fig. 4. Ruggedness (mean and interquartile range across $n_{\text{segments}} = 1000$ fitness profiles with ASR and random $\boldsymbol{x}_i$ sampling) on the robot navigation problem, one plot row for each problem factor, one plot column for each ruggedness metric, one line for each fitness function.

All to all, these experiments suggest that the intrinsic complexity of control problems, where the mapping between the search space (here a point $\boldsymbol{\theta} \in \mathbb{R}^p$ representing the parameters of an ANN) and the fitness space (here an aggregate of the behaviour resulting from a simulation) is very indirect, cannot be easily captured relying on the notion of fitness landscape. Further investigation is hence needed to fully understand when and how evolutionary optimisation is able to solve this kind of problems.

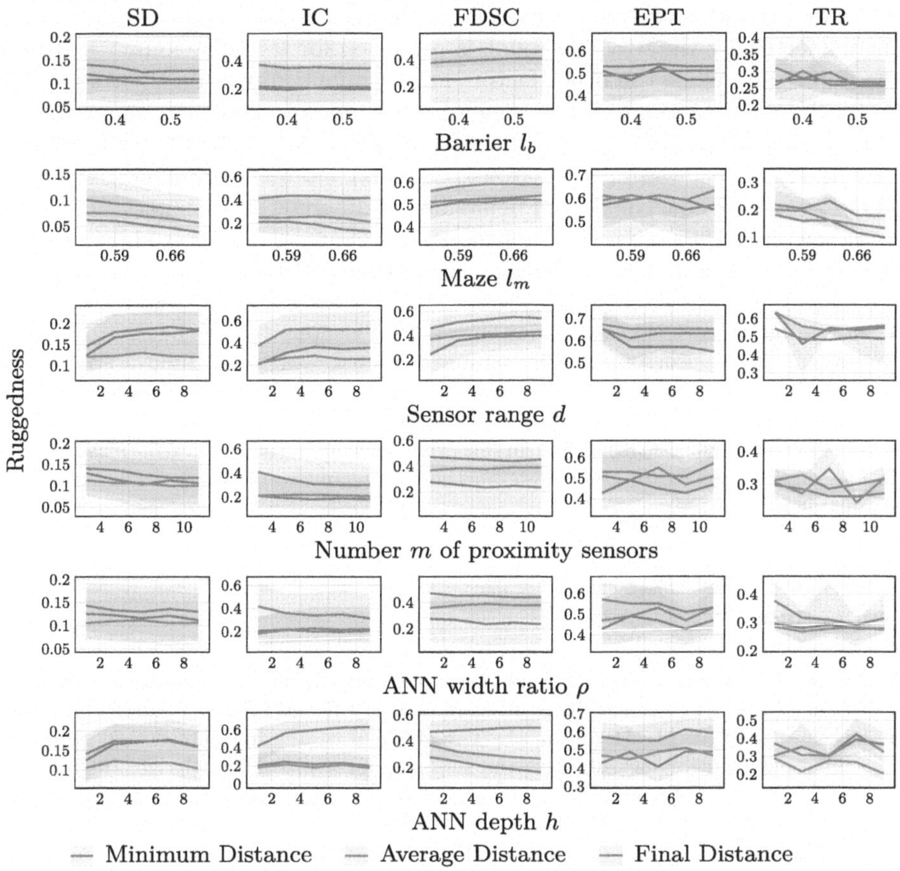

Fig. 5. Ruggedness (mean and interquartile range across $n_{\text{segments}} = 1000$ fitness profiles with ASR and $\boldsymbol{x}_i$ coming from optimisation) on the robot navigation problem, one plot row for each problem factor, one plot column for each ruggedness metric, one line for each fitness function.

4 Conclusion

In this study, we performed several experiments aimed at understanding what factors shape the fitness landscape, namely, which ones make it more or less rugged, for a simple ER problem. Our approach involved gathering a comprehensive set of five different metrics to assess fitness landscape ruggedness. We validated these measures on a benchmark problem with a known landscape and, subsequently, applied them to a more complex robotic navigation task. By systematically varying individual factors, we were able to isolate their specific effects on landscape ruggedness, attempting to highlight how each element contributes to the overall fitness landscape.

We discovered that the ruggedness of the fitness landscape is most significantly affected by the way the robot is rewarded for its behaviour, *i.e.*, by the fitness function, while the other factors have a minor impact on the ruggedness. Our work offers a more detailed view of the shape of the fitness landscape for ER problems. In particular, it confirms that the choice of the fitness function is a key design decision. Moreover, it suggests that more effort should be devoted to understanding how the search behaviour of different evolutionary algorithms (EAs), *i.e.*, how they navigate the fitness landscape, affects their ability to cope with complex optimisation problems arising from ER tasks.

Acknowledgements. This work has been partially supported by the PNRR research activities of the consortium iNEST (Interconnected North-Est Innovation Ecosystem) funded by the European Union Next-GenerationEU (Piano Nazionale di Ripresa e Resilienza (PNRR) – Missione 4, Componente 2, Investimento 1.5 – D.D. 1058 23/06/2022, ECS_00000043). This manuscript reflects only the authors' views and opinions, neither the European Union nor the European Commission can be considered responsible for them.

References

1. Auerbach, J., et al.: Robogen: robot generation through artificial evolution. In: Artificial Life Conference Proceedings, pp. 136–137, MIT Press, Cambridge (2014)
2. Divband Soorati, M., Hamann, H.: The effect of fitness function design on performance in evolutionary robotics: the influence of a priori knowledge. In: Proceedings of the 2015 Annual Conference on Genetic and Evolutionary Computation, pp. 153–160 (2015)
3. El Saliby, M., Nadizar, G., Salvato, E., Medvet, E.: Eventually, all you need is a simple evolutionary algorithm (for neuroevolution of continuous control policies). In: Proceedings of the Genetic and Evolutionary Computation Conference Companion, Association for Computing Machinery (2024)
4. Hoshino, T., Mitsumoto, D., Nagano, T.: Fractal fitness landscape and loss of robustness in evolutionary robot navigation. Auton. Robot. **5**, 199–213 (1998)
5. Katada, Y., Handa, Y.: Measurements of features of fitness landscapes on competitive co-evolutionary robotics. In: 2010 World Automation Congress, pp. 1–6, IEEE (2010)
6. Li, W., et al.: Evaluation of frameworks that combine evolution and learning to design robots in complex morphological spaces. IEEE Trans. Evol. Comput. (2023)
7. Lipson, H., Sunspiral, V., Bongard, J., Cheney, N.: On the difficulty of co-optimizing morphology and control in evolved virtual creatures. In: Artificial Life Conference Proceedings, pp. 226–233, MIT Press, Cambridge (2016)
8. Malan, K.M.: A survey of advances in landscape analysis for optimisation. Algorithms **14**(2), 40 (2021)
9. Malan, K.M., Engelbrecht, A.P.: Quantifying ruggedness of continuous landscapes using entropy. In: 2009 IEEE Congress on Evolutionary Computation, pp. 1440–1447, IEEE (2009)
10. Malan, K.M., Engelbrecht, A.P.: A survey of techniques for characterising fitness landscapes and some possible ways forward. Inf. Sci. **241**, 148–163 (2013)

11. Medvet, E., Nadizar, G., Manzoni, L.: JGEA: a modular java framework for experimenting with evolutionary computation. In: Proceedings of the Genetic and Evolutionary Computation Conference Companion, pp. 2009–2018 (2022)
12. Mühlenbein, H., Schomisch, M., Born, J.: The parallel genetic algorithm as function optimizer. Parallel Comput. **17**(6–7), 619–632 (1991)
13. Muñoz, M.A., Kirley, M., Halgamuge, S.K.: Exploratory landscape analysis of continuous space optimization problems using information content. IEEE Trans. Evol. Comput. **19**(1), 74–87 (2014)
14. Nadizar, G., , Medvet, E., Wilson, D.G.: Searching for a diversity of interpretable graph control policies. In: Proceedings of the Genetic and Evolutionary Computation Conference. Association for Computing Machinery (2024)
15. Neri, F., Turner, A.: A fitness landscape analysis approach for reinforcement learning in the control of the coupled inverted pendulum task. In: International Conference on the Applications of Evolutionary Computation (Part of EvoStar), pp. 69–85. Springer, Heidelberg (2023). https://doi.org/10.1007/978-3-031-30229-9_5
16. Pitzer, E., Affenzeller, M.: A comprehensive survey on fitness landscape analysis. In: Recent Advances in Intelligent Engineering Systems, pp. 161–191 (2012)
17. Riley, S.J., DeGloria, S.D., Elliot, R.: Index that quantifies topographic heterogeneity. intermountain J. Scie. **5**(1-4), 23–27 (1999)
18. Smith, T., Husbands, P., O'Shea, M.: Not measuring evolvability: initial investigation of an evolutionary robotics search space. In: Proceedings of the 2001 Congress on Evolutionary Computation (IEEE Cat. No. 01TH8546), vol. 1, pp. 9–16, IEEE (2001)
19. Smith, T., Philippides, A., Husbands, P., O'Shea, M.: Neutrality and ruggedness in robot landscapes. In: Proceedings of the 2002 Congress on Evolutionary Computation. CEC'02 (Cat. No. 02TH8600), vol. 2, pp. 1348–1353. IEEE (2002)
20. Storn, R., Price, K.: Differential evolution-a simple and efficient heuristic for global optimization over continuous spaces. J. Global Optim. **11**, 341–359 (1997)
21. Vassilev, V.K., Fogarty, T.C., Miller, J.F.: Information characteristics and the structure of landscapes. Evol. Comput. **8**(1), 31–60 (2000)
22. Vassiliades, V., Chatzilygeroudis, K., Mouret, J.B.: Using centroidal voronoi tessellations to scale up the multidimensional archive of phenotypic elites algorithm. IEEE Trans. Evol. Comput. **22**(4), 623–630 (2017)
23. Weinberger, E.D., Stadler, P.F.: Why some fitness landscapes are fractal. J. Theor. Biol. **163**(2), 255–275 (1993)

Steps to Chemically Embodied Cognition

Sergio Rubin[1(✉)], Pasquale Stano[2], Andrea Roli[3], and Luisa Damiano[1]

[1] IULM University, Milan, Italy
{sergio.rubin,luisa.damiano}@iulm.it
[2] Department of Biological and Environmental Sciences and Technologies (DiSTeBA), University of Salento, Lecce, Italy
[3] University of Bologna – Campus of Cesena, Cesena, Italy

Abstract. Attempts to implement autopoietic organization – essential for distinguishing the living from the non-living – in software and hardware models often miss key aspects, such as the realization of metabolic closure, which enable the system to interpret and interact with its physical environment. To address this, we aim to integrate chemical organization theory and active inference – both developed to encode the metabolism and cognition of autopoietic systems, respectively – to create a new generation of wetware models within the "Synthetic Biology (SB) and Embodied AI" research program. This approach may allow us to test critical aspects of synthetic life-like systems, including the implementation of sensory-motor loops and the anticipatory and predictive decision-making that arises from self-production and agent-environment interactions. Synthetic systems that exhibit operations beyond cybernetic (reactive) feedback loops and information processing could be considered minimal forms of chemically embodied cognition.

Keywords: Autopoiesis · metabolic closure · minimal cognition · active inference · chemical organization theory

1 Introduction

Embodied AI (EAI) emerged in the early 1990s as a response to the limitations of classical AI, based on the idea that artificial cognitive systems cannot express their full potential without a body capable of interacting with a physical and social environment. EAI emphasizes the role of the body in cognition. Instead of relying only on computer programs, EAI research programs explore "embodied agents", that is, biologically inspired robots capable of learning about their environment and performing cognitive tasks through their physical bodies. Based on scientific insights on natural cognition, this approach attempts at creating, in robotic agents, increasingly complex cognitive abilities starting from (combinations of) sensory-motor processes. The goal is to model in robots the full spectrum of natural cognitive processes, capturing the functioning of "the embodied mind". Since the late 1990s, EAI approaches have shown that significant advancements can be generated, in terms of useful applications. However, at

T. Carletti et al. (Eds.): WIVACE 2024, CCIS 2532, pp. 84–97, 2025.
https://doi.org/10.1007/978-3-031-93631-9_7

the same time, its pioneers question the possibility, for their embodied agents, to match the cognitive abilities of living organisms. Already in 1997 Rodney Brooks formulated this difficulty in terms of a lack of grasp on the principles of the organization of life [1], triggering a (still ongoing) epistemological debate on the realizability of the modeling ambitions of EAI [2]. "Organismically-inspired robotics" [3] and "Enactive AI" [4] are research programs that attempt to fill the gaps of EAI by drawing inspiration from well-defined theoretical models of biological organization based on autopoiesis and related concepts [4–6]. But, despite their high theoretical significance, these programs have yet to yield concrete organizational approaches in EAI that effectively model natural cognition by artificially implementing the organizational features of life. We think that the main reason for the limitations of classical robotic concepts, plans, and constructions—when referred to cognitive and mind-like properties—is that they are developed in a domain substantially different from the domain wherein biological organisms exist. Impressive advancements are still under development, guided by the "Brooksian paradigm", which advocates for robots to rely on sensory-motor coupling with the environment combined with proprioception, rather than on mere abstract symbolic reasoning.

The challenging goal of synthetically modeling natural cognition and related biological processes can be more effectively achieved within the chemical domain. This is because the essential features of this domain, which are crucial for biological cognition and life, in general, cannot be transferred in the hardware and software domains. Based on this thesis, a few research programs have been proposed (e.g., "Synthetic Biology (SB) and Embodied AI"), inspired by the second-order cybernetic ambition of a fruitful convergence between the engineering problem of constructing intelligent devices for practical applications and the scientific issue of understanding the origin(s) of cognition and life. The aim is twofold: (i) ideating and implementing wetware models of minimal, archetypical living and cognitive systems; (ii) developing these models into useful engineering applications. Here we develop these research axes and, in particular, the programmatic idea that modeling synthetically the organizational features of biological cognitive systems (and not just their behavior) is the crucial guideline for a deep chemical approach to EAI.

In Sect. 2, we introduce the concepts of molecular autopoiesis, embodied cognition, and AI. Section 3 explores an autopoietic-based approach to synthetic biology within the context of embodied AI. Section 4 delves into sense-making in embodied AI, drawing on Chemical Organization Theory (COT) and the Free Energy Principle (FEP). Section 5 presents experimental perspectives on chemically embodied cognition within the Synthetic Biology and Embodied AI Research Program. Finally, Sect. 6 offers key conclusions and future perspectives.

2 Molecular Autopoiesis, Embodied Cognition and AI

Experimental and theoretical investigations on chemical and biochemical networks play a key role in understanding abiotic, prebiotic, and biotic systems because of the complex phenomenology emerging from their dynamics. The diversity of chemical species involved in these networks generates different network topologies and dynamic stability. For example, while evolved (current) enzymes, working as operators, exert strong selection (constraints) toward their substrates, it is expected that primitive catalysts would have been more promiscuous. Despite the still existing mystery about primitive chemical networks and structures, it is generally agreed that, focusing on metabolism, several core carbon-based metabolic pathways drove the origin of primitive cells, and that such core minimal metabolism recapitulates primitive biogenesis [7]. Such sort of networks exist in a dynamical form, in the continuous processes of production and consumption of their components, due to a circular organization of operators. When these self-maintaining systems become able to self-define (versus their environment) via operational closure – i.e., the endogenous production of a boundary throughout a circular organization of operators – they meet the criteria to be defined "autopoietic". In line with the biology of cognition developed by Maturana around the notion of autopoiesis, the operational closure is always molecular and autopoietic systems are considered both living and cognitive [5]. An autopoietic system continuously specifies its own organization through its operations of production of its own components, i.e., by operational closure (a.k.a. metabolic closure/closure of constraints) [8]. Such an organization involves compensating continuous environmental perturbations (e.g. variations in the concentration of chemicals, changes in the temperature, etc.) to remain at low entropy.

For Maturana and Varela [5], cognition is an operational effective action in their structural coupling with the environment without involving information processing. A cognitive system is capable of both action and perception, where the coupling of the two ensures that (i) the system exhibits operational closure, and (ii) its emergent behavior in the environment fulfills a constraint. This autopoietic organizational basis of cognition has later been elaborated as "sense-making", which can be described as the endogenous generation of operational meanings for perceived environmental perturbations. This semantic interpretation of the environmental variations can enable predictive behavior in terms of anticipation [9], allowing the biological system to infer perturbative conditions for the maintenance of its organizational continuity. Moreover, the sense-making processes become inherently intertwined with the system's self-production. In our view, this theoretical framework can promote new and promising approaches to EAI, when developed through frontier wetware modeling methods from SB.

Noticeably, current software and/or hardware EAI systems, while displaying computational processes useful to solve some cognitive tasks, lack cognition in terms of structural coupling and sense-making, since they lack metabolic-like dynamics of self-production [10]. That is: to address the current open issues in EAI, the still missing move is the introduction in the field of wetware chemical models, which are particularly suitable for the artificial realization of the biological organization. In other words, there is a need of a wetware approach to EAI based on autopoietic organizational. We provide here a road map to this modeling, which would also help to re-address questions about the relation of continuity between life and cognition, and to clarify whether minimal (synthetic) biological systems can implement cognitive processes at the fundamental (bio)chemical level.

3 An Autopoietic-Based Approach of Synthetic Biology to Embodied AI

Synthetic Biology can be widely defined as those bottom-up approaches to the construction (synthesis) of complex (bio)chemical systems that share with biological organisms essential traits of their structural and/or dynamical organization. The SB we refer to, therefore, includes investigations on the origins of life, chemistry of complex dynamical systems (systems chemistry), wetware Artificial Life (AL), bio-organic and chemo-enzymatic strategies, etc. These research areas have a common playground: the chemical domain, which we believe is indeed the right space for investigating various kinds of organizational approach to EAI. Our main claim is that, thanks to their chemical peculiarities, SB approaches can provide concepts and tools for an effective implementation of the biological organization with respect to life and cognition modeling, overcoming the current limitations. More specifically, we think that effective models of metabolic cognitive processes have not been fruitful because the capacity of material transformations, a specific feature of chemistry, simply cannot be emulated by hardware systems, while software systems cannot even "perform" in the physical space – they just produce simulations of physical processes in a virtual space. Chemical and biochemical networks, approached both experimentally and theoretically, are essential for understanding abiotic, prebiotic, and biotic systems due to the complex behaviors emerging from their dynamics and to the actual possibility of realizing the above-mentioned "organizational closure". The diversity of chemical species and their transformations generates varying topologies and dynamic stability in chemical networks. An example comes from primitive chemical networks, e.g., the core carbon-based metabolic pathways which have been crucial for biogenesis and the origin of primitive cells [7]. These sorts of network operate dynamically, produce and consume their components in a cyclically organized manner. When such self-maintaining systems can self-delimitate themselves, so that their "unity" is proved because the systems components relate to each other by a net of causal relations, they are termed "autopoietic".

Maturana and Varela's characterized living systems as those "machines" that have an autopoietic organization [5], continuously specifying their organization through the production of their components (metabolic closure/constraint closure, [8]). These complex mechanisms involve (require) the compensation for environmental disturbances in order to remain organized. Maturana and Varela [5] viewed this last aspect as a fundamental cognitive process, described as "structural coupling" with the environment. In other words, from their theory it follows that autopoietic systems are necessarily living *and* cognitive. It is possible to recognize such important dynamics from a semantic perspective and call it "sense-making". With this term we refer to the generation of operational meanings for environmental perturbations. The latter are managed by the system by means of self-regulation schemes, intertwined with self-production, and assure that the system can continue to exist, despite the perturbations. This theoretical framework offers the possibility of exploring new directions in EAI because there are experimental landscapes that can allow, in principle, the construction of artificial systems capable of being engaged into such autopoietic sense-making processes. Artificial systems based on software and hardware, in contrary, cannot be extended to the (necessary) self-production dynamics of metabolic chemical systems [10]. We maintain that only wetware approaches allow the generation of biological-like EAI, because these are the only systems capable of metabolic self-regulation: we aim, therefore, at exploring EAI by grounding it into a life-like organization. Creating a detailed plan for this new modeling method could help answer questions about the link between life and cognition.

4 Sense-Making Embodied AI Through Chemical Organization Theory (COT) and Free Energy Principle (FEP)

A framework that allows SB-EAI modeling of cognitive processes can be conceived by combining Chemical Organization Theory (COT) [11,12] and active inference, a corollary of Free Energy Principle (FEP) [13]. COT is a formalization of the metabolic closure of autopoietic systems that links the structure of reaction networks to their dynamics. COT establishes a formal criterion for determining a reaction network to be operationally closed and stoichiometrically self-maintaining. The parts of the reaction network that fulfill these two conditions, called organizations, contain all stable states that the reaction network can reach in the long term: For any long-term steady state that the reaction network reaches, the molecular species whose concentration is larger than zero-that is, the active part of the network-will be a chemical organization unit [11]. However, depending on the kinetic constraints, such as reaction rates or boundary conditions, some organizations become feasible, but others do not. Organizations can hence be characterized in terms of the possible ways, given particular kinetic constraints, that the consumed species (molecular components) are produced by reaction pathways. On the other hand FEP is a formalization of embodied cognition of autopoietic systems that can be read as a physics of sentience. By sentience we mean that the system is equipped with sensory apparatus – that in the FEP framework represents sensory states – which allows the system to interpret perturbations and signals from the environment. It works as a normative account of sentient behavior that describes it as maximizing Bayesian model evidence and decision-making. It entails (i) establishing a particular partition of states of the system based upon conditional independencies of internal and external states through blanket (sensory and active) states, (ii) unpacking the implications of this partition in terms of Bayesian inference, and (iii) describing the paths of system's states with variational principles [13].

Thus, our main working hypothesis is that a minimal self-maintaining metabolic system can arise, by active inference, from a set of chemical reactions, based on the primitive core metabolism. When the dynamics of the system is evaluated, it must be out-of-equilibrium dynamical state. The requirements for its implementations are: (i) identify a (small) number of key chemical reactions able to self-maintain in COT terms; (ii) determine whether some of chemical reactions or products can act as internal and sensory-active states; (iii) let the system interact with a fluctuating environment to determine whether it remains in out-of-equilibrium dynamics in terms of active inference (Fig. 1). That is, whether the system self-maintains by actively inferring a fluctuating environment. Such forms of minimal chemically embodied cognition should be understood in terms of sensory-active loops that act selectively on environmental molecules and transform them for the benefit of the system's own operation of self-maintaining. In other words, through active states the system will plan and select environmental molecules for the preservation its own identity. In this sense, this autopoietic-like chemical system would be capable of sense-making, based upon a continuous chemical self-producing organization.

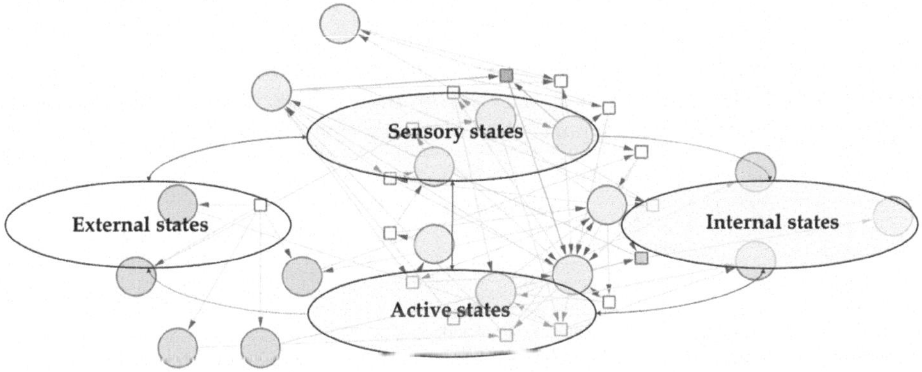

Fig. 1. Minimal chemical reaction network capable of self-maintenance with partition states for active inference.

This is quite decisive, since EAI emphasizes the importance of sensorimotor coupling and embodied cognition. A reverse engineering -deconstructing the system to understand how it functions- could be applied seeking to establish what are the key molecular components -specific molecules that are essential for the system to function- upon which the dynamical stability of the system organization depends. This means to study the ability of the system to maintain its overall structure and function over time, despite changes or perturbations. Such a determination could be based on measurement of the production rates of chemical components that the system cycles to achieve self-maintenance in a fluctuating environment (e.g., under changing the concentration of chemical components) such that the assimilation rates for different chemical components are determined by active inference and thus by the dynamic loop of sensory, active and internal states. It is important to consider that the time dynamics of such autopoietic-like chemical organization would avoid dissipation in an analytically determined timescale. In essence, this means that the autopoietic-like chemical organization must be analyzed to understand how it sustains itself within a given timeframe, avoiding breakdown or instability.

5 Experimental Perspectives Towards Chemically Embodied Cognition

Based on the above-mentioned proposal, we call for a novel turn in experimental SB (and systems chemistry) that explicitly includes the mechanisms underlying cognition and sense-making based on the mechanisms derived from our analysis. Based on the current technical progress, investigations on the so-called "synthetic cells" (SCs) are ideal candidates for developing experimental approaches to minimal cognition, and so contributing to the "experimental epistemology" program pioneered by McCulloch decades ago. SC technology, indeed, is advancing quite well, promoted by the significant efforts of a growing international community. Important consortia have proven to be effective in boosting SC research [14,15], which is often presented as one of the most innovative directions in biotechnology. To date, studies on cell-like systems cover distinct interests, from the origins-of-life to the construction of SC, from molecular communication research to possible applications as "smart" drug delivery systems (Fig. 2).

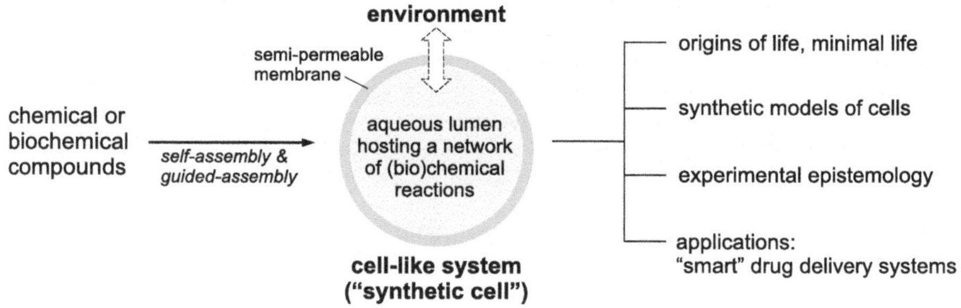

Fig. 2. Cell-like systems, often called "Synthetic Cells" (SCs), can be built in the laboratory by means of procedures based on self-assembly and guided-assembly. The resulting structures resemble biological cells in simple manner, and – to date – are not alive. Importantly, SCs exchange matter with their environment thanks to a semi-permeable membrane, possibly functionalized with transporters or pores. The membrane itself, as well as the other components of the boundary, should be parts of the dynamical reaction "network" that constitute the entire cell-like system. On the right, fields of application of SCs. See text for more details.

Systems made of allegedly primitive molecules are employed to model primitive cells. In this context, the self-replication of nucleic acids such as primitive ribozyme or the self-reproduction of the whole cell-like systems becomes important experimental targets. The usage of primitive (or anyway simple) molecules serves to demonstrate that some cell-like behaviors are achievable in conditions that are far from modern biological ones. In the same manner, searching for the minimal complexity associated with the chemical networks that constitute such artificial systems is a task that can help to elucidate the early steps of primitive life. A similar approach can be developed by using contemporary molecules (DNA, ribosomes, enzymes, etc.). In this case, the focus is shifted from "primitiveness" in the chemical sense, to simplified versions of cellular life. For example, the identification of the irreducible genetic-metabolic core networks provides useful guidance to the construction of SCs [16]. Such an approach looks at the problem of "minimal life" from the bottom-up, while others are devoted to simplifying extant biological cells by removing unnecessary genes (top-down) [17]. The possibility of constructing a simplified version of cells helps to understand the dynamics of complex multi-molecular and often hierarchical systems in the absence of the complicated background present in biological cells. In this way, it is possible to study complex mechanisms from a privileged viewpoint. These research directions are widespread and are attracting the attention of an increasing number of scientists. They pertain to basic science but already promise possible developments in applied research. Indeed SCs can be developed as tools in nanomedicine, tissue engineering, drug- and gene-delivery "smart" agents. There are already interesting reports showing how SCs could produce therapeutic proteins, acting in situ, near the damaged cell [18].

Although it can be easily recognized that a decisive contribution to the birth of bottom-up minimal life and origins-of-life investigations based on the construction of synthetic cells actually comes from the autopoietic tradition – in particular from pioneer studies on 'chemical autopoiesis' in the 1990s [19] – the central theme has always been focused on the material problem of realizing a self-bounded system capable of producing its own components. This is evident when research lines in origins-of-life and bottom-up synthetic biology are analyzed. A possible reason is that early and current researchers generally concentrate their attention on the chemical problem of generating a self-producing network, which is far from being trivial (and not yet realized).

On the other hand, autopoiesis and cognition are indissolubly bound to each other, since they both arise from operational closure, yet as different domains, that of the organization itself and of the relation with its environment [10]. However, investigations that explicitly aim to consider the cognitive capability of minimal chemical systems are rather rare. We consider that a chemical system capable of maintaining operational closure and a relationship with its environment at least as prescribed in COT and FEP respectively, can effectively be considered a proto-cognitive system. These would contribute, as anticipated above, to an "experimental epistemology" program, in the tradition of early scholars in cybernetics (e.g., the already mentioned Warren S. McCulloch), and in second-order cybernetics (Heinz von Foerster), but also to an "experimental ontology" program, in the tradition of relational biology (Robert Rosen).

5.1 Synthetic Biology and Embodied AI Research Program

Our research program seeks to explore, in a preliminary and possibly rudimentary phase, how biology and cognition (in minimal versions) can emerge from physics and chemistry. It is not difficult to reconstruct the genealogy of the modern experimental attempts, as a seminal publication is available and well known to specialists, i.e., the Varela-Luisi programmatic paper on the use of reverse micelles for constructing an autopoietic system in the chemical laboratory [20]. The article resulted from the encounter of the two authors at a 1983 "Other Realities" meeting in Alpbach (Austria) [21]. Some years later, the first experimental results were published, giving rise to the so-called "chemical autopoiesis", which was mainly developed by Pier Luigi Luisi and collaborators at the ETH Zürich in the 1990s, using reverse micelles, regular micelles, and – importantly – liposomes [22]. The first steps toward SCs originated, slowly, from the attempts that followed from these early papers.

Our research program starts from the rediscovery of these early studies and of their recent developments, emphasizing those aspects that are firmly grounded in autopoietic constructive dynamics. Maturana and Varela's autopoiesis offers a privileged possibility for a dialogue between epistemology and the sciences of the artificial [23], mainly developed, till now, in the software domain (e.g., in the field of "Artificial Life").

Chemical approaches represent the first, and most evident, shift in the practice of autopoiesis-inspired modeling (develop wetware models instead of software models). However, our research program further exploits autopoietic principles to tailor a realistic experimental route based on autonomy [6]. In particular, instead of insisting on the full set of transformations required to make an autopoietic network, we will target a narrower set of processes capable of adaptive behavior. More specifically, a possible goal for implementing EAI in wetware is the construction of a network of processes that recursively depend on each other and that can be recognized as a unity in the domain in which they exist ([6], p. 55). To this aim we need to (i) identify, (ii) model in silico, and later (iii) realize in vitro a minimal set of self-constructive processes in SCs. These processes must operate in a mutual (and overall-circular) dependence, and on a timescale that is faster than the metabolic timescale, so that the required condition of material production, transformation, and destruction can be partially or completely omitted (or outsourced to an auxiliary non-autopoietic network). Although we know that this "middle way" approach departs from an authentic autopoietic organization, it will allow us to explore adaptive behavior in a manner overtly inspired to autopoietic autonomy.

With these pragmatic considerations in mind, a preliminary selection of plausible mechanisms for future investigations has been carried out, leading to the following conceptual layout.

Identification. By taking inspiration from known (bio)chemical systems, we can think of at least three categories of possible self-constructive networks. The first is based on coupling of physical and chemical processes, such as those involving sensitive microenvironments (particles, membranes, and other assemblies) and chemical reactions. For example, morphological or functional changes in vesicles driven by internal production of chemicals and a correlated variation of internalized reactions (rates, selectivity, etc.) due to morphological changes. A second category can be represented by chemical networks with or without enzyme catalysis, that operate out-of-equilibrium. They can be perturbed by variations in one or more of their "nodes", in a manner that a new state is reached. In turn, the new state should lead to a modification of the perturbation or of its effect. A third category could be related to biochemical mechanisms of signaling and gene regulation (examples come from systems and synthetic biology, especially when such networks have been already made in vitro). To keep such networks active, a continuous consumption of energy and matter is required –making evident the need of an auxiliary system that provides just the necessary chemicals without being itself directly involved in the regulatory dynamics. That role can be played by a transcription-translation machinery, included in the physical body of the agent of interest, but not including –as it already happens– self-generative processes. In a previous publication, we have depicted a possible scheme based on bacterial two-component systems, which are able to sense environmental variables and regulate gene expression. In particular, we have hypothesized a recurrent network based on the increase or on the decrease of the concentration of regulators (synthesis/degradation), or of their activity (via allosteric effects) [24].

In Silico Modeling. The systems of interest will first be simulated, by in silico modeling, for their capability of displaying adaptive behavior. Our initial plan is to apply periodic environmental conditions (without and with noise), acting both as perturbations and as material supplies for sustaining the network, and explore the conditions for existence and adaptation (a form of learning). While rate constants and binding constants are the equivalents of neural network parameters (i.e., weights, biases, thresholds), chemical networks have as additional parameters the concentrations of the participating species. In turn, these concentrations might depend on reactions within the network, and thus generate a recurrent network. In addition to this kind of modeling, we also intend to investigate chemical networks from the viewpoint of semantic information theories. The first conceptual shift we propose is to move away from considering information as a static entity, and instead understanding it as a dynamic process –specifically– the process of being in-formed. This perspective aligns with the early discussions on the nature of information, mainly debated in the early days of cybernetics and not further developed at that time. Notably, while Claude Shannon's definition of information [25] does not address meaning [26], others – such as Gregory Bateson and Donald M. MacKay – emphasized its semantic aspects (e.g., Bateson defined information as "a difference that makes a difference" [27]). Similarly, as remarked by Gordon Pask in his monography on cybernetics, MacKay's view suggests that information carries meaning when there is a selective function enacted by an organism's behavior [28]. For a system to undergo in-formation, it must be capable of altering its structure and dynamics in response to internal or external events, which act as perturbations or, in the FEP framework as 'belief priors' to its pre-existing behavior. For example, in a life-like chemical network, the relationships between elements (chemical interactions or reactions) must change both qualitatively and quantitatively even with antecedence to perturbation. The ability of a network to accommodate perturbations while maintaining its overall organization (as occurs in autopoiesis) becomes a key feature in developing new systems in SB and systems chemistry that are both cognitive and capable of sense-making.

In Vitro Realization. The current approaches typically developed as "Systems Chemistry" [29] and SB are perfectly suitable for the wetware implementation of the artificial systems we refer to. Our design will always involve particles of various types as cellular mimics in order to fulfill a primary need of distinction, confinement, and individuation (versus the environment) of the reactive networks we will explore. Particles of various sizes and structures are currently explored in the field, such as micelles, reverse micelles, vesicles (lipid and fatty acid vesicles), coacervate, water-in-oil droplets, and similar. A peculiarity of such systems is that their structure can be easily tuned by varying the nature or the amount of their components, and this offers the possibility of reversible (when possible) perturbations. Chemical and biochemical networks have also been intensively explored, especially in the case of enzyme-catalyzed systems. Enzyme-free reaction networks appear more challenging to control, but being more "prebiotic-like", are certainly intriguing targets for exploration. However, achieving autonomous and autopoietic dynamics is a challenging goal.

The summarized perspective represents an innovative shift –at least in the realm of wetware implementations– from seeing autopoiesis merely as the self-production of system components to acknowledging its original and broader conception, which includes action, perception, and structural coupling to environmental perturbations that the system has selected based on their compatibility with its ongoing dynamics. In 2004 Luisi and Bitbol discussed, in a commentary, the concept of autopoietic cognition in an artificial 'metabolic' network [30]. We believe that the prototypes we intend to explore (in silico and later in vitro) can be fruitfully guided by the theoretical frameworks of COT and FEP taking advantage of the recent advancements in systems chemistry. Such novel approaches can help move research in the desired direction and offer a promising path toward developing minimal cognitive systems in the laboratory.

6 Conclusion

The autopoietic-based steps to chemically embodied cognition through the integration of COT and active inference (a.k.a. FEP), outlined in this paper, aims to create a new form of bioinspired AI, specifically embodied AI in synthetic cells. Our vision diverges from the traditional computational approach, as our focus is not on building programmable chemical robots, though some of the modeling and experimental tools may overlap.

Investigating how cognition as sense-making emerges in certain chemically dynamic systems that exhibit an autopoietic-like organization is not trivial. These processes can occur on both short and long timescales. The theory and experiments, but also the conceptual and numerical models of COT and active inference has never been applied to minimal self-sustaining metabolic systems operating out of equilibrium.

From a practical standpoint, SB approaches, including systems chemistry, provide promising experimental platforms for this research. Although our primary focus is on fundamental questions regarding the origins of cognition in minimally complex chemical systems, we believe that this work can also contribute to applied research, such as molecular robotics and "chemical AI" [31,32]. This could be particularly valuable in scenarios where artificial agents, such as synthetic cells, establish communication with biological cells, as in the field of nanomedicine – an exciting area with significant potential for development in the near future.

Acknowledgement. The research has been developed within the PRIN-2022 project n. 20222HHXAX (Ministero dell'Università e della Ricerca, Italy), entitled "Org(SB-EAI)—An Organizational Approach to the Synthetic Modeling of Cognition based on Synthetic Biology and Embodied AI", funded by the European Union, Next Generation EU, Mission 4, Component 1, CUP F53D23004560006. A similar report on the topics covered by the above-mentioned project has been recently published in a conference proceedings volume [33].

Disclosure of Interests. The authors have no competing interests.

References

1. Brooks, R.A.: From earwigs to humans. Robot. Auton. Syst. **20**, 291–304 (1997)
2. Ziemke, T.: Are robots embodied? In: Proceedings of the First International Workshop on Epigenetic Robotics. Modeling Cognitive Development in Robotic Systems, pp. 75-83. Lund University Press, Lund (2001)
3. Di Paolo, E.A.: Organismically-inspired robotics: homeostatic adaptation and teleology beyond the closed sensorimotor loop. In: Murase, K., Asakura, T. (eds.) Dynamical Systems Approach to Embodiment and Sociality, pp. 19–42. Advanced Knowledge International, Adelaide, Australia (2003)
4. Froese, T., Ziemke, T.: Enactive artificial intelligence: investigating the systemic organization of life and mind. Artif. Intell. **174**, 466–500 (2009)
5. Maturana, H.R., Varela, F.J.: Autopoiesis and Cognition: The Realization of the Living. D. Reidel Publishing Company, Dordrecht (1980)
6. Varela, F.J.: Principles of Biological Autonomy. Elsevier North-Holland Inc., New York (1979)
7. Morowitz, H.J.: Beginnings of Cellular Life: Metabolism Recapitulates Biogenesis. Yale University Press, New Haven (1992)
8. Kauffman, S.A., Roli, A.: A third transition in science? Interface Focus **13**, 20220063 (2023)
9. Rosen, R.: Anticipatory Systems. Philosophical, Mathematical and Methodological Foundations. Elsevier, New York (1985)

10. Rubin, S.: Cartography of the multiple formal systems of molecular autopoiesis: from the biology of cognition and enaction to anticipation and active inference. Biosystems **230**, 104955 (2023)
11. Dittrich, P., Speroni Di Fenizio, P.: Chemical organisation theory. In: Al-Rubeai, M., Fussenegger, M. (Eds.) Systems Biology, pp. 361–393. Springer, Dordrecht (2007)
12. Dittrich, P., Speroni Di Fenizio, P.: Chemical organisation theory. Bull. Math. Biol. **69**, 1199–1231 (2007)
13. Friston, K.: Life as we know it. J. Roy. Soc. Interface **10**, 20130475 (2013)
14. Schwille, P., et al.: MaxSynBio: avenues towards creating cells from the bottom up. Angew. Chem. Int. Ed. **57**, 13382–13392 (2018)
15. Frischmon, C., Sorenson, C., Winikoff, M., Adamala, K.P.: Build-a-cell: engineering a synthetic cell community. Life **11**, 1176 (2021)
16. Luisi, P.L., Oberholzer, T., Lazcano, A.: The notion of a DNA minimal cell: a general discourse and some guidelines for an experimental approach. Helv. Chim. Acta **85**, 1759–1777 (2002)
17. Gibson, D.G., et al.: Creation of a bacterial cell controlled by a chemically synthesized genome. Science **329**, 52–56 (2010)
18. Krinsky, N., et al.: Synthetic cells synthesize therapeutic proteins inside tumors. Adv. Healthc. Mater. **7**, e1701163 (2018)
19. Stano, P.: The birth of liposome-based synthetic biology: a brief account. In: Pearson, B.R. (ed.) Liposomes: Historical, Clinical and Molecular Perspectives, pp. 37–52. Nova Science Publishers Inc, Hauppauge (2017)
20. Luisi, P.L., Varela, F.: Self-replicating micelles: a chemical version of minimal autopoietic systems. Origins Life Evol. Biosphere **19**, 633–643 (1989)
21. Luisi, P.L.: My encounters with Francisco Varela. Syst. Res. **28**, 689–695 (2011)
22. Walde, P., Wick, R., Fresta, M., Mangone, A., Luisi, P.L.: Autopoietic self-reproduction of fatty-acid vesicles. J. Am. Chem. Soc. **116**, 11649–11654 (1994)
23. Ceruti, M., Damiano, L.: Il Passato nel Futuro. Mente, Reti e Alberi della Conoscenza. An essay published in: Maturana H. R., Varela F. J.: L'Albero della Conoscenza. Le Radici Biologiche della Conoscenza Umana (Italian translation of "Maturana, H. R.; Varela, F. J.: El Árbol Del Conocimiento"). Mimesis Edizioni, Milano (2024)
24. Braccini, M., Collinson, E., Roli, A., Fellermann, H., Stano, P.: Recurrent neural networks in synthetic cells: a route to autonomous molecular agents? Front. Bioeng. Biotechnol. **11**, 1210334 (2023)
25. Shannon, C. E.: A mathematical theory of communication. Bell Syst. Tech. J. **27**, 379–423, 623–656 (1948)
26. Logan, R.K.: What is information?: Why is it relativistic and what is its relationship to materiality, meaning and organization. Information **3**, 68–91 (2012)
27. Bateson, G.: Steps to an Ecology of Mind. Collected Essays in Anthropology, Psychiatry, Evolution, and Epistemology. University of Chicago Press, Chicago (1972)
28. Pask, G.: An Approach to Cybernetics. Hutchinson and Co., London (1961)
29. Ashkenasy, G., Hermans, T.M., Otto, S., Taylor, A.F.: Systems chemistry. Chem. Soc. Rev. **46**, 2543–2554 (2017)
30. Bitbol, M., Luisi, P.L.: Autopoiesis with or without cognition: defining life at its edge. J. Royal Soc. Interface **1**, 99–107 (2004)
31. Murata, S., Toyota, T., Nomura, S.M., Nakakuki, T., Kuzuya, A.: Molecular cybernetics: challenges toward cellular chemical artificial intelligence. Adv. Func. Mater. **32**, 2201866 (2022)

32. Gentili, P.L., Stano, P.: Living cells and biological mechanisms as prototypes for developing chemical artificial intelligence. Biochem. Biophys. Res. Commun. **720**, 150060 (2024)
33. Rubin, S., Stano, P., Roli, A., Damiano, L.: Sense-making in embodied AI. Towards autopoietic chemical AI. In: Callaos, N., Gaile-Sarkane, E., Lace, N., Sánchez, B., Savoie, M. (eds.) Proceedings of the 28th World Multi-Conference on Systemics, Cybernetics and Informatics (WMSCI 2024); 10–13 September 2024, Virtual Conference, pp. 85–91. The International Institute of Informatics and Cybernetics: Winter Garden, FL (2024)

A Comparative Study on Decision-Making Mechanisms in a Site Selection Task

Ahmed Almansoori[1,2(✉)], Dari Trendafilov[1], Muhanad Alkilabi[1], and Elio Tuci[1]

[1] Faculty of Computer Science, University of Namur, Namur, Belgium
ahmed.almansoori@unamur.be
[2] Faculty of Engineering, University of Kerbala, Kerbala, Iraq

Abstract. In swarm robotics, it is crucial for swarm members to reach a consensus on a single option from a set of alternatives to complete complex tasks autonomously. Typically, individual mechanisms underpinning such collective behaviour are designed using either hand-coded or automatic approaches. In this paper, we aim to compare the performance of robotic swarms controlled by mechanisms designed using both types of techniques in a site-selection task. The evaluated hand-coded mechanisms are based on the voter model and majority rule, while the automatic design approach involves an evolved dynamic neural network mechanism. The evaluation is conducted in a simulated environment that represents different operating conditions and swarm sizes and follows the same protocol for all decision-making mechanisms. The results reveal that the evolved neural network controller demonstrates better behavioural responses, including more accurate decision-making and increased resilience to varying environments and group sizes, compared to traditional hand-coded approaches.

1 Introduction

In swarm robotics, collective behaviour is a global response arising, through a self-organisation process, from the local interaction of multiple simple agents with their physical and social environment. This emergent behaviour allows robotic swarms to achieve complex tasks beyond the capabilities of individual robots. While self-organisation offers undeniable advantages for designing adaptive and resilient systems, its inherent unpredictability and decentralised nature pose significant hurdles for system engineers [3]. To address the design problem, researchers have investigated a variety of controller design approaches, such as hand-coded and automatic design techniques. Hand-coded mechanisms produce behavioural strategies that can be easily described in operational terms. However, their capability to adapt to different sources of variability tends to be limited to the circumstances clearly predicted by the designer, which leaves the robots potentially unprepared to tackle unexpected events encountered in real-world settings [3]. In contrast, evolutionary robotics (ER), an automatic design

T. Carletti et al. (Eds.): WIVACE 2024, CCIS 2532, pp. 98–110, 2025.
https://doi.org/10.1007/978-3-031-93631-9_8

strategy, employs evolutionary computation techniques to develop controllers based on artificial neural networks [10]. The primary advantage of ER lies in its use of evaluation functions that prioritise the performance of the swarm as a whole over the behaviour of individual robots. This approach allows for an automated design process that is adaptable to the specific requirements, however poses challenges for the analysis and the interpretation of such controllers [7,9,12,15].

Collective decision-making, one of the most studied collective behaviours in swarm robotics, refers to a decision problem in which natural or artificial agents collectively make a choice among two or more alternatives, with the final decision being a group outcome rather than an individual one [16]. Within the domain of collective decision-making, robot swarms demonstrated two fundamental abilities: task allocation and consensus achievement. Task allocation involves improving overall performance by splitting the swarm into subgroups focused on different tasks. While consensus achievement allows the swarm to adopt the same opinion about different options. The individual opinion formation process is influenced by direct sampling of the physical environment and by social interaction based on local communication.

In this paper, we present a comprehensive comparative analysis of swarm performance in the site selection task, building on the paradigm illustrated in [1], and evaluating three different decision-making mechanisms for opinion formation - two hand-coded ones based on the voter model and the majority rule and one evolved dynamic neural network as described in [1]. The evaluation of these approaches across a range of operating conditions and swarm sizes aims to provide deeper insights into their relative strengths and weaknesses, eventually contributing to the development of more effective and adaptable collective decision-making strategies for robotic swarms.

2 Related Work

To date, site selection research is based predominantly on hand-coded strategies. [20] studies a binary site-selection task using the voter model, which makes a robot switch to the opinion of a randomly chosen spatially close neighbour. [18,19] explore the majority rule, which makes a robot switch to the most frequent opinion among n spatially close neighbours. In these studies, robots alternate between exploring options and disseminating their current opinion for a time proportional to the option's quality. This modulation of the dissemination time is the main factor driving consensus achievement on the best quality option. Related research [4,11] on a binary site selection scenario investigates the effect of different opinion formation strategies on the decision dynamics. [14] explores the relationship between the maximum communication distance and the accuracy of the collective decision-making process in dynamic environments with three sites of different quality using the voter model with cross-inhibition, where conflicting information leads robots to reset their opinions and seek others.

Diverging from the previous body of research, a recent study demonstrates the potential of evolved dynamic neural networks as a more flexible and adaptable alternative for site selection [1]. By leveraging evolutionary computation techniques, the authors synthesised controllers enabling robotic swarms to effectively solve a ternary site selection task, without explicit assumptions on individual responses or group dynamics. The proposed approach does not assume a correlation between environmental features and robot behaviour, such as the positive correlation between option quality and dissemination time used in [4,11,18–20] or between option quality and the frequency of the individual's opinion update process used in [13,14]. Furthermore, this work avoids built-in assumptions on how robots process perceptual cues and communication signals to form their opinions, as seen in the voter or majority models. Instead, the evolutionary process develops individual mechanisms for integrating these signals and creating the necessary relationships for generating modulations that break the initial equilibrium and push the group toward consensus on the best option.

As demonstrated in Sect. 4, our approach generates effective, robust and scalable group strategies. As discussed in Sect. 3, in spite of the simplifications dictated by the necessity to limit the complexity of this initial study, the task not only possesses all the relevant characteristics of classic site selection scenarios, but also features some properties that make it particularly challenging. For example, outside the social context, single robots do not have the means to systematically make correct decisions on which is the best quality site. This makes it particularly hard for the swarm designer to predict and potentially to hand-code the actions underpinning virtuous collective responses. In view of the promising results that we obtain in terms of accuracy and robustness of the decision-making strategies, we discuss possible further developments in Sect. 5.

3 Methods

The site-selection scenario investigated in this study closely replicates the one described in [1], and is implemented using a single square arena of 2 m × 2 m in which 21 simulated robots pseudo-randomly move for the entire duration of the evaluation while avoiding collisions with other robots and arena walls.

In both, hand-coded and neural network-controlled swarms, the development of individual opinions and the subsequent decision-making process follow a similar structure. The robots alternate between two phases: i) exploration phase, in which the robots perceive the cue signalling the respective site's quality, but they cannot communicate; and ii) dissemination phase, in which they are allowed to communicate their respective opinions on the best quality site to spatially close robots, but they cannot sample the site's quality. At the end of the dissemination state, each robot updates its opinion using either the VM or MM for the swarms controlled by hand-coded mechanism or the neural network controller for the swarm controlled by evolved mechanism. Finally, each robot moves to the exploration state that corresponds to its new opinion (see Fig. 2b).

The main differences between hand-coded mechanisms (VM and MM) and the evolved neural network are in the dissemination time and opinion update mechanisms. In VM and MM, the dissemination time is directly proportional to the robot's estimate of site quality, promoting robots with higher site quality to propagate their opinions for longer duration. Following hand-coded mechanisms, the robots update their opinions based only on social information received from their neighbours. In the VM, this involves adopting a random neighbour's opinion, while in the MM, it involves adopting the majority opinion. Conversely, the evolved neural network does not impose a direct relationship between quality estimation and dissemination time. Instead, during the evolution phase, the network learns how to use the environmental bias among the available options to create both the indirect feedback modulation and opinion update mechanism needed to complete the task. This is done through a complex interaction of sensory inputs and communication signals.

Alike other studies on site selection with simulated agents (e.g., see [11]), our experimental setup is simplified with respect to those elements that can be omitted while preserving the distinctive challenges related to the development of a consensus to the best quality option through a process based on a collective decision. In particular, in our scenario depicted in Fig. 1a, the sites are not distinctive locations, which the robots join and leave based on the experimental phase. The robots never actually leave the square arena during evaluation. The back-and-forth movement between the nest and sites is implemented through modifications of the robots perceptual system, for which either communication is allowed (this consists of the robot being in a dissemination phase and metaphorically placed in the nest) or the site quality sampling is allowed (this consists of the robot being in an exploration phase and metaphorically placed in the site corresponding to its current opinion). Note that, this way of instantiating the nest/sites is meant to simplify the scenario with respect to the following: i) it eliminates the requirement to develop control mechanisms underpinning the back-and-forth navigation between nest and sites; ii) it reduces the evaluation time by eliminating the nest-sites transfer time. This is particularly helpful to reduce the design time of decision mechanisms which tends to be quite large with the use of evolutionary computation techniques.

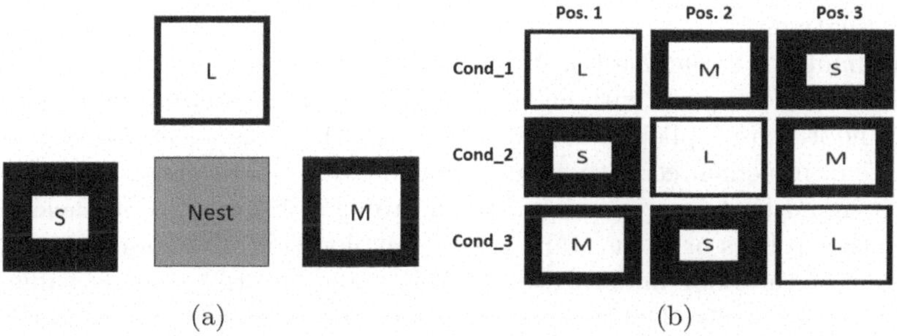

Fig. 1. a) The site-selection scenario with nest and three sites – L, M, and S [1]. b) The position of each site in the three experimental conditions [1].

The scenario features three different sites (or options) – large (L), medium (M), and small (S) – with varying white floor proportions (80%, 50%, and 20%, respectively) and a nest (see Fig. 1a). The higher the proportion of white the better the site's quality. The nest is a location in which the colour of the floor is gray (see Fig. 1a). The swarm is evaluated in three experimental conditions (Cond_1, Cond_2, and Cond_3) that differ in the position of the best quality site (see Fig. 1b). The positions of the sites in the metaphorical space change based on the experimental condition: in Cond_1, site L is in position 1, site M is in position 2, and site S in position 3; in Cond_2, site S is in position 1, site L is in position 2, and site M in position 3; in Cond_3, site M is in position 1, site S is in position 2, and site L in position 3 (see Fig. 1b). Evaluation takes place in all three conditions. The task of the swarm is to determine which position hosts the highest quality site in all three conditions.

At the beginning of each trial, 21 robots are placed randomly in the arena and are initialized uniformly with the three possible opinions (seven each) – i.e., the best quality site is located in position 1, 2 or 3, respectively. Each robot starts the trial in the exploration phase by sampling the site's quality corresponding to its current opinion. For each robot, the duration of the exploration phase is sampled from an exponential distribution with 10 s mean. At the end, of its exploration phase, each robot switches to the dissemination phase in which it has no access to the cue signaling the quality of the site, but is allowed to disseminate through communication its opinion to spatially close neighbors. The duration of the dissemination phase is also sampled from an exponential distribution with 10 s mean. At the end of its dissemination phase, each robot returns to an exploration phase that takes place in the site corresponding to the robot's opinion. The latter is set at the end of the dissemination phase by choosing the opinion most frequently expressed during that phase. The metaphorical back-and-forth movement between the nest and the sites of the robots lasts 800 s – this is the trial time within which the group has to reach a consensus. Note that, the random variations in exploration and dissemination durations progressively bring the swarm into an "asynchronous" state, in which robots operate in different phases simultaneously. A trial is considered successful whenever all the robots hold the same opinion concerning the position of the best quality site for at least 10 s consecutively.

The simulated e-puck2 robot [8], a platform frequently used in swarm robotic experiments, is equipped with eight infrared sensors located around the robot's body, a floor sensor placed on the bottom of the chassis, and a communication system implemented with a range and bearing board. The infrared sensors return a signal that is proportional to the robot's distance to obstacles, which can be other robots or arena walls. During exploration, the floor sensor detects the colour of the floor underneath the robot and returns a single value, 0 for black and 1 for white. During dissemination, the floor sensor constantly reads 0.5, indicating that the robot is on a gray floor in the nest. The robot opinion is a three-bit signal, i.e., $\{1,0,0\}$ if the best quality site is considered located in position 1 (Pos. 1); $\{0,1,0\}$ – position 2 (Pos. 2); and $\{0,0,1\}$ – position 3 (Pos.

3), respectively. For the entire duration of the dissemination phase, each robot emits a three-bits signal corresponding to its opinion. Every robot in the dissemination phase receives, for the entire duration of this phase, a signal from the spatially closest neighbor among those in the dissemination phase and located at less than 50 cm distance. If there is no disseminating robot within the communication range (<50 cm), the communication signal of a robot receiver is set to $\{0, 0, 0\}$. When a robot is in the exploration phase, the communication signal is constantly set to $\{0.5, 0.5, 0.5\}$. To compensate for the simulation-reality gap, uniform noise is added to all sensor readings, motor outputs and robot position (see [6] for a similar approach).

The robot controller is made of two modules. The "walk" module generates the pseudo-random walk and obstacle avoidance behaviour. The "decision" module generates the robot's opinion. The walk module makes the robot move according to an isotropic random walk consisting of straight motion for 5 s at a speed of 20 cm/s and rotation with turning angles sampled from a wrapped Cauchy distribution [5], defined by:

$$f_\omega(\theta, \mu, \rho) = \frac{1}{2\pi} \frac{1 - \rho^2}{1 + \rho^2 - 2\rho\cos(\theta - \mu)}, \quad 0 < \rho < 1; \tag{1}$$

with skewness $\rho = 0.5$. Upon obstacle detection (wall or other robots), the robot stops and rotates with a random angle generated uniformly in the interval $[-\pi, \pi]$. The robot resumes its random walk after turning, provided that there is no obstruction, otherwise it repeats the maneuver.

The decision module is a three-layer dynamic neural network (see Fig. 2a and [2] for more details), with a four-neuron fully recurrent hidden layer (i.e., $N_{H,1}$, $N_{H,2}$, $N_{H,3}$ and $N_{H,4}$). The four-neuron input layer relays sensory activation into the network, with the activation of $N_{I,1}$ corresponding to the reading of the floor sensor, and the activation of neurons $N_{I,2}$, $N_{I,3}$, and $N_{I,4}$ corresponding to the three-bit communication signal. The three-neuron output layer ($N_{O,1}$,

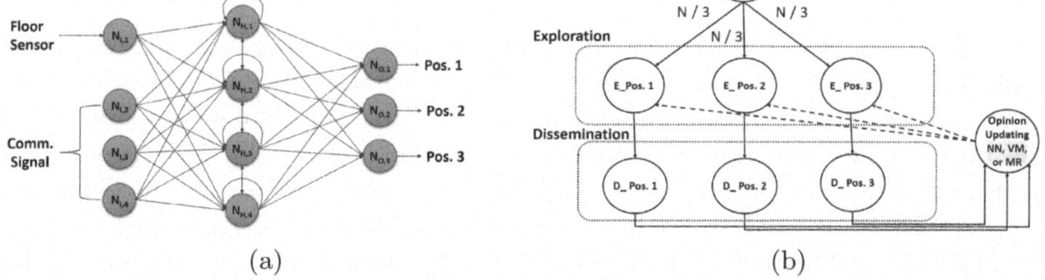

Fig. 2. (a) The decision module made of a dynamic neural network [1]. (b) The probabilistic finite-state machine. E_Pos 1, E_Pos 2, E_Pos 3, D_Pos 1, D_Pos 2 and D_Pos 3 represent the exploration and the dissemination state, respectively. Solid lines represent deterministic transitions, while dotted lines represent stochastic transitions.

$N_{O,2}$, and $N_{O,3}$) generates the robot opinion. The activation of each hidden unit (H_j) is governed by the following differential equation:

$$\tau_j \dot{H}_j = -H_j + \sum_{h=1}^{4} W_{hj}^{H} \sigma(H_h + \beta^{H_h}) + \sum_{i=1}^{4} W_{ij}^{I} I_i; \quad \sigma(x) = \frac{1}{1+e^{-x}}; \tag{2}$$

where τ_j is the decay constant, W_{hj}^{H} is the strength of the synaptic connection from hidden neuron h to hidden neuron j, and W_{ij}^{I} is the strength of the connection from input neuron i to hidden neuron j, and I_i is the activation of input neuron N_{I_i}. The activation of each output neuron (O_j) is computed by the following:

$$O_j = \sum_{i=1}^{4} W_{ij}^{O} \, \sigma(H_i + \beta^{H_i}) + \beta^{O_j}; \tag{3}$$

where $j \in [1, 2, 3]$. H_i is the activation of the hidden neuron i, β^{O_j} and β^{H_i} are bias terms, and W_{ij}^{O} is the strength of the synaptic connection from hidden neuron i to output neuron j. The weights of the connections between neurons, the bias terms and the decay constants are genetically encoded parameters. Cell potentials are set to zero each time the network is reset. State equations are integrated using the forward Euler method with a step size of 0.1 s.

At every update cycle, the robot opinion (OP) is derived by normalising the activation of the output neurons (O_j) with the following softmax function:

$$Softmax(\vec{O})_j = \frac{e^{O_j}}{\sum_{n=1}^{3} e^{O_n}}, j \in 1, 2, 3; \quad \vec{O} = (O_n) \in \mathbb{R}^3; \tag{4}$$

The highest activated output defines the robot's opinion according to:

$$\text{OP} = \begin{cases} \text{Pos. 1} & \text{if } O_1 = max_j O_j, \\ \text{Pos. 2} & \text{if } O_2 = max_j O_j, \\ \text{Pos. 3} & \text{if } O_3 = max_j O_j, \\ \text{no change} & \text{if } O_1 = O_2 = O_3. \end{cases} \tag{5}$$

A tournament selection evolutionary algorithm using linear ranking is employed to set the network parameters. The population contains 64 genotypes. Each genotype vector comprises 44 connections, four decay constants, and seven bias terms, which are randomly initialised in the range $[0, 1]$. New genotypes, except the elite, are produced by applying mutation with a random Gaussian $N(0, 0.1)$ offset applied to each component encoded in the genotype, with a probability of 0.03. At each new generation the highest scoring individual (the elite) is retained unchanged from the previous generation. The remainder of the new population is generated by binary tournament selection from the 45 best individuals of the old population.

At the beginning of each differently seeded evaluation trial, the decision module is reset (i.e., robots have no memories from previous evaluations) and cloned on each of the 21 robots forming a homogeneous swarm. The robots are randomly positioned in the arena with a randomly chosen orientation in $[0, 2\pi]$. Each robot performs a pseudo-random walk as illustrated above for the entire duration of the trial (i.e., 800 s, $T = 8000$ simulation cycles). The fitness of a genotype F is defined as the average group evaluation score after it has been assessed once in each condition.

The opinion OP_{rt} of robot r at time step t, being $OP_{rt} = 1$ if the robot holds the correct opinion and $OP_{rt} = 0$ otherwise, contributes to calculating the average group evaluation score F as follows:

$$F_e = \frac{2}{TR} \sum_{t=T/2}^{T} \sum_{r=1}^{R} OP_{rt}, \quad F = \frac{1}{E} \sum_{e=1}^{E} F_e, \tag{6}$$

where F_e in trial e is computed from trial half-time to trial end.

4 Results

In order to compare the performance of our neural-network-based decision-making mechanism (NNM) with the extensively studied voter model (VM) and majority model (MM) (both implemented as described by [17]), we conducted three tests with simulated swarms of robots in the site selection scenario, focusing on the robustness of collective decision-making strategies.

In all tests, unless otherwise indicated, the swarm size is 21 robots, with white floor coverage set at 80% for site L (best site), 50% for site M, and 20% for site S. Each randomly seeded trial lasts 800 s, during which robots operate following the procedure described in Sect. 3 The robots' communication range in all tests is 50 cm and the objective is to collectively identify the location of the site with the largest proportion of white floor (best option). Consensus is achieved when all robots hold the correct opinion for at least 10 s. For the NNM-controlled swarm, trials are repeated for all three conditions (Cond_1, Cond_2, Cond_3) outlined in Sect. 3, due to evolutionary bias. For VM and MM-controlled swarms, trials are conducted only in Cond_1, as these strategies are functionally symmetric with respect to the location of the best option. The applied performance measures are accuracy (number of trials with correct consensus) and time to consensus.

The results reveal the pronounced superiority of the NNM over the VM and MM controllers in terms of performance accuracy over 50 experimental trials (see Fig. 3a). The NNM maintains a high accuracy rate across all three conditions with minor deviations, which demonstrates its ability to collectively capture the environmental bias, represented by the variation of the proportion of the white floor, and exploit it to generate the required feedback modulation and opinion updating mechanism to push the swarm to achieve correct consensus. The significantly lower accuracy rates of the VM and MM suggest that these models struggled with the correct estimation of the sites' qualities, which is crucial for generating an ample distinction in the dissemination periods that break the initial equilibrium toward the correct opinion. The consensus time of

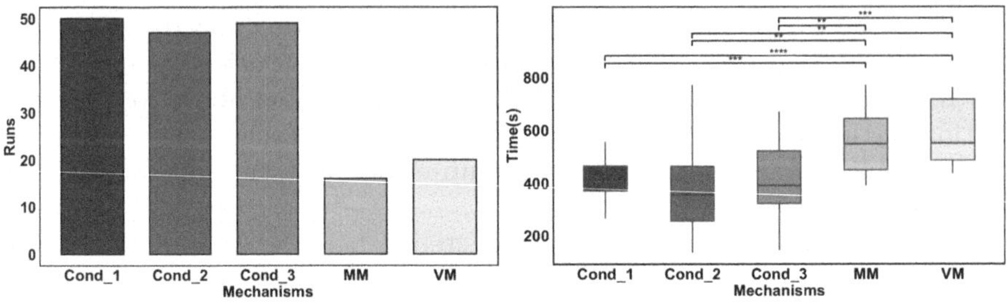

(a) (b)

Fig. 3. (a) Accuracy (i.e., consensus for at least 10 s) and (b) consensus time (successful trials only) over 50 trials of the NNM (all three conditions), MM and VM controllers. Pairwise Mann-Whitney U-test p-values – $*$ for $p < 0.05$, $**$ for $p < 0.01$, $***$ for $p < 0.001$, and $****$ for $p < 0.0001$.

the NNM is significantly lower than the VM and MM, corroborated by a pairwise Mann-Whitney U-test (see Fig. 3b). In this experiment, the initial opinions of the robots were uniformly and randomly distributed over the swarm (seven robots each), which follows a common practice in swarm robotics for site selection studies and is also applied during the evolution phase of the NNM.

In order to explore the robustness of the controllers under conditions that diverge from the standard setup, we performed series of tests with varying the initial number of robots holding the correct opinion from zero to 21. The results (see Fig. 4) reveal that the NNM controller remains resilient across diverse initial configurations, performing reliably with just a minor deviation in Cond_3, which can be attributed to evolutionary bias [1]. In contrast, the accuracy of the VM and MM controllers declines significantly when the initial number of

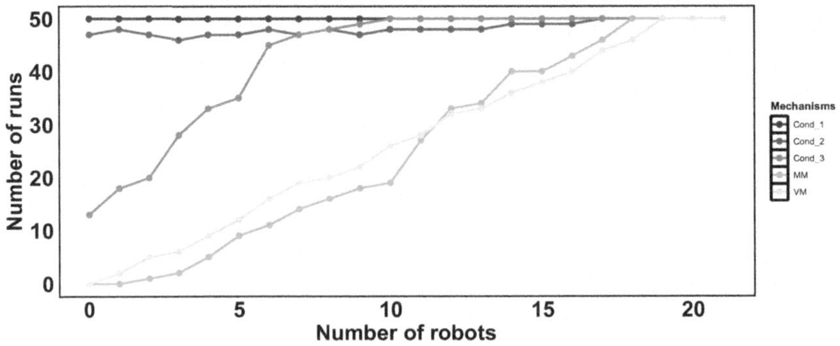

Fig. 4. The number of successful trials (i.e., with correct consensus) out of 50 runs, presented as a function of the initial number of robots holding the correct opinion (ranging from 0 to 21).

robots holding the correct opinion is lower than in the standard setup. Notably, consensus is never reached in cases where no robots initially hold the correct opinion.

We have investigated the scalability of the studied decision-making controllers with swarms of 42, 84, and 126 robots while keeping the robot density in the arena unchanged – thus the arena size has been modified to $2\sqrt{2}$ m × $2\sqrt{2}$ m (42 robots), 4 m × 4 m (84 robots), and $2\sqrt{6}$ m × $2\sqrt{6}$ m (126 robots). Figure 5a reveals that the accuracy generally drops with increasing the swarm size, however the performance of the NNM controller is the least affected among all. Considering our strict definition of consensus, which requires all robots to share the correct opinion for at least 10 s, we have investigated the evolution of opinions in more detail and found that the NNM controller often fails by a margin only of a few robots at the end of the trial (see Fig. 5b), which is not the case for the MM and VM controllers. This evidence indicates that the NNM controller is less affected by increasing swarm size than the VM and MM controllers.

The results of our study reveal a significant performance drop for the voter model and the majority rule controllers with respect to reported previous work [4,11,20], which is arguably due to the lack of global information typically used for positive feedback modulation. Earlier research has relied on the assumption that robots possess sensors capable of precisely estimating options' quality and employed these estimates for regulating dissemination time or frequency of communication events, thus explicitly reinforcing the positive feedback allowing to break the initial equilibrium and incite the swarm toward the correct option. In our experimental setup, the quality estimation occurs during exploration, where robots move randomly across the arena and sample the floor to estimate the site quality. This random exploration, however, leads to unstable (noisy) quality estimates that seem to have a severe detrimental effect on the VM and MM performance, which rely largely om a consistent distinction between options' qualities for generating the required bias in dissemination times that facilitates consensus towards the correct option.

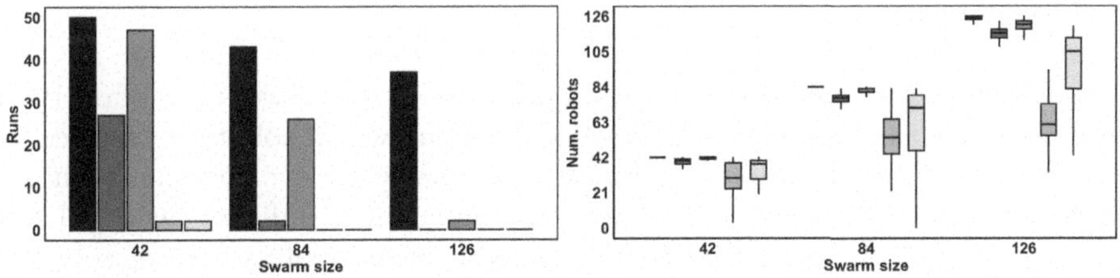

(a) (b)

Fig. 5. Results of the scalability test of the NNM (all three conditions), MM and VM controllers with swarm size of 42, 84, and 126 over 50 trials. (a) The number of successful trials (i.e., with correct consensus). (b) The number of robots with correct opinion at the end of each trial.

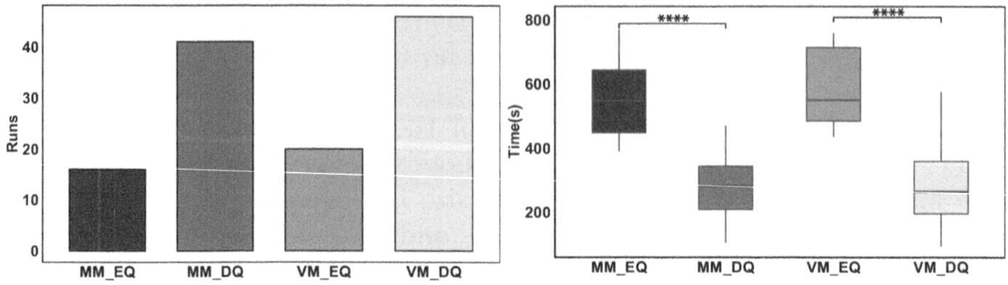

(a) (b)

Fig. 6. (a) Accuracy and (b) consensus time of the VM and MM controllers regulated with (estimated) noisy (MM_EQ and VM_EQ) and (direct) accurate option quality information (MM_DQ and VM_DQ). Pairwise Mann-Whitney U-test p-values – $*$ for $p < 0.05$, $**$ for $p < 0.01$, $***$ for $p < 0.001$, and $****$ for $p < 0.0001$.

To verify that this performance drop was indeed due to the lack of reliable quality estimation and not an implementation artifact in our solution, we conducted an additional test in which the robots were able to detect the correct quality of the sites for direct modulation, as reported in previous studies [4,19,20].

The results (see Fig. 6) reveal a pronounced significant difference in MM and VM performance when the robots are equipped with reliable (direct) and unreliable (estimated) option quality information. Our finding supports the hypothesis that accurate quality estimation is paramount for the effectiveness of this family of controllers. In this light, our results emphasize the strength and benefits of the evolutionary approach, which can yield high performance even in environments for which the option quality cannot be inferred directly. Hence, the automatic evolutionary process appears to be the most effective known solution to date in this scenario.

5 Conclusions

This paper presents the insights of a comparative study of two types of decision-making mechanisms for the site selection task in swarm robotics: hand-coded (i.e., voter model and majority rule) and automatically evolved (i.e., dynamic neural network). The comparisons are conducted in a uniform fashion based on standard performance criteria. The scalability and robustness of solutions to environmental variability are evaluated in a range of conditions. The results demonstrate that the evolved neural network controller offers better behavioural responses, higher accuracy and stronger resilience to varying environmental conditions with respect to traditional hand-coded solutions. Future work will focus on: i) more comprehensive analysis of the operational principles underlying opinion formation in single robots; ii) progressive elimination of simplifications in order to port the solution to physical robots; iii) modification of the site selection scenario by distributing the cues non-homogeneously; iv) development of integrated neuro-controllers supporting simultaneously opinion formation as well as environmental exploration.

Acknowledgements. A.A. thanks the CERUNA doctoral fellowship by the University of Namur for financial support. D.T. is funded by the European Union's Horizon 2020 research and innovation programme under the Marie Curie grant agreement No 101034383. Computational resources have been provided by the Consortium des Équipements de Calcul Intensif (CÉCI), funded by the Fonds de la Recherche Scientifique de Belgique (F.R.S.-FNRS) Grant No. 2.5020.11 and by the Walloon Region.

References

1. Almansoori, A., Trendafilov, D., Alkilabi, M., Tuci, E.: On the design of control mechanisms for a site selection task in a simulated swarm of robots. In: Hamann, H., et al. (eds.) Swarm Intelligence. ANTS 2024. Lecture Notes in Computer Science, vol. 14987. Springer (2024). https://doi.org/10.1007/978-3-031-70932-6_18
2. Beer, R.D.: A dynamical systems perspective on agent-environment interaction. Art. Intell. **72**, 173–215 (1995)
3. Brambilla, M., Ferrante, E., Birattari, M., Dorigo, M.: Swarm robotics: a review from the swarm engineering perspective. Swarm Intell. **7**(1), 1–41 (2013)
4. Masi, G.D., Prasetyo, J., Zakir, R., Mankovskii, N., Ferrante, E., Tuci, E.: Robot swarm democracy: the importance of informed individuals against zealots. Swarm Intell. **15**(4), 315–338 (2021). https://doi.org/10.1007/s11721-021-00197-3
5. Kato, S., Jones, M.: An extended family of circular distributions related to wrapped Cauchy distributions via Brownian motion. Bernoulli **19**(1), 154–171 (2013)
6. Ligot, A., Birattari, M.: Simulation-only experiments to mimic the effects of the reality gap in the automatic design of robot swarms. Swarm Intell. **14**(1), 1–24 (2020)
7. Mendiburu, F.J., Ramos, D.G., Morais, M.R., Lima, A.M., Birattari, M.: Automode-mate: automatic off-line design of spatially-organizing behaviors for robot swarms. Swarm Evol. Comput. **74**, 101118 (2022)
8. Mondada, F., et al.: The e-puck, a robot designed for education in engineering. In: Proceedings of the 9th International Conference on Autonomous Robot Systems and Competitions, vol. 1, pp. 59–65 (2009)
9. Nelson, A.L., Barlow, G.J., Doitsidis, L.: Fitness functions in evolutionary robotics: a survey and analysis. Robot. Auton. Syst. **57**(4), 345–370 (2009)
10. Nolfi, S., Floreano, D.: Evolutionary Robotics: The Biology, Intelligence, and Technology of Self-Organizing Machines. MIT Press (2000)
11. Prasetyo, J., De Masi, G., Ferrante, E.: Collective decision making in dynamic environments. Swarm Intell. (1), 217–243 (2019). https://doi.org/10.1007/s11721-019-00169-8
12. Salman, M., Garzón Ramos, D., Birattari, M.: Automatic design of stigmergy-based behaviours for robot swarms. Commun. Eng. **3**(1), 30 (2024)
13. Talamali, M.S., Marshall, J.A., Bose, T., Reina, A.: Improving collective decision accuracy via time-varying cross-inhibition. In: 2019 International Conference on Robotics and Automation (ICRA), pp. 9652–9659. IEEE (2019)

14. Talamali, M.S., Saha, A., Marshall, J.A., Reina, A.: When less is more: robot swarms adapt better to changes with constrained communication. Sci. Robot. **6**(56), eabf1416 (2021)
15. Tuci, E., Rabérin, A.: On the design of generalist strategies for swarms of simulated robots engaged in a task-allocation scenario. Swarm Intell. **9**(4), 267–290 (2015). https://doi.org/10.1007/s11721-015-0113-y
16. Valentini, G., Ferrante, E., Dorigo, M.: The best-of-n problem in robot swarms: formalization, state of the art, and novel perspectives. Front. Robot. AI **4**, 9 (2017). https://doi.org/10.3389/frobt.2017.00009
17. Valentini, G., Brambilla, D., Hamann, H., Dorigo, M.: Collective perception of environmental features in a robot swarm. In: Dorigo, M., et al. (eds.) ANTS 2016. LNCS, vol. 9882, pp. 65–76. Springer, Cham (2016). https://doi.org/10.1007/978-3-319-44427-7_6
18. Valentini, G., Ferrante, E., Hamann, H., Dorigo, M.: Collective decision with 100 kilobots: speed versus accuracy in binary discrimination problems. Auton. Agents Multi-Agent Syst. **30**(3) (2016). https://doi.org/10.1007/s10458-015-9323-3
19. Valentini, G., Hamann, H., Dorigo, M.: Efficient decision-making in a self-organizing robot swarm: on the speed versus accuracy trade-off. In: Proceedings of the 2015 International Conference on Autonomous Agents and Multiagent Systems, pp. 1305–1314 (2015)
20. Valentini, G., Hamann, H., Dorigo, M., et al.: Self-organized collective decision making: the weighted voter model. In: AAMAS, vol. 14, pp. 45–52. Citeseer (2014)

Emergent Naming System in an Unknown Dynamic Environment: A Collective Perception Scenario

Dat Nguyen Tien Hoang(✉) and Nicolas Cambier

Vrije Universiteit Amsterdam, Amsterdam, The Netherlands
datnguyen11.2003@gmail.com, n.cambier.research@gmail.com

Abstract. Best-of-n problems have been rigorously studied in the context of swarm intelligence. However, most solutions to this problem rely on a series of assumptions such as prior knowledge about the number or features of available options. In this study, we investigate a collective perception task with no prior knowledge about option features, in a dynamic environment. The added constraint renders previous approaches impractical, as robots are no longer able to use option features to identify them when communicating. To resolve this problem, we propose a novel probabilistic Naming Game routine that allows agents to converge on words that uniquely identify such features. We also propose a new quality estimation method that is compatible with the proposed Naming Game routine and easier to optimise. We compare the proposed approach with two baselines: decision convergence without communication, and decision convergence with pre-existing identifiers. We perform scalability tests for larger swarm sizes in larger environments. Our findings indicate that swarms utilising the probabilistic Naming Game to interpret neighbours' communication signals for the purpose of quality estimation generally performed better than swarms that do not communicate and that this behaviour is relatively scalable. We also found that swarms using communication suffer from social inertia, that impacts the probabilistic Naming Game swarms less than swarms with prior environmental information.

Keywords: Swarm Intelligence · Collective Perception · Unknown environment · Naming Game · Best-of-n · Collective Decision Making

1 Introduction

Collective decision-making is a decentralised process in which a population of agents come to a consensus on a particular decision. This decision should emerge as the consequence of the population interacting and communicating locally, and should not be attributable to any individual in the swarm [14]. In other words, as there exists no "coordinator" to aggregate the opinions of individuals, the collective decision emerges as a function of swarm individual behaviours. This

T. Carletti et al. (Eds.): WIVACE 2024, CCIS 2532, pp. 111–124, 2025.
https://doi.org/10.1007/978-3-031-93631-9_9

not only allows for trivial scaling of the system but also increases the tolerance of the system in terms of the number of agents. Collective decision-making has regularly been explored and formulated as instances of best-of-n problems: In an environment with n options of varying qualities, determine which option is of the highest quality.

One specific instance of the best-of-n problem is the *Collective Perception* task, whereby the agents have to determine which of $n = 2$ (sometimes extended to $n > 2$ [5,9]) options in a checked environment is most prevalent [13], as presented in Fig. 1. This task is useful to analyse the generality of a model as any given option is scattered in the environment, thus its quality (i.e., the proportion of the option in the environment) can not be determined from a single vantage point. Various approaches to the task have been proposed, some of which were hermetic to environmental changes [8,13], while others were adaptable to a dynamic environment [1,10,11]. These works however all assume prior knowledge of the environment: unique identifiers of options are shared across the population, hence individuals' opinions regarding the options can be trivially communicated by using a signal pre-associated with a given option's features (e.g. 0 for black and 1 for white).

Fig. 1. A (55%) white-dominant environment with 20 agents. Agents' colours indicate their current opinion (green-white, red-black). (Color figure online)

Consider now scenarios in which environmental features are unknown and agents' sensors are realistically noisy; identifying options by their sensor values becomes impossible as we do not know in advance what threshold to use to distinguish between two options, and that threshold might vary from agent to agent depending on the sensor noise. Moreover, pointing an option or guiding other agents toward it is non-trivial and time-consuming for swarm robots. In such a scenario, successful communication therefore requires agents to agree on some symbols (e.g. values) referencing each option. In other words, the swarm must learn to "name" environmental options. For this purpose, language games are useful, as demonstrated by a previous work on collective navigation using virtual pheromones, that assumed a similar set of constraints as the one we have described above [6] (i.e. no preexisting values associated with each option).

Language games emerged as a study of how symbolic language evolves in a population. In particular, consider the naming game: At any time, 2 individuals are randomly selected from the population, with one being the speaker and the other the hearer. The speaker chooses a topic and a word associated to that topic from its memory, and communicates it to the hearer. If the hearer knows the word and associates it to the same topic as the speaker, then the naming game is considered a success and the association is reinforced. Otherwise, it is considered a failure and the association is weakened. A population playing this naming game will eventually converge to a set of word-topic associations that is shared across the whole population [12]. Various versions of the naming game, as well as numerous repair strategies for when the naming game fails have been analysed [3,7]. However, to the authors' best knowledge, the naming game research so far had all assumed, at least partly, definitive topic disambiguation from the hearer (e.g. via pointing), which requires some adaptation given realistic constraints (as we outlined above, referring to options are not possible).

In this study, we consider a collective perception scenario [13] in which the binary environment is unknown and changes after some time. We present a novel approach to the task by playing a Probabilistic Naming Game (PNG) on top of a simple yet original quality estimation method. We then show how effective the communication induced by the naming game is, compared to the cases in which the population 1) does not communicate and 2) communicates with prior environment knowledge. We tuned all relevant parameter settings on a small swarm and environment size, then the same parameters were tested for scalability. We found that the PNG swarms show a significant performance improvement compared to swarms that do not communicate, despite being less scalable than swarms with prior knowledge. PNG swarms even outperform prior knowledge swarms after the environment change in terms of our defined Performance metric. We also discuss the implications of these results, the limitations of this study and our approach, and potential future research directions. In this paper, we focus on the best-of-2 case, but we propose a method to extend to best-of-n ($n > 2$) in Sect. 4.

2 Methodology

We consider a collective perception task in which the binary options of the environment are *a priori* unknown and change after some time; this implies that the options present in the environment could not be directly referred to from the outset, and that the "correct decision" of the swarm is not static. It should be noted that even though the agents could not communicate the options, they are still able to differentiate any two options features with their sensor. The environment being unknown entails a constraint that previous approaches did not take into account, rendering them unsuitable: inherently agents could not globally assign identifiers to the options, and consequently are unable to communicate the options and by extension opinions regarding the options. To resolve this, we propose an approach consisting of a homogeneous swarm playing a probabilistic version of the naming game. Swarm individuals utilizing this method would be able to communicate effectively and converge on the decision of the highest-quality option (i.e., option making up the highest proportion of the environment).

2.1 Approach

At any timestep, the agent performs the four following routines: random walk (and obstacle avoidance), quality-selection, opinion signal matching, and probabilistic naming game. The approach was motivated by the following rationale:

- Agents explore the environment "individually", forming an idea of which option is more/less dominant.
- Agents play the naming game with that knowledge and come to a consensus on the identifiers for all options using Naming Game messages.
- Agents should be able to use the identifiers to (sustain and/or) converge to a consensus faster (especially after the environmental changes). For this purpose, they communicate with quality-related messages, using the names from the Naming Game as identifiers.

The main obstacle to this approach is that the agents are only able to converge to some option-identifier pairs if they agree on which option is more/less dominant; yet coming to a consensus on the dominant option depends on interpreting the communication signals. To resolve this issue, we propose a probabilistic version of the naming game with lateral inhibition strategy [12]. That is, for each word, either increase or decrease their association strength with each option according to whether the word-option association corresponds to the agent's experiences; such that the naming system converges as the swarm's decision converges, forming a positive feedback loop. We hypothesize that this allows the population to 1) sustain the consensus once it is achieved and 2) make use of the "naming" identifiers to converge faster after the environment changes.

Random Walk Routine. We implemented a random walk routine similar to that of NNM [1]. Every 5 s, the agent uniformly samples an angle $\omega \in [-\pi, \pi]$, rotates itself by ω, then continues to move forward for 5 s. At any timestep, if the agent senses an obstacle ahead (i.e., the border or another agent), then it performs the sampling and rotating sub-routine until no obstacles are sensed.

Quality-Selection Routine. Swarm individuals have mappings {option: quality}, indicating the estimated quality for some encountered options. When the agent encounters a new option, they initialise that option's quality to the mean of other known option's qualities; if this is the first option encountered, the initialised value is inconsequential as it is normalized every timestep. At any timestep where the individual senses a given option, the quality corresponding to that option is incremented by ϵ_{sens}. Similarly, an option's quality is increased by ϵ_{comm} when the individual receives an opinion signal for that particular option (the signal is matched in the *Opinion signal matching routine*). The qualities of the options are normalized after this incrementation. We also put an upper limit on the quality of any option at 0.55 for a similar reason.

The routine not only gives more weight to the more recently sensed/received options, but also makes further increments/decrements to qualities smaller when the qualities are higher/lower, increasing sensitivity to environment changes.[1]

At any time T, agents have some probability P_{comm} to broadcast a word that represents their opinion. An agent's opinion is the dominant option with the higher quality score while the word corresponding to the opinion is the one with the highest word-option association score (2.1). If no words are present in the lexicon for the more dominant option i, then the agent generates a random word $w_i \in [0, 255]$ and associates w_i with the option with score s.

Opinion Signal Matching Routine. At any time T, suppose that the agent received the word w_{opi} in a quality-related message from its neighbour, it would choose an option that is likely to be represented by this word. This option is chosen probabilistically, proportionally to the normalized scores of the word for each option (Eq. 1). Let us recall that this matched option's quality is then increased by ϵ_{comm} in the *Quality-Selection Routine.*

$$P(matchedO = o) \sim score[o][w_{opi}] \tag{1}$$

Probabilistic Naming Game Routine. As a hearer, this routine is triggered at any time T, if the agent has received Naming Game message as $[w_0, w_1]$ (2.1), and the agent sensed option o_S in the environment. The goal of the routine is to match w_0 and w_1 with some options o_0 and o_1. We introduce the parameter c_{bias} governing this bias term. Figure 2 describes the word interpretation subroutine for the Probabilistic Naming Game. It should be noted that the algorithm utilises $quality[opinion]$ as a proxy for quality difference; this is only possible since the environment is binary: certainty for one option translates to certainty for the other. Also note that in Fig. 2, $score[o][w]$ refers to the normalized scores, as we want the relative certainty between options for each word.

The matching gives more weight to the quality of the option if the agent is unsure which option the word should be matched with, and less when they are certain: At the onset, agents rely on the qualities as the options do not have reliable "names"; conversely, agents relying on the changing qualities (due to the dynamic environment) late in the experiment could cause the naming system to diverge. Furthermore, as the absolute quality value of the option is dependent on ϵ_{sens} and ϵ_{comm}, setting the two parameters high to increase sensitivity compelled the naming to converge prematurely. This is the rationale for the parameter c_{bias}: one option is certainly more frequent than the other; if the agent is sensing o_S, it is more likely that the option is the dominant one.

[1] https://github.com/DaBo111/Wivace2024-EmergentNaming/blob/main/Wivace2024_appendix.pdf.

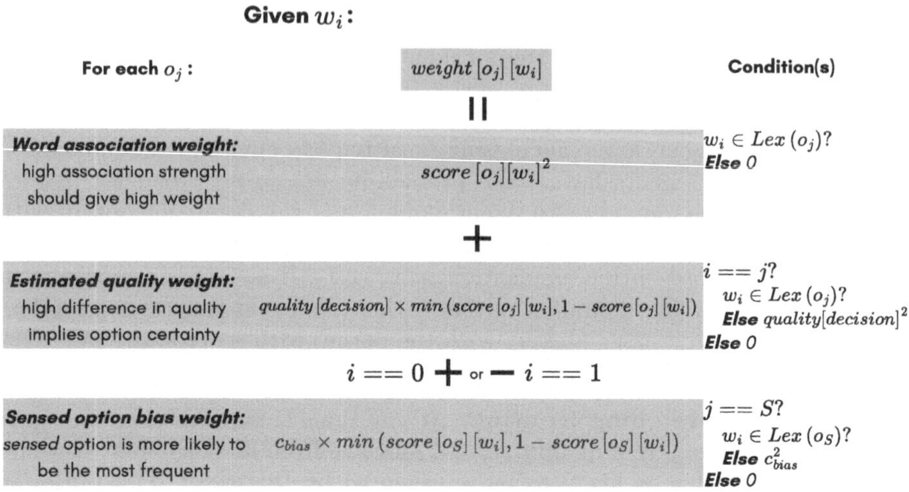

Fig. 2. Options weighting for each word. The final weight of an option is the summation of three components. They are calculated by the illustrated formulas given conditions on the right are satisfied, otherwise the "*Else*" value is used for each of the components instead. $Lex(o)$ refers to the lexicon of o (i.e., {all words w} that is associated with o).

For every word w_i, the matching (Fig. 2) returns the unnormalized probability distribution for all options o. The matched o_i is then chosen with some probability (Eq. 2).

$$P(o_i = o) \sim weight[o][w_i] \tag{2}$$

If w_i is not in the lexicon of o_i, initialise the score of w_i in o_i with s, otherwise increase the score with δ_{inc} (Eq. 3). All other words $w^* \neq w_i$ in the lexicon get their scores inhibited with δ_{inh} (Eq. 4). For memory efficiency purposes, if the association score for any option-word pair decreases below ϵ_{min}, the agent removes the association from its memory. In the following equations, $score[o][w]$ refers to the unnormalized scores.

$$score[o_i][w_i]_{T+1} = score[o_i][w_i]_T + \delta_{inc} * (1 - score[o_i][w_i]_T) \tag{3}$$

$$score[o_i][w^*]_{T+1} = score[o_i][w^*]_T - \delta_{inh} * score[o_i][w^*]_T, \forall w^* \neq w_i \tag{4}$$

The agents also naively deal with homonyms (i.e., same words but different meanings) by just removing them from the lexicon. Homonyms are identified by getting the most probable word for each option; if the words are the same, then they are homonyms.

As a speaker, at any time T, agents have some probability P_{NG} to broadcast some words $[w_0, w_1]$. w_0 refers to the word associated with the agent's current opinion (dominant option) and w_1 refers to the word associated with the non-dominant option.

2.2 Experimental Setup

We propose a novel approach to the collective perception task with *a priori* unknown about environmental features (in this case, the colour of the options). We designed a set of experiments and metrics to investigate to what extent this proposed approach assists in the swarm's decision-making process in terms of convergence speed, convergence accuracy, resilience to environmental changes (i.e. option swap), and scalability.

Environment. The environment is a 200 cm × 200 cm continuous plane bounded by a barrier. The surface of this plane consists of 10 cm × 10 cm cells, representing options of the environment; here, the cells can either take the value 0 (white) or 1 (black). One of the options is dominant, covering 55% of the plane. To simulate the experimental conditions of both [1,13], we implemented our model in the violet-simulator[2], and let the agent's shape be a circle of diameter 7 cm × 7 cm (the cross-section of an e-puck robot), moving at 20 cm/s. All the specifications, including the ones mentioned henceforth, are scaled in terms of "pixels" (p) and "ticks" (T) (instead of "centimetres" (cm) and "seconds" (s)) with the ratio of $10T = 1\,\text{s}, 1\,\text{cm} = 5p$. To simulate an unknown environment, the sensor values of random agents are swapped. After some time T_{change}, the environment changes such that the dominant option becomes non-dominant and vice versa; this change occurs not simply by flipping the options' feature but instead by providing the environment with a new option distribution. For the simplicity of evaluation, we keep track of the true (pre-swapped) sensor values of each agent and evaluate their opinion on that true sensor value. In both the duration before and after the environment change, the swarm has to come to a consensus on the current dominant option of the environment. Hereafter, we refer to "phase 1" as the part of the simulation before the environment change and "phase 2" as the section after the change.

Experimental Parameters. We performed experiments with 20, 40 and 100 agents to evaluate the scalability of our approach; for this test, we also scaled the environment dimensions to preserve the swarm density, and all other settings were kept identical. The parameters of our model were manually tuned for the setting with 20 agents as shown in Table 1:

Table 1. Parameters

ϵ_{sens}	ϵ_{comm}	ϵ_{min}	δ_{inh}	δ_{inc}	s	P_{NG}	P_{comm}	c_{bias}
5e−04	2.5e−04	0.01	1.5e−03	1e−03	0.5	0.15	0.05	0.75

[2] https://github.com/m-rots/violet.

Evaluation Metrics. We define a swarm as having converged on a decision if they can sustain the consensus for at least $50s$ ($500T$) both pre and post environment change. The value of $50s$ was chosen to be in the same scale as [1]'s experiments relative to the duration of each simulation run. As we also want to investigate the effect of our approach on the decision convergence of the swarm in a dynamic environment, we propose a "Performance metric" (5).

$$Performance = \frac{\Sigma_{t=T}^{D^*}((t-T) * e^{R_{opi_T}})}{\Sigma_{t=T}^{D^*}((t-T)}, \tag{5}$$

where $T = 0, D^* = T_{change}$ if measured in phase 1 and $T = T_{change}, D^* = Duration$ if measured in phase 2. R_{opi_T} is the ratio of population that holds the correct opinion in timestep T.

R_{opi_T} is exponentiated as we place more importance on whether the population actually converges; this is then weighted by the timestep value to further emphasize convergence in the later timesteps (as opposed to high R_{opi_T} at the beginning but do not converge throughout the run).

3 Results

We performed 100 seeded runs with the optimised parameters in an environment with 20, 40 and 100 agents for 3 different communication strategies, the proposed strategies and 2 baselines: one wherein the agents do not communicate and only estimate the options' qualities via the quality estimation method (No Communication, NC); the other wherein the agents communicate with prior information (Prior Information, PI), i.e., directly communicating the pre-swapped sensor values. Each seeded run lasted for $2000s$ ($20000T$) and the environment changes were scheduled at $6000T$, $10000T$ and $14000T$ (in distinct runs).

3.1 Decision Convergence

We present the evaluation results demonstrating how the 20-agent swarm's decisions converge in the duration of the simulation in Fig. 3 (Left). First and foremost, we observe that the communication routine does in fact impact convergence; in fact, the routine enables the sustenance of the consensus. This is illustrated by the fact that swarms that do not estimate quality via communication are more likely to fail to converge to a consensus both before and after the environment change. Secondly, in phase 1, swarms with prior information of the environment can converge much faster than the ones that do not.

This difference in speed, however, is not observed in phase 2 of the simulation: the convergence speeds of the two settings are relatively similar. In fact, if the PNG swarm is allowed sufficient time to converge on both the decision and the language (see Sect. 3.2), the behaviour of the swarm after the swap is almost identical to that of the PI swarm. Furthermore, one can observe a slight correlation between the duration of time from the point of convergence to the

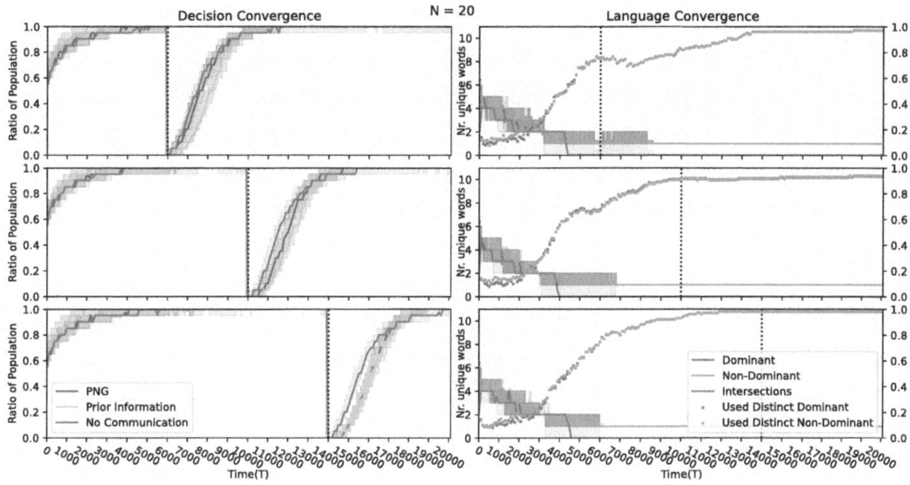

Fig. 3. Swarm of 20 agents. **Upper, Middle, Lower:** $T_{change} = 6000, 10000$ and 14000. **Left:** Lines with shaded percentiles represent the ratio of agents having the correct opinion. **Right:** Lines with shaded percentiles represent the number of unique words the population used for a specific option (Blue: dominant, Yellow: non-dominant); the Green line represents the number of unique intersections between the two lexicons. Scatter dots represent the mean ratio of the population using a specific distinct word (words that are not in the intersection of the lexicon) for each of the options (Blue: dominant, Yellow: non-dominant). (Color figure online)

swap and the time it takes for the swarm to start deviating from the consensus after the swap for swarms that did use communication. This behaviour can be attributed to the quality estimation method of our approach: as the swarm spends more time in an environment, individuals are allowed to further reinforce the quality associated with the dominant option in that environment. We call this phenomenon "social inertia", the longer the swarm is exposed to an environment the less sensitive it is to changes; this is somewhat alleviated by our decision to cap the estimated quality at 0.55 in Sect. 2.

3.2 Language Convergence

We can observe, in Fig. 3 (Right), that, during all three environment change schedules, the 20-agent swarm can converge to a unique word for each option in most simulation runs. Interestingly, the decision convergence of the swarm does not require the whole population to converge on the language; this is most obvious in the case with $T_{change} = 10000$, wherein the swarm reaches a sustained consensus in most of the simulation runs at around $T = 6000$ while only around 0.8 of the population has reached a language consensus. Figure 3, $T_{change} = 6000$ also demonstrates the case when the change occurs before the swarms reach a decision consensus: there is a decrease in the population's meaningful language

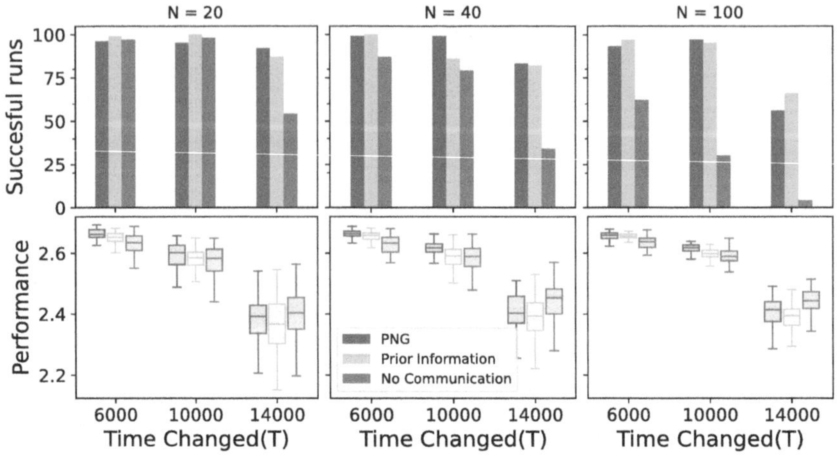

Fig. 4. Upper: Number of runs in which the swarm converged to a decision consensus in *phase* 2 and **Lower:** the "Performance" of the swarm in phase 2 for swarms of 20 (**Left**), 40 (**Middle**) and 100 agents (**Right**) in 100 simulation runs.

usage shortly after the change, yet the population is still able to converge on a meaningful language system.

3.3 Scalability

Figure 4 (Upper) shows the number of simulation runs in which the swarm was able to converge to the correct decision in phase 2. Unexpectedly, in the basic setting (swarm size $N = 20$), swarms that do not interact could still somewhat consistently converge to a decision; this implies that the task can still be achieved without requiring collective behaviour. However, this behaviour is neither scalable nor robust without the effect of communication (PI and PNG swarms). This is evidenced by results when the swarm size is scaled to 40 and 100 agents and/or when the change schedule is delayed. Table 2 shows that, while PNG performs significantly worse than or on par with PI (in terms of successful runs), it also significantly outperforms NC for both phases in most settings.

Figure 4 (Lower) demonstrates the Performance metric defined in Sect. 2.2. We observe an interesting pattern here: across all three scales, PNG is more likely to, and converge faster than both PI and NC for $T_{change} = 6000$ and 10000 while for $T_{change} = 14000$, PNG performs better than PI but worse than NC in phase 2. Table 2 shows that, in phase 1, the Performance of PNG is significantly worse than that of PI while being comparatively similar to NC. This performance difference is significant in most cases. The difference in performance between PNG and PI in phase 2 can be attributed to the fact that PI swarms take longer to start switching opinions after the environment changes (see Sect. 3.1). This (lack of) social inertia also plays a role in the sudden increase in NC's performance relative to PNG and PI when $T_{change} = 14000$. The gradual degradation in

Table 2. Mann-Whitney U test for performance and binomial test for number of successful runs for various swarm sizes. PNG was tested against the two baselines in 100 simulation runs. * indicates whether H_0 was rejected to α = 5e−02. A conclusion is drawn for each phase if PNG is either significantly better/worse in both metrics, or in one metric and similar in the other metric. P1 refers to "Phase 1" and P2 refers to "Phase 2".

N	T_{change}	Against	Performance				Successful runs				Conclusions
			Phase 1		Phase 2		Phase 1		Phase 2		
			H_1	p-value	H_1	p-value	H_1	p-value	H_1	p-value	
20	6000	PriorInfo	<*	4e−13	>*	2e−05	<*	0e−00	<*	1e−02	P1 worse
		NoComm	≠	1e−01	>*	1e−11	≠	5e−01	≠	5e−01	P2 better
	10000	PriorInfo	<*	5e−15	>*	2e−03	<*	0e−00	<*	0e−00	P1 worse
		NoComm	≠	6e−01	>*	3e−03	≠	3e−01	≠	5e−01	P2 better
	14000	PriorInfo	<*	5e−16	≠	3e−01	<*	0e−00	≠	1e−01	P1 worse
		NoComm	≠	3e−01	≠	1e−01	<*	0e−00	>*	9e−17	P1 worse, P2 better
40	6000	PriorInfo	<*	3e−19	>*	2e−03	<*	2e−30	<*	0e−00	P1 worse
		NoComm	≠	7e−01	>*	1e−16	>*	1e−02	>*	1e−05	P1, P2 better
	10000	PriorInfo	≠	4e−01	>*	1e−10	≠	8e−01	>*	4e−06	P2 better
		NoComm	≠	7e−01	>*	1e−08	≠	3e−01	>*	1e−09	P2 better
	14000	PriorInfo	<*	1e−22	>*	3e−02	≠	6e−01	≠	8e−01	P1 worse, P2 better
		NoComm	≠	6e−01	<*	5e−05	>*	5e−03	>*	8e−24	P1 better
100	6000	PriorInfo	<*	1e−29	≠	3e−01	<*	3e−115	<*	3e−02	P1, P2 worse
		NoComm	≠	4e−01	>*	2e−15	>*	2e−02	>*	1e−12	P1, P2 better
	10000	PriorInfo	<*	4e−33	>*	5e−13	<*	1e−35	≠	4e−01	P1 worse, P2 better
		NoComm	≠	7e−01	>*	3e−15	>*	4e−09	>*	1e−46	P1, P2 better
	14000	PriorInfo	<*	4e−33	>*	7e−03	≠	1e−00	<*	2e−02	P1 worse
		NoComm	≠	1e−01	<*	3e−09	>*	1e−17	>*	4e−51	P1 better

Performance as T_{change} increases is caused by the same dynamic (for NC, this degradation is due to not social inertia, but purely self-reinforcement inertia). Furthermore, this Performance metric does not fluctuate much as the setting is scaled.

All in all, our results show that PNG generally outperforms NC in phase 2 while being overshadowed by PI in phase 1.

4 Conclusions

In this study, we presented a novel approach to the Collective Perception scenario [13] in a dynamic environment and with additional constraints, namely the absence of any prior knowledge about the features of available options. Our contribution is two-fold: on the one hand, we proposed a simple quality estimation strategy that is capable of dealing with a dynamic environment; on the other hand, we showed how a probabilistic naming game routine could be adapted to the task, enabling meaningful communication between individuals in the swarm while also offering a better trade-off between resilience to environmental changes and achieving opinion convergence (compared to swarms with prior environment knowledge).

We tested our Naming Game approach against two baselines, one with known prior (Prior Information, PI) and one wherein agents do not communicate (No Communication, NC), and found that, while the results of our Probabilistic Naming Game (PNG) approach do not reach that of PI in phase 1, it outperformed NC in phase 2 in most cases; this phenomenon is observed across all simulation scales. This means that the swarm is able to create meaningful communication about the available options. We also found that, in terms of the Performance metric, PNG is statistically significantly better than both PI and NC in phase 2 if the environment changes shortly after the beginning of the experiment, and worse than NC as the change happens later, demonstrating the effect of social inertia. Our communication approach enables the swarm to be relatively scalable: larger swarms using PNG and PI are still able to converge somewhat consistently in phase 2 when the environment change schedule is early, while NC's convergence rate plummets. PNG's results could be seen as the intermediate between PI and NC, having less social inertia than the former (while also obeying harder constraints) but converging more consistently than the latter. Unexpectedly, we found that our quality estimation method performs well enough, such that the task can be achieved relatively consistently without interaction in the "easy" setting; to the best of the authors' knowledge, this was neither mentioned nor demonstrated in any prior related work.

Although the study focused specifically on the Collective Perception task, our probabilistic Naming Game approach can be adapted to other best-of-2 problems, either coupled with our proposed quality estimation method or with another suitable strategy (that expresses agents' certainty regarding an option e.g., NMM [1]). Extending to best-of-n, however, requires modification in the Naming Game word interpretation sub-routine as an assumption does not immediately generalize to $n > 2$: in a binary environment, certainty of one option implies certainty of the other option; a solution to this could be for the hearer to match a received word to an option by a "one-vs-all" method. We conjecture that this enables language convergence, as agents treat i^{th} option as dominant when matching i^{th} word, supposing all other "more dominant" options $j < i$ do not exist, essentially reducing the task to be equivalent to the setting studied in this paper. Another interesting extension could be to adapt our approach to continuous environments in which the naming game routine could be motivated by a model studied for colour naming patterns [4]. This could also tackle the assumption that agents can individually differentiate different options: handling sensor noise might be non-trivial and agents have to learn to categorize their own sensor values. Another assumption we implicitly made was that the environment options are "somewhat" uniformly distributed, as this allows individuals to form their own opinions to match the words of the probabilistic Naming Game; future research could either address this problem or show that this requirement is strictly necessary for the task to be feasible. Another immediate next step from this study is to determine whether the probabilistic Naming Game is asymmetric when spoken non-deterministically and interpreted deterministically (as opposed to our approach); analogously, the Naming Game's feedback requirement is asymmetric: only updates in the hearers' lexicon is required for the naming system to converge [3].

Acknowledgments. Violet-simulator was developed by Storm Timmermans. The seeded runs were performed on the Distributed ASCI Supercomputer DAS5 [2].

Disclosure of Interests. The authors have no competing interests to declare that are relevant to the content of this article.

References

1. Almansoori, A., Alkilabi, M., Tuci, E.: On the evolution of adaptable and scalable mechanisms for collective decision-making in a swarm of robots. Swarm Intell. 1–21 (2024)
2. Bal, H., et al.: A medium-scale distributed system for computer science research: infrastructure for the long term. Computer **49**(05), 54–63 (2016). https://doi.org/10.1109/MC.2016.127
3. Baronchelli, A.: Role of feedback and broadcasting in the naming game. Phys. Rev. E **83**(4), 046103 (2011)
4. Baronchelli, A., Gong, T., Puglisi, A., Loreto, V.: Modeling the emergence of universality in color naming patterns. Proc. Natl. Acad. Sci. **107**(6), 2403–2407 (2010)
5. Bartashevich, P., Mostaghim, S.: Multi-featured collective perception with evidence theory: tackling spatial correlations. Swarm Intell. **15**(1), 83–110 (2021)
6. Cambier, N., Eiben, A., Ferrante, E.: Emergent naming system in an unstructured environment: a shortest-path discovery case study. In: Artificial Life Conference Proceedings 35, vol. 2023, p. 47. MIT Press, Cambridge (2023)
7. Chen, G., Lou, Y.: Naming Game. Springer, Cham (2019)
8. Ebert, J.T., Gauci, M., Mallmann-Trenn, F., Nagpal, R.: Bayes bots: collective Bayesian decision-making in decentralized robot swarms. In: 2020 IEEE International Conference on Robotics and Automation (ICRA), pp. 7186–7192 (2020). https://doi.org/10.1109/ICRA40945.2020.9196584
9. Ebert, J.T., Gauci, M., Nagpal, R.: Multi-feature collective decision making in robot swarms. In: Proceedings of the 17th International Conference on Autonomous Agents and MultiAgent Systems, pp. 1711–1719 (2018)
10. Pfister, K., Hamann, H.: Collective decision-making with Bayesian robots in dynamic environments. In: 2022 IEEE/RSJ International Conference on Intelligent Robots and Systems (IROS), pp. 7245–7250. IEEE (2022)
11. Soorati, M.D., Krome, M., Mora-Mendoza, M., Ghofrani, J., Hamann, H.: Plasticity in collective decision-making for robots: creating global reference frames, detecting dynamic environments, and preventing lock-ins. In: 2019 IEEE/RSJ International Conference on Intelligent Robots and Systems (IROS), pp. 4100–4105 (2019). https://doi.org/10.1109/IROS40897.2019.8967777
12. Steels, L., Loetzsch, M., et al.: The grounded naming game. Exp. Cult. Lang. Evol. **3**, 41–59 (2012)

13. Valentini, G., Brambilla, D., Hamann, H., Dorigo, M.: Collective perception of environmental features in a robot swarm. In: Dorigo, M., et al. (eds.) ANTS 2016. LNCS, vol. 9882, pp. 65–76. Springer, Cham (2016). https://doi.org/10.1007/978-3-319-44427-7_6
14. Valentini, G., Ferrante, E., Dorigo, M.: The best-of-n problem in robot swarms: formalization, state of the art, and novel perspectives. Front. Robot. AI **4**, 9 (2017)

Mixed-Variable Surrogate-Based Optimization Using Affinity Features to Improve the Distance for Categorical Variables

Charlotte Beauthier[1](✉), Rajan Filomeno Coelho[1], Sylverio Pool Marquez[2,3], Caroline Sainvitu[1], and Annick Sartenaer[2,3]

[1] Cenaero, rue des Frères Wright 29, 6041 Charleroi, Belgium
{charlotte.beauthier,rajan.coelho,caroline.sainvitu}@cenaero.be
[2] University of Namur, rue Joseph Grafé 2, 5000 Namur, Belgium
{sylverio.poolmarquez,annick.sartenaer}@unamur.be
[3] Namur Institute for Complex Systems (naXys), rue Joseph Grafé 2, 5000 Namur, Belgium

Abstract. In practice, various industrial design problems have continuous, discrete, and categorical variables. Our objective is to manage efficiently these different types of variables within a surrogate-based optimization process. In this work, we propose to redefine the notion of distance between the possible values of a categorical variable (named "attributes"), through the concept of "affinity". The notion of affinities between attributes can be interpreted as a weighted relationship between attributes. These affinities are usually defined based on a physical intuition of the designer. Indeed, affinities are generally implicitly associated to the behavior of one or several outputs that behave(s) similarly for various attributes. In order to study the impact of the use of affinities, numerical results are presented on specific test problems coming from structural and mechanical design frameworks.

Keywords: Surrogate-based optimization · Mixed-variable optimization · Evolutionary algorithm · Derivative-free optimization · Categorical variables

1 Introduction

The objective of this work is the development of methods combining numerical simulation tools and advanced optimization algorithms, which play a crucial role in exploring new conceptual designs. Nowadays the exploitation of multidisciplinary optimization using high-fidelity simulation models is common to many engineering design problems. A globally effective approach to high-fidelity optimization problems based on computationally expensive analysis lies in the exploitation of surrogate models. They act as cheap-to-evaluate alternatives to

T. Carletti et al. (Eds.): WIVACE 2024, CCIS 2532, pp. 125–136, 2025.
https://doi.org/10.1007/978-3-031-93631-9_10

the original high-fidelity models reducing the computational cost, while still providing improved designs. Furthermore, in practice, various industrial design problems have continuous, discrete, and categorical variables. From an engineering point of view, the specific case of categorical variables is of great practical interest by their ability to represent the choice of a material, the type of engine architecture, the shape of a cross-section for a beam profile, etc.

This work takes place in a development environment provided in [7]. More precisely, the developments are carried out in MINAMO, Cenaero's in-house design space exploration and multi-disciplinary optimization platform. The underlying principle of Surrogate-Based Optimization (SBO) consists in accelerating the optimization process by essentially exploiting surrogate models for the objectives and constraints evaluation, with a minimal number of function calls to the high-fidelity models in order to keep the computational cost within affordable limits [4]. In more details, this online SBO, carried out by MINAMO, consists on several steps, as illustrated in Fig. 1. First of all, a Design of Experiments (DoE) is defined using an a priori space filling technique LCVT (Latinized Centroidal Voronoi Tessellations). Here, the objective is to extract as much information as possible from a minimum number of experiments. After the evaluation of the DoE by the high-fidelity models, the next step is then the online SBO process. The surrogate models (a tuned version of Radial Basis Functions (RBF) networks here) are trained based on the available information in the database, and an evolutionary optimization step is launched to generate new best candidates for the given optimization problem. These candidates are then evaluated by the high-fidelity models and their accurate performance is checked afterwards. Finally, the new candidates are added to the database and the online SBO is repeated until a satisfactory performance is achieved.

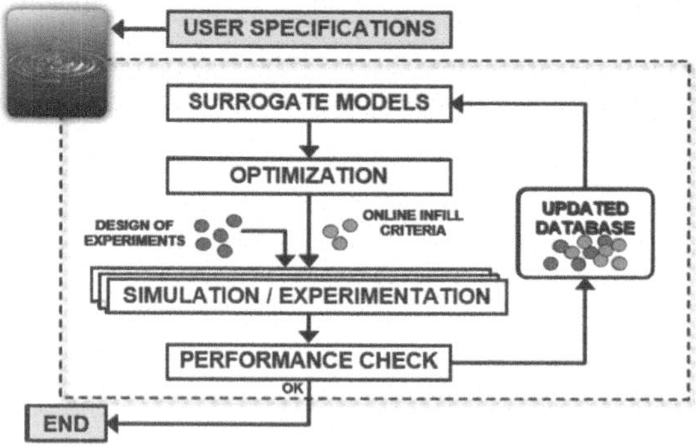

Fig. 1. Online surrogate-based optimization scheme process within MINAMO.

The contributions of this paper are focused on the management of mixed variables in this surrogate-based optimization process. For this purpose, a method is proposed to redefine the notion of distance between two attributes of a categorical variable, through the concept of affinity.

In this paper, Sect. 2 introduces and investigates the concept of affinity for categorical variables while Sect. 3.2 describes the test problems (coming from mechanical and structural design), explains the experimental setup, and presents the numerical results obtained. Finally, Sect. 4 summarizes the current contributions and discusses the next steps for future research.

2 Affinity-Based Distance

2.1 Mixed-Variable Distance

Our objective is to manage efficiently and concurrently the different types of variables in MINAMO, within a surrogate-based optimization process. Since the construction of the surrogate models and the optimization search require the comparison of individuals with variables of different types, the best approach is to define a specific distance, as mentioned in [10]. One way to handle applications with both continuous and categorical variables is by means of a heterogeneous distance function that uses different distance functions on the different kinds of variables. One approach that has been used in MINAMO is to use the overlap metric for categorical variables and the normalized Euclidean distance for continuous variables (real, integer, and discrete ones) (see [11] and [10]).

The authors in [11] and [10] define a heterogeneous distance function which sets the distance[1] between two values or attributes x_i and y_i of a given component i of a vector of variables as:

$$\text{dist}(x_i, y_i) = \begin{cases} \text{norm\_diff}(x_i, y_i) & \text{if } x_i, y_i \text{ are real values} \\ \text{overlap}(x_i, y_i) & \text{if } x_i, y_i \text{ are attributes} \end{cases} \tag{1}$$

where

$$\text{norm\_diff}(x_i, y_i) = \frac{|x_i - y_i|}{\max_i - \min_i} \tag{2}$$

and

$$\text{overlap}(x_i, y_i) = \begin{cases} 0 \text{ if } x_i = y_i \\ 1 \text{ otherwise} \end{cases}, \tag{3}$$

with $\max_i$ and $\min_i$ being the upper and lower bound for variable i, respectively. The overall distance is therefore given by the Heterogeneous Euclidean-Overlap Metric function $\text{HEOM}(x, y)$:

$$\text{HEOM}(x, y) = \sqrt{\frac{\sum_{i=1}^{n} \text{dist}(x_i, y_i)^2}{n}}, \tag{4}$$

where n is the number of variables.

[1] For the sake of clarity, the term "distance" will be used both for the distance between two vectors and the distance between two components of a vector.

2.2 Notion of Affinity

The overlap distance can be slightly modified by introducing the notion of affinities between attributes for a given categorical variable. The rationale behind the role of affinities consists in providing a better representation of the intrinsic topology of the design space. Intuitively, the notion of affinities can be interpreted as a link between attributes. A typical example in mechanical design is the choice of a material:

$$X = \{\texttt{steel\_variant\_1}, \texttt{steel\_variant\_2}, \texttt{steel\_variant\_3}, \texttt{wood\_variant\_1}, \texttt{wood\_variant\_2}, \texttt{wood\_variant\_3}, \texttt{aluminum}\}.$$

We can define affinities between all the steel variants and between all the wood variants, while there is no link between aluminum and the other materials, hence the set X is partitioned into three groups. These affinities are usually defined based on a physical intuition of the designer. It is generally implicitly associated to the behavior of one or several responses that behave(s) similarly for several attributes. Some attributes can therefore be defined as close because they have a similar impact on one or more physical quantities. This is the reason why, in the remainder of this study, we will define affinities between attributes, depending on the way attributes are grouped together following a sensible (or not) physical or engineering intuition.

Based on the definition of these affinities, the overlap distance formulation written in Eq. (3) becomes:

$$\text{overlap}(x_i, y_i) = \begin{cases} 0 & \text{if } x_i = y_i \\ 1 - \texttt{affinity}(x_i, y_i) & \text{otherwise} \end{cases}, \tag{5}$$

where $\texttt{affinity}(x_i, y_i) \in [0, 1]$ is a numerical value defining the affinity/link between two attributes x_i and y_i, with $\texttt{affinity}(x_i, y_i) = \texttt{affinity}(y_i, x_i)$ and $\texttt{affinity}(x_i, y_i) = 0$ for two attributes belonging to two different groups. Note that, as soon as the set of attributes is partitioned into distinct groups, Eq. (5) remains a distance in mathematical sense if $\texttt{affinity}(x_i, y_i) \leq 0.5\ \forall i$. Indeed, in that case, the overlap distance between two different attributes within the same group is always greater than or equal to 0.5 and the triangular inequality is still satisfied since $\texttt{overlap}(x_i, y_i) \in [0, 1]$. For larger values than 0.5, the triangular inequality may not be respected, unless a same numerical value of affinity is used for each group of the partition. This is due to the fact that the overlap distance between attributes of two different groups remains equal to 1. Based on the previous example, suppose that we define:

$$\texttt{affinity}(\texttt{steel\_variant\_i}, \texttt{steel\_variant\_j}) = 0.3, \text{ for } 1 \leq i, j \leq 3,$$

such that the overlap distances between these attributes are equal to 0.7 and no more 1 for all steel materials. We can apply the same operation for wood materials. The affinity value 0.3 is at this stage chosen arbitrarily to give an importance to the defined affinities. Figure 2 represents a simple example based on a categorical variable $X = \{\texttt{steel\_variant\_1}, \texttt{steel\_variant\_2}, \texttt{wood}\}$ with an affinity of 0.3 between steel_variant_1 and steel_variant_2. Now, since the distance and affinities between attributes have an influence in the construction of the surrogate models and in the search process, the impact of the refinement of the distance on the surrogate-based optimization convergence results is presented in the following section.

Fig. 2. Illustration of an updated overlap distance when affinities are defined between attributes.

3 Numerical Results

3.1 Test Problems

In order to benchmark surrogate-based optimization algorithms, we use a set of test problems coming from several contexts (engineering, design, etc.)

Welded Beam Design Problem. In this optimization problem presented in [8], a rectangular bar welded at one end is used to support a load at the other end. The goal is to minimize the manufacturing cost by playing on a total of six optimization variables. In particular, the first variable x_1 sets the weld configuration used between two possible ones, and x_2 gives the material used among four possibilities for the beam and its support ($x_2 \in \{\texttt{steel}, \texttt{cast\_iron}, \texttt{aluminum}, \texttt{brass}\}$). The last four variables are related to geometrical variables (h, t, b, l illustrated in Fig. 3). The constraints are related to the bending stress, the buckling load, the beam deflection, and the weld shear stress. The outputs (objective and constraints) of this test problem can be expressed analytically.

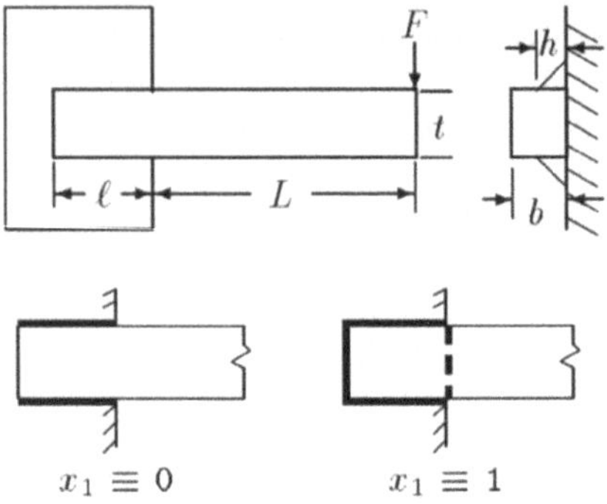

Fig. 3. Representation of the welded beam problem with the two possible weld configurations [8].

Structural Design Problems. A series of examples taken from the structural optimization literature (and initially defined for continuous variables only) have been adapted to include categorical variables. The first structure (`t10b`) is the **10-bar truss**, widely used as benchmark in structural optimization. The geometry and loads are defined in [6]. The boundary conditions impose zero horizontal and vertical displacements on the two nodes located on the left of the structure (see Fig. 4). The structure is a truss, i.e., all bar elements undergo tension or compression only. Inspired by [6], the continuous design variables include, in addition to the cross-section variables, the three vertical positions of the upper nodes, leading to a parametric and shape design set.

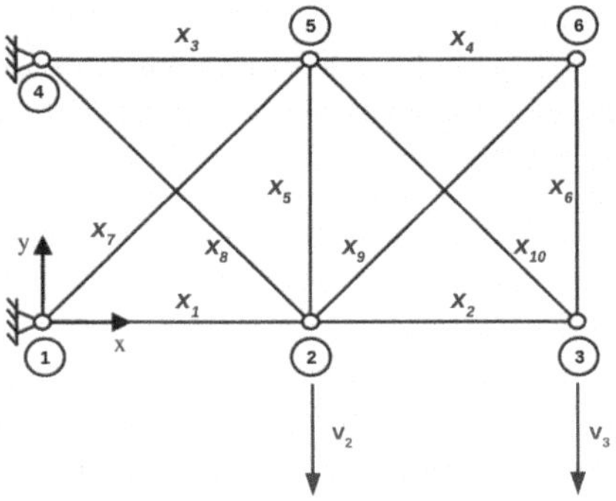

Fig. 4. 10-bar truss (`t10b`) example.

The second structure (`dome`) is a **3D reticulated dome**, investigated in detail in [2], see Fig. 5. It is composed of 241 nodes and 686 elements, and should be modeled as a frame (beam elements). A vertical load of 1 kN is applied on the top of the dome.

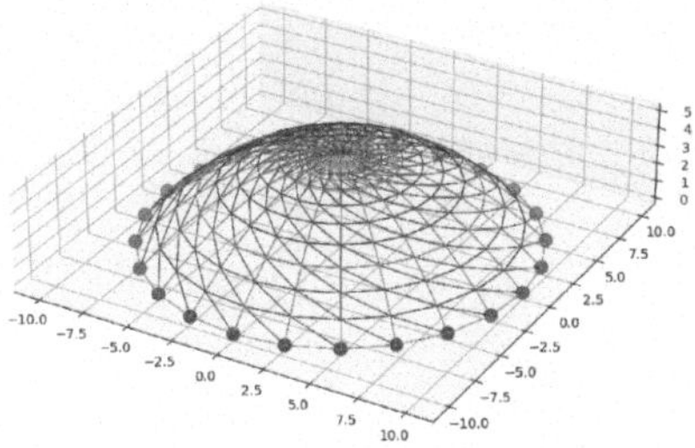

Fig. 5. Reticulated dome example: plane view of the ground structure.

The third structure (`bldg`) is a **35-storey skyscraper** investigated in [5], see Fig. 6. This building is composed of 244 nodes and 942 elements. Horizontal wind loads are applied on one side of the building, with a nodal value of 1 kN. The foundation nodes are fixed.

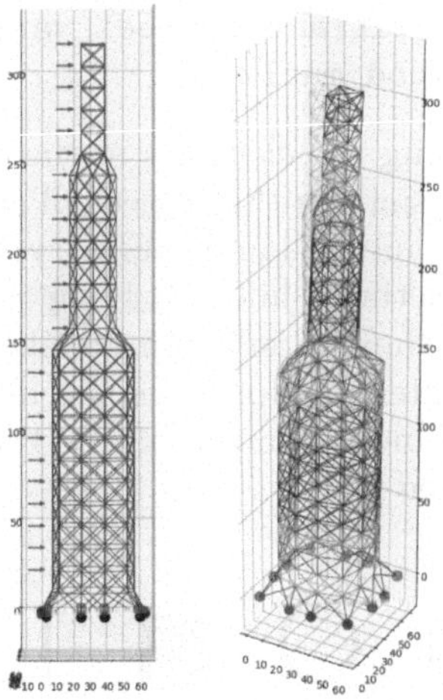

Fig. 6. 35-storey skyscraper (`bldg`) example.

The last example (`wing`) is a **simplified wing box structure** inspired by the model used in [1] (see Fig. 7), with 100 nodes and 162 elements. The 2D wing is loaded by a uniformly distributed lift loading.

The parametrization of these structural test cases are defined as follows: the variables are the diameters of the bar or beam cross-sections (except for `t10b`, where the three nodal positions are also tackled through continuous variables); additionally, two types of categorical variables are used, namely the cross-section shape and the material. The shape must be taken in the {`Circular_full`, `Circular_tubular`, `Square_full`, `Square_tubular`} set (see Fig. 8). Regarding the material, two options are available:

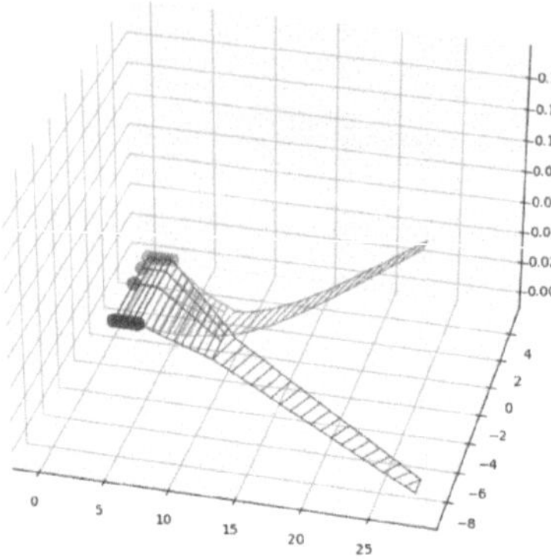

Fig. 7. Simplified wing box structure (`wing`) example.

- `light`: only three materials are available: {`steel`, `aluminum`, `timber`};
- `all`: an extended set of ten materials is available: {`steel`, `aluminum`, `timber`, `steel_variant_1`, `steel_variant_2`, `steel_variant_3`, `steel_variant_4`, `wood_beech`, `wood_elm`, `wood_maple`}.

We use a strategy for grouping the elements in the `dome`, `bldg` and `wing` cases in order to reduce the dimension. In that cases, all elements from a same group share the same cross-section material and geometry. The goal in these problems is to obtain the lightest structure while satisfying resistance and deformability constraints. The core of the simulation is a finite element analysis assessing the structural responses (mass, internal efforts, displacements) for a given parametric design. This analysis is performed by means of FEAPpv, a finite element program developed at UC Berkeley, see [9].

Fig. 8. Cross-section shape defined as a (categorical) variable.

Table 1 summarizes the test problems: n, n_c, and n_{cat} are, respectively, the total number of variables, and the number of continuous and categorical variables. The column n_{attr} contains the number of possible attributes for each categorical variable, while the n_{constr} column shows the number of constraints considered in the optimization specification.

Table 1. Summary of the structural and mechanical test problems.

Problem	Material	n	n_c	n_{cat}	n_{attr}		n_{constr}
t10b	all	15	13	2	4	10	3
	light	15	13	2	4	3	3
dome	all	14	12	2	4	10	3
	light	14	12	2	4	3	3
bldg	all	34	32	2	4	10	3
	light	34	32	2	4	3	3
wing	all	5	3	2	4	10	3
	light	5	3	2	4	3	3
welded beam	/	6	4	2	2	4	4

3.2 Numerical Results Using Affinities

In this section, we analyze the optimization results obtained on the mechanic and structural examples described in the previous section. In the sequel, we would like to test the concept of affinities, based on good choices of the designer (will defining "well-grouped" attributes improve the results?) but also based on wrong choices, as seen as mistake of the designer or bad intuition (will defining "ill-grouped" attributes deteriorate the results?). The following "well-grouped" and "ill-grouped" attributes can be defined intuitively based on the material, through the similarity in Young's modulus and density values and based on the definition of the cross-sections (only for the structural problems), see Tables 2 and 3.

Table 2. Definition of the "well-grouped" attributes for the considered test problems.

Problem	Cross-sections	Materials
welded beam	/	{cast_iron, brass}
structural cases	{●, ■}	{steel, steel_variant_1, steel_variant_2, steel_variant_3, steel_variant_4}
	{○, □}	{timber, wood_beech, wood_elm, wood_maple}

Table 3. Definition of the "ill-grouped" attributes for the considered test problems.

Problem	Cross-sections	Materials
welded beam	/	{steel, aluminum}
structural cases	{●, □} {○, ■}	{steel, timber} {steel_variant_1, wood_beech} {steel_variant_2, wood_elm} {steel_variant_3, wood_maple}

Surrogate-based optimization is studied here on 100 independent runs based on $3n$ points in the inital DoE. We performed $200-3n$ iterations of the surrogate-based optimization (except for bldg cases: $400-3n$) in order to have a total of 200 function evaluations (400 for bldg), with a population size of 200 points and 200 generations in the evolutionary optimization step (on the surrogate models) on three different versions:

- `sbo_dft`: default SBO version with no defined affinities;
- `sbo_affXX`: SBO version with well-grouped attributes, where `XX` specifies the affinity numerical value $\in \{0.1, 0.2, \ldots, 0.8, 0.9\}$, identical for each defined group;
- `sbo_ill_affXX`: SBO version with ill-grouped attributes, where `XX` specifies the affinity numerical value $\in \{0.3, 0.8\}$, identical for each defined group.[2]

In order to give a global view of the performances of the different versions, the performance profiles [3] are used which represent the percentage of solved problems by each version according to τ, where τ is the proportion of the budget used by a version to solve a problem, compared to the minimal budget required by the best version(s) to solve the same problem. The performance profiles comparing the default version and the version using ill-grouped attributes are shown in Fig. 9. These results confirm that defining ill-grouped attributes deteriorates the optimization results. The performance profiles comparing the default version and the version using well-grouped attributes with different values $\{0.1, 0.2, \ldots, 0.8, 0.9\}$ are shown in Fig. 10, where we can observe that good results can be obtained with values less than 0.4. This is why we are showing in Fig. 11 the performance profiles comparing the default version with only versions with well-grouped attributes of values $\{0.1, 0.2, 0.3\}$. It illustrates clearly

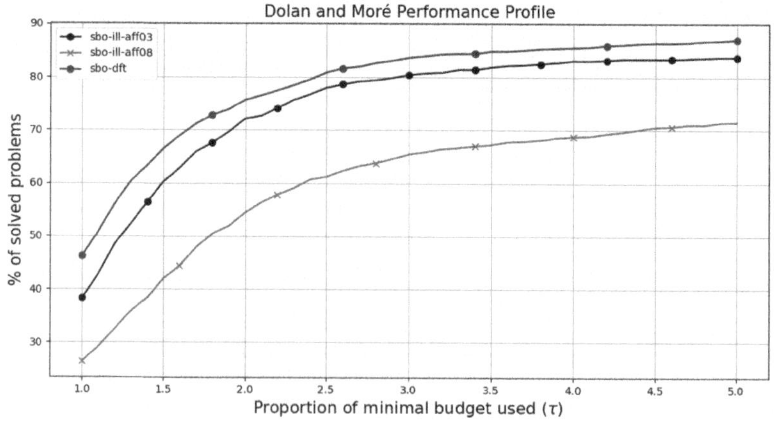

Fig. 9. Performance profiles comparing default and ill-grouped attributes versions.

[2] All the values from 0.1 to 0.9 are not tested here since the idea is only to confirm that using ill-grouped attributes deteriorates the optimization results.

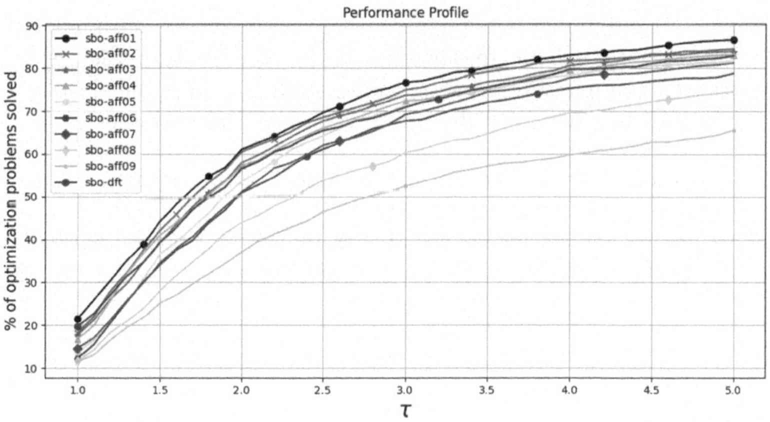

Fig. 10. Performance profiles comparing default and well-grouped attributes versions (with all values).

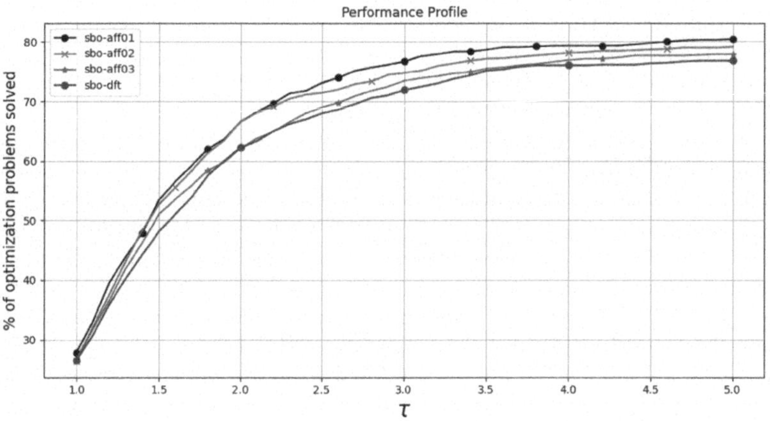

Fig. 11. Performance profiles comparing default and well-grouped attributes versions (with best values).

that the use of affinities for well-grouped attributes can help obtaining better performances.

4 Conclusions

The concept of affinities between attributes of categorical variables has been introduced. In order to study the impact of the use of affinities, specific test problems on which these affinities can be defined have been studied. The first results obtained are promising, showing how they can impact the quality of the results in a surrogate-based optimization process. A deeper analysis is required to better understand how to determine them, potentially with an automatical tool. Future work will be undertaken to further analyze their role on the surrogate-based optimization process, along with the possible development of other formulations of distance well suited to categorical variables.

Acknowledgments. The present research is supported by the Walloon Region via the "Win2wal 2021 - n°2110089 - MASTERMIND" research project and benefited from computational resources made available on Lucia, the Tier-1 supercomputer of the Walloon Region, infrastructure funded by the Walloon Region under the grant agreement n°1910247.

References

1. Anhalt, C., Monner, H., Breitbach, E.: Interdisciplinary Wing Design – Structural Aspects. Technical report, DLR (2003), SAE Technical Paper 2003-01-3026
2. Descamps, B., Filomeno Coelho, R.: The nominal force method for truss geometry and topology optimization incorporating stability considerations. Int. J. Solids Struct. **51**(13), 2390–2399 (2014)
3. Dolan, E.D., Moré, J.J.: Benchmarking optimization software with performance profiles. Math. Program. **91**(2), 201–213 (2002)
4. Forrester, A., Keane, A.: Recent advances in surrogate-based optimization. Prog. Aerosp. Sci. **45**(1–3), 50–79 (2009)
5. Kaveh, A., Talatahari, S.: Optimization of large-scale truss structures using modified charged system search. Int. J. Optim. Civil Eng. **1**, 15–28 (2011)
6. Prayogo, D., Harsono, K., Eko Prasetyo, K., Tjong Wong, F., Tjandra, D.: Size, topology, and shape optimization of truss structures using symbiotic organisms search. In: 2019 International Conference on Advanced Mechatronics, Intelligent Manufacture and Industrial Automation (ICAMIMIA), pp. 197–201 (2019)
7. Sainvitu, C., Iliopoulou, V., Lepot, I.: Global optimization with expensive functions - sample turbomachinery design application. In: Diehl, M., Glineur, F., Jarlebring, E., Michiels, W. (eds.) Recent Advances in Optimization and its Applications in Engineering, pp. 499–509. Springer, Heidelberg (2010)
8. Shahan, D.W., Backlund, P.B., Seepersad, C.C.: Classifier-guided sampling for discrete variable, discontinuous design space exploration. In: Volume 3B: 39th Design Automation Conference. American Society of Mechanical Engineers, Portland, Oregon, USA (2013). https://doi.org/10.1115/DETC2013-13138
9. Taylor, R.L.: FEAP – A Finite Element Analysis Program (2008), Version 8.2 User Manual
10. Wilson, D.R., Martinez, T.R.: Improved heterogeneous distance functions. J. Artif. Intell. Res. **6**, 1–34 (1997). https://doi.org/10.1613/jair.346
11. Wilson, D., Martinez, T.: Heterogeneous radial basis function networks. In: Proceedings of International Conference on Neural Networks (ICNN 1996), vol. 2, pp. 1263–1267. IEEE, Washington, DC, USA (1996). https://doi.org/10.1109/ICNN.1996.549079

Using Pseudo-attractors of Genetic Regulatory Networks to Analyze Biological Data

Gianluca D'Addese[1], Carlo Crovetti[1], Roberto Serra[1,2,3], and Marco Villani[1,2(✉)]

[1] Department of Physics, Informatics and Mathematics, Modena and Reggio Emilia University, 41121 Modena, Italy
marco.villani@unimore.it
[2] European Centre for Living Technology, 30123 Venice, Italy
[3] Institute of Advanced Studies, University of Amsterdam, 1012 WX Amsterdam, The Netherlands

Abstract. Using models to interpret data is a fundamental aspect of data analysis, enabling us to understand patterns, make predictions, and derive insights from raw information. In particular, using models transform raw or otherwise mute data into valuable knowledge. In this work we show how an intuition concerning Random Boolean Networks (RBNs) - a well-known model in the field of complex systems - allows the introduction of a new type of observable, useful for the analysis of real biological data. We then apply this idea to single-cell data on humans and mice.

Keyword: Random Boolean Networks · models · single cell data

1 Introduction

1.1 Random Boolean Networks and Generic Properties

Random Boolean Networks are simplified representations of gene regulatory networks (GRNs), which have been extensively studied as abstract models of complex systems due to the flexibility of their dynamic behavior, which can be adjusted from ordered to disordered by changing a few key parameters [1–4]. RBNs have been applied to various biological phenomena, such as cell differentiation [5–7], and to several fields including robotics [8–10], the study of evolutionary processes [11–14], and the simulation of social systems [15–18].

The behavior of RBNs can be compared to that of GRNs, allowing for inferences about their dynamics. RBNs exhibit dissipative dynamics, leading finite networks to eventually settle into a limited number of attractor states, which can be either constant (fixed points) or periodic (limit cycles) after an initial transient phase. Under the assumption that these attractors correspond to distinct cell types in multicellular organisms [2], interesting scaling relationships between the number of genes and the number of different cell types have been proposed: in the case of ensembles of RBNs having critical or slightly ordered dynamics such relationships are somewhat similar to those observed in

T. Carletti et al. (Eds.): WIVACE 2024, CCIS 2532, pp. 137–147, 2025.
https://doi.org/10.1007/978-3-031-93631-9_11

nature [19, 20]. These results support the idea that the RBN ensembles could effectively simulate some characteristics of the genetic networks of living beings. This hypothesis is further supported by studies showing that RBNs can reproduce data from HeLa cells [21] and the distribution of gene perturbation avalanches in yeast [22]. Nowadays, advancements in genomic and related technologies enable the prediction of the structure of various subsets of real genetic networks [4]. However, the number of thoroughly understood cases remains limited, and significant uncertainties exist regarding the larger networks that contain these partially known circuits. Consequently, investigating generic properties using random networks continues to be an important area of research.

1.2 The "Common Sea"

Due to their significance, critical networks have been extensively studied and discussed in numerous papers. However, some important features have been largely overlooked. This work aims to explore the existence and properties of a relatively large set of nodes that take the same value across all attractors, called "common sea" (CS) [23]. Nodes that do not belong to the common sea, which take different "specific" values in different attractors, are called specific nodes, and their collection is referred to as "specific part" (SP).

As mentioned above, RBN attractors can be either constant (fixed points) or periodic (cycles). However, for reasons discussed in [23], it makes sense to define a (constant) pseudo-attractor for each dynamic attractor, whose components are identical to the time averages (binarized by comparison with a threshold) of the corresponding components of the true attractor. The mapping from true dynamic attractors to pseudo-attractors is therefore not injective, and different attractors can be represented by the same pseudo-attractor, which therefore becomes the representation of a class of attractors all having similar average activities once compared with a threshold (Fig. 1).

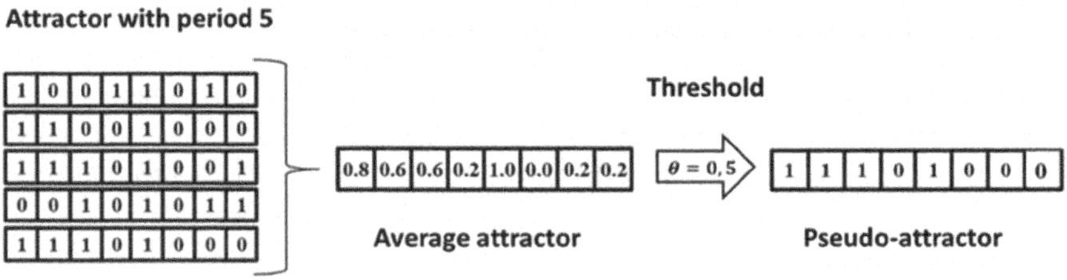

Fig. 1. The construction of a pseudo-attractor. Starting from an attractor of the RBN, the component-by-component averages are obtained and compared with a threshold: the value of the pseudo-attractor components are "0" if the average is less than the threshold, and "1" otherwise

We can therefore define the common sea (CS) as the set of nodes assuming the same values in all the pseudo-attractors of a given network realization (see Fig. 2). By examining different realizations with different parameter values it is possible to understand how the CS is influenced by the characteristics of the network, see [23, 24].

It is important to note that the concept of "common sea" is new and distinct from previous definitions. In the RBN literature, fixed points are sometimes called "frozen nodes" and entire network segments are referred to as "frozen sea" [19]. However, the common sea definition is different because (i) it is based on pseudo-attractors, meaning that nodes oscillating in the "true" dynamic attractor can still belong to the CS, and (ii) it requires that nodes take the same value in all attractors, so some nodes that are "frozen" in some attractors may not be part of the CS. Property (i) also differentiates the common sea from the "stable cores" defined in previous studies [3, 25].

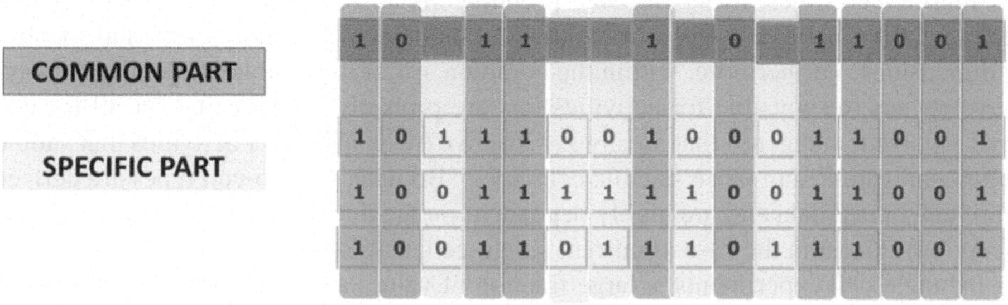

Fig. 2. The identification of the common sea (the set of genes that have the same value in all pseudo-attractors) and consequently of the specific parts. Genes that are always "off" correspond to genes that are normally never expressed by cell types (but which, for example, can perform a function in transition periods). In the following the part of the common sea always expressed is called "UNI".

In [23, 24] it is shown that the dimensions of the common sea of RBNs belonging to the same ensemble are very similar, and interestingly that these dimensions (i) are already relevant for small systems (a fraction always greater than 80% of the genes of the systems used) and that (ii) grow with the number of nodes of the system, so that already with a few hundred genes the fraction exceeds the value of 0.95. Half of this common sea is represented by always active genes. These considerations are more or less similar for most of the threshold values used to binarize the pseudo-attractors.

2 Common Sea in Biological Data

2.1 Adult and Fetal Cell Types

The same approach (identification of pseudo-attractors, search of the common part) can be adopted for analyzing and give meaningful interpretation of real data. In particular, in this work we are interested in single cell data concerning two living species, namely humans and mice.

In the data sets we analyze here [26, 27], the expression levels of numerous genes in single cells are measured (>700,000 single cells from >50 human tissues, and >400,000 single cells from >10 mouse tissues). These data are very noisy, so to obtain a reliable average profile for a given cell type, it is useful to aggregate the contributions of multiple instances of the same type (the variance of the average of several instances of the same type decreasing with the square root of the number of instances). The result is a vector representing the cell type, very similar to the "average attractor" mentioned by studying

RBNs. In the biological literature, these typical expression profiles are referred to as "pseudo-cells".

We propose to binarize these profiles by comparing the value of each component to a threshold: each gene is considered active only if its activation exceeds that threshold. In this way, Boolean pseudo-cell types can be defined by assigning a value of 0 to each gene if its activation level is below the threshold and a value of 1 otherwise.

This work therefore presents the result of the analysis of experimental data of human and murine cell types, using Boolean pseudo-cells as analogues of Boolean pseudo-attractors. This approach helps to identify the biological "common sea" and identify its dimensions. Furthermore, within the common sea, it is possible to identify "always active" genes (responsible for activities that are probably fundamental for all the cell types involved) and genes that are always inactive (responsible for activities that should be kept unexpressed in those specific cell types). Boolean pseudo-cell types are derived from real-valued data and are highly dependent on the threshold you choose - with one exception, at least in the case of current single-cell analyses.

In single-cell experiments, a large fraction of values are often null (they are below any threshold). Many of these values are probably measurement errors, given the small amount of starting materials. The averaging procedure should correct for such errors: but what is the meaning of the case in which all measurements of a gene in all attractors are exactly null? This situation is not uncommon: we are particularly interested in the case where it occurs if we analyze adult tissue cell types and fetal cell types separately.

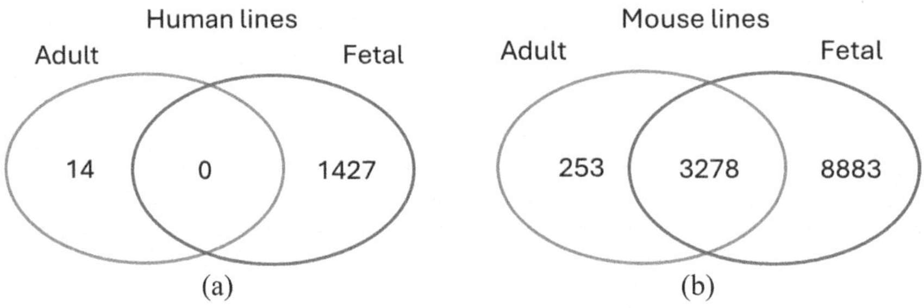

Fig. 3. (a) In human lines (27341 genes in total): the number of genes that show no activity in any single cell of any adult cell type, the number of genes that show no activity in any single cell of any fetal cell type, and the number of genes that show no activity in any single cell of any cell type, whether adult or fetal (in this case, there are only genes that are always inactive in the adult types but active in the fetal types, or genes that are always inactive in the fetal types but active in the adult types). (b) The same measurement, performed in the mouse single cell database (34947 genes in total).

A first interesting observation is that in the case of human lines there is an alternation: genes inactive for adult cell types are active in fetal cell types, or vice versa (and there are no permanently inactive genes – Fig. 3a). The total inactivity therefore seems to be at least partly related to these large groupings (with an imbalance towards fetal types, in which 5.2% of the genes (1427/27341*100 are kept inactive – Fig. 3a). These genes are part of the inactive "common sea" of all fetal cell types, while this is not the case for adult cell types.

A similar situation is also present in mouse cell lines (Fig. 3b), where however a large number of inactive genes is also present in both groups. Using the idea presented here, however, it is possible to interpret this number in different ways: (i) we are not taking into account all possible cell types, (ii) there are genes whose function is to regulate transient situations not present at the time of the measurements or (iii) there are numerous systematic measurement errors.[1] The first two hypotheses lead to decidedly relevant consequences. It is very interesting to note, eventually, that in both cell lines there is a notable increase in genes never expressed in the case of fetal cell types. This parallelism encourages reflection on whether there are inherent qualitative differences between adult cell types and fetal cell types.

2.2 Two Different Thresholding Approaches

Is it possible to use the idea of a "common sea" more comprehensively, in order to better characterize these two groups of cell types? One difficulty is in identifying a correct value for the threshold to use: in fact, at least up to now there is no definitive reason to set a specific threshold value.[2] In the following, however, we will show that there are features of the common seas of adult cell types and fetal cell types that are largely independent of specific threshold values - thus getting rid of this problem.

Furthermore, in order to obtain the Boolean pseudo-cells we can use two different types of thresholding: an absolute threshold equal for all genes (method S1) and a mechanism in which for each gene we use a potentially different threshold (method S2) [23]. The first approach allows us to make some general considerations, while the second one addresses the hypothesis that the same concentrations of chemically different compounds can have different impacts on the regulatory network. The results from both approaches support the idea that fetal cell types are much more similar to each other (are less "differentiated") than are adult cell types.

The first approach is to simply examine the profiles of the available cell types and, gene by gene, compare the average of the counts with a threshold: if the average is less than or equal to the threshold the corresponding gene is to be considered "inactive" - otherwise it is "active".

We then obtain a Boolean pseudo-cell for each profile, and it is therefore possible to identify the size of the "common sea" of a group of Boolean pseudo-cells corresponding to a specific threshold. It is then possible to repeat the operation for many thresholds and observe the corresponding variation of this size, both for the group of adult cell types and for the group of fetal cell types. In fact, we must Keep in mind that the two dimensions are not directly comparable. The observations regarding the RBN model remind us that the size of the common sea depends on the number of pseudo-attractors: the greater the

[1] The probability that a purely random distribution of so many null values would give such high numbers of genes that are always inactive (or such high numbers of genes that are completely inactive in adult cell types, or in fetal cell types) is completely negligible.

[2] We can note that the identification of pseudo-attractors in synthetic RBNs shows little dependence on the threshold value (data not shown).

number of pseudo-attractors, the smaller the size of the common sea - a relationship that is not linear and tends to saturate [23].

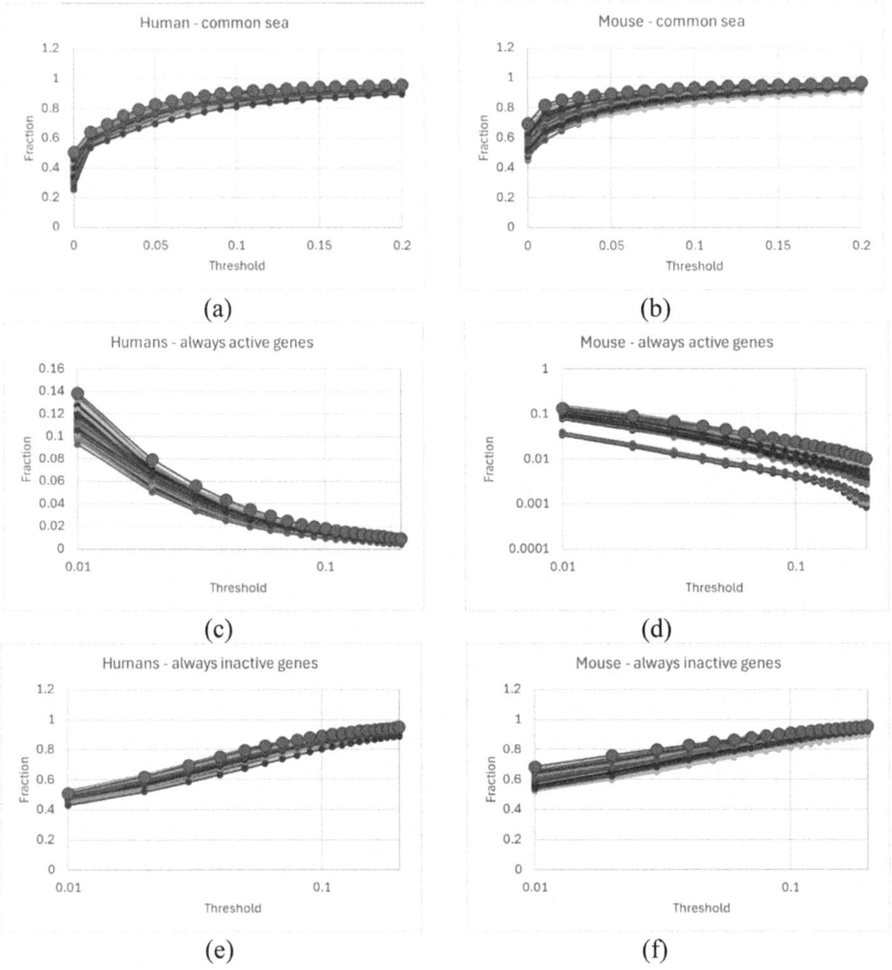

Fig. 4. S1 threshold method (a) The fraction of genes belonging to the common sea is greater if only human fetal cell types are taken into consideration (the plot with big red spheres), while random extractions of 12 types carried out from the 51 human mature types always present a smaller common sea (the other lines): this happens at every threshold value used. The same happens if we compare the fraction of genes that are always active (c), and the fraction of genes that are always inactive (e). The same situation holds for the common sea of fetal mouse cell types (b), for the fraction of genes always active in fetal mouse cell types (d), and for the fraction of genes always inactive in fetal mouse cell types (f). Threshold values are shown until reaching a situation where more than 95% of the genes are detected as always inactive. (Color figure online)

The human database contains 51 adult and 12 fetal cell types, while the mouse database contains 45 adult and 7 fetal cell types. In the first case we then proceeded to extract 100 random groups of adult cell types, each of size 12: what we will compare are the trends in the size of the common sea of these clusters (100 trends), against the trend in the size of the common sea of the group composed by the 12 fetal types. The same procedure was adopted for the murine cell lines.

In human cell types, no ensemble of 12 adult cell types shows a common sea having a size comparable with the common sea present when considering fetal cell types: the same occurs for the fraction of always-active genes, and for the fraction of always-inactive genes (Fig. 4). Two orders of considerations can therefore be made. The first underlines the fact that fetal cell types show a very high mutual similarity (a large common sea). The second highlights that a larger fraction of always-active genes could indicate a lower degree of differentiation compared to adult cell types: in fact, many genes with variegated activity in different adult cell types are indifferently always-active in fetal cell types. The same observation applies when repeating the analysis with murine cell types.

This observation is consistent with the idea, common to many researchers but so far never evaluated quantitatively (at least to the best of our knowledge), that fetal cells are "less differentiated" than the cells of mature individuals.

It is interesting to verify if and which of these observations are maintained using the second thresholding method, in which for each gene we can potentially use different thresholds.

In fact, both in human and mouse lines the maximum counts corresponding to various genes can differ even by two orders of magnitude. Furthermore, for the same gene the maximum counts on fetal cells only and on adult cells only can also differ significantly. The fact that - gene by gene - most of the "F/A" quotients between the maximum count of fetal types and the maximum count in adult types are less than 1.0 supports the idea that fetal types are less differentiated than adult types: fetal genes, in addition to having a widespread activity among many cell types (the previous considerations), also have lower activation levels (see Fig. 5).

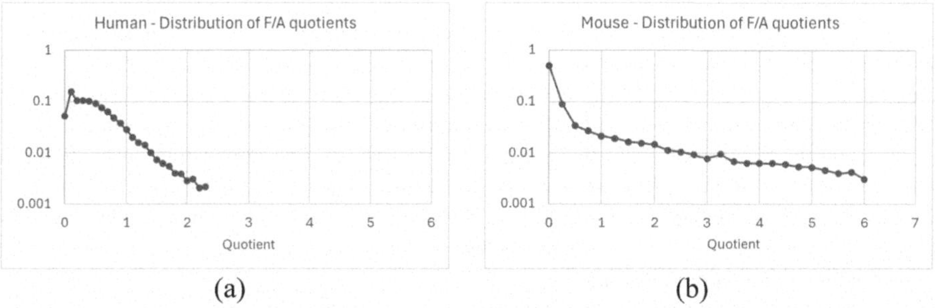

Fig. 5. (a) The distribution of quotients - gene by gene - between the maximum count of human fetal types and the maximum count of human adult types. 86% of the measurements are less than or equal to 1.0. (b) The same for mouse cell lines. 67% of the measurements are less than or equal to 1.0.

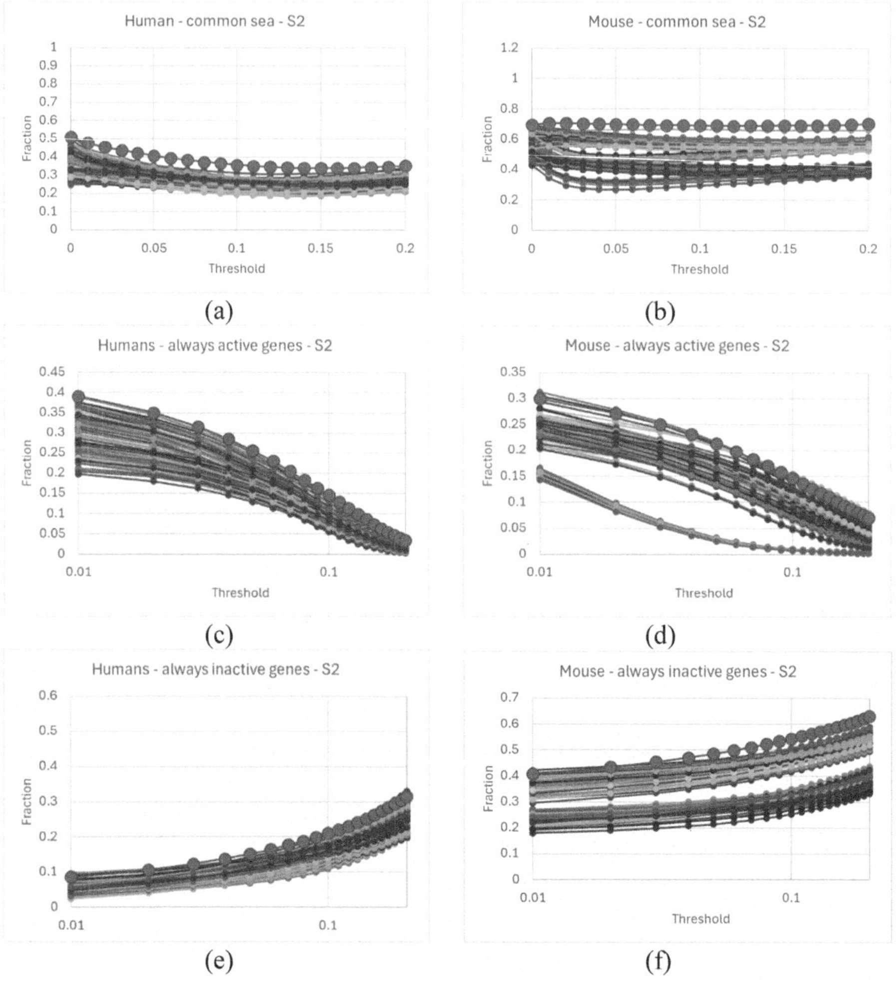

Fig. 6. S2 threshold method (a) The fraction of genes belonging to the common sea is greater if only human fetal cell types are taken into consideration (the plot with big red spheres) vs. random extractions of 12 types from the 51 human mature types (the other lines): this happens at almost all the threshold values used. The same happens if we compare the fraction of genes that are always active (c), and the fraction of genes that are always inactive (e). The same situation holds for the common sea of fetal mouse cell types (b), for the fraction of genes always active in fetal mouse cell types (d), and for the fraction of genes always inactive in fetal mouse cell types (f). Threshold values are shown until reaching a situation where more than 95% of the genes are detected as always inactive. (Color figure online)

The second approach then divides each gene's count value by its maximum value before comparing it to the threshold: the meaning of the threshold therefore derives from a relative comparison, where for example "0.05" indicates a threshold equal to one twentieth of the maximum possible value. The maximum value of each gene is calculated on all cell types, regardless of whether adult or fetal: it is therefore possible that, for a given threshold, a gene is detected as inactive in all fetal cell types, while it could be active in some adult cell types (or vice versa).

Despite this different approach, the second thresholding method also identifies a large common sea for fetal cell lines, with many genes always active and many genes always inactive, both for human and mouse lines. The fetal cell type plots are always above the corresponding extraction ensemble plots (Fig. 6).

2.3 Genes Expressed Only by a Single Cell Type

The argument that fetal cell types are less differentiated than adult ones suggests that the number of active genes in a single fetal cell type (an indicator of specific diversity) should be lower than that found in adult cell types. This is indeed the case: the fraction of genes expressed only by a single cell type is greater in mature types than in fetal types (Fig. 7).

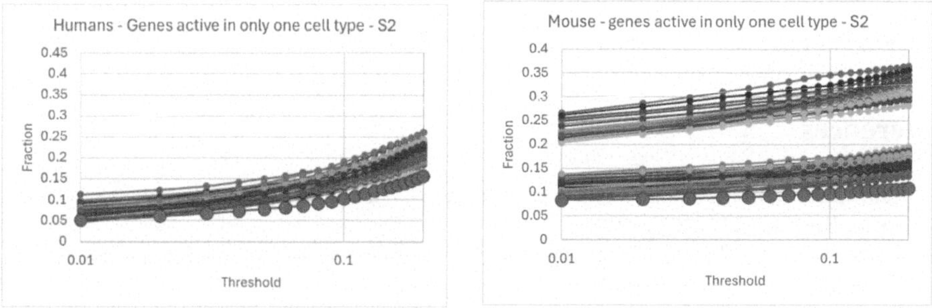

(a) (b)

Fig. 7. S2 threshold method (a) The fraction of genes expressed by a single human cell type is greater in mature types than in fetal types (the plot with big red spheres, with respect to the other lines). (b) The same observation applies to murine cell types. (Color figure online)

3 Discussion

The use of generic models (which capture the correct dynamic regime even if they do not share all the details of the system under consideration) allows to suggest new types of observables, useful to extract new information from biological data.

In this work we show how the identification of a "common sea" in generic regulatory network models (the RBN framework) has allowed a new use of the data contained in the so-called pseudo-cell, highlighting interesting differences between adult and fetal cell types. In particular, fetal cell types show a large common sea, composed of both genes active in all fetal cell types and genes that do not show activity in each of these cell types. The number of genes active in all fetal types is higher than the number of genes active in adult types: the same occurs for the number of inactive genes. This observation appears to be correct in the databases we used on humans and mice: it will be interesting to see whether it remains valid also in data from other databases, and in different living beings.

The results presented here could be a clue to support the hypothesis that fetal cell types are at a less differentiated stage than adult cell types. A further clue in this direction is given by the observation that the number of genes active in a single cell type (an indicator of specificity) is lower in fetal cell types than in adult cell types.

Furthermore, unless you use particularly high (and implausible) thresholds, it can be noted that the fraction of active genes belonging to the common part of real systems (human, mouse – Figs. 4a and 5a) seems to be lower than that detected in current models (a remark which applies to RBNs, but also to other models). This observation highlights that our current understanding of gene regulatory networks is largely incomplete and that new research approaches are needed.

Acknowledgments. M.V. acknowledges support by Università degli Studi di Modena e Reggio Emilia (FAR2023 project of the Department of Physics, Informatics and Mathematics).

Disclosure of Interests. The authors have no competing interests.

References

1. Kauffman, S.A.: Metabolic stability and epigenesis in randomly constructed genetic nets. J. Theor. Biol. **22**, 437–467 (1969)
2. Kauffman, S.A., Kauffman, M.: The Origins of Order: Self-Organization and Selection in Evolution. Oxford University Press, Oxford (1993)
3. Bastolla, U., Parisi, G.: Relevant elements, magnetization and dynamical properties in Kauffman networks: a numerical study. Phys. D. Nonlinear Phenom. **115**, 203–218 (1998)
4. Schwab, J.D., Kühlwein, S.D., Ikonomi, N., Kühl, M., Kestler, H.A.: Concepts in Boolean network modeling: what do they all mean? Comput. Struct. Biotechnol. J. **18**, 571–582 (2020)
5. Villani, M., Barbieri, A., Serra, R.: A dynamical model of genetic networks for cell differentiation. PLoS ONE **6**, e17703 (2011)
6. Villani, M., Serra, R.: On the dynamical properties of a model of cell differentiation. EURASIP J. Bioinform. Syst. Biol. **2013**, 4 (2013)
7. Braccini, M., Roli, A., Villani, M., Serra, R.: Dynamical properties and path dependence in a gene-network model of cell differentiation. Soft. Comput. **25**, 6775–6787 (2021)
8. Roli, A., Manfroni, M., Pinciroli, C., Birattari, M.: On the design of Boolean network robots. In: European Conference on the Applications of Evolutionary Computation, pp. 43–52. Springer, Heidelberg (2011)
9. Roli, A., Villani, M., Serra, R., Garattoni, L., Pinciroli, C., Birattari, M.: Identification of dynamical structures in artificial brains: an analysis of boolean network controlled robots. In: Baldoni, M., Baroglio, C., Boella, G., Micalizio, R. (eds.) Proceedings of the AI*IA 2013: Advances in Artificial Intelligence, Torino, Italy, 4–6 December 2013, pp. 324–335. Springer, Cham (2013)
10. Roli, A., Villani, M., Serra, R., Benedettini, S., Pinciroli, C., Birattari, M.: Dynamical properties of artificially evolved boolean network robots. In: Proceedings of the AI*IA 2015: Advances in Artificial Intelligence, Ferrara, Italy, 23–25 September 2015

11. Aldana, M., Balleza, E., Kauffman, S., Resendiz, O.: Robustness and evolvability in genetic regulatory networks. J. Theor. Biol. **245**, 433–448 (2007)
12. Torres-Sosa, C., Huang, S., Aldana, M.: Criticality is an emergent property of genetic networks that exhibit evolvability. PLoS Comput. Biol. **8**, e1002669 (2012)
13. Villani, M., Magrì, S., Roli, A., Serra, R.: Evolving always-critical networks. Life **10**, 22 (2020)
14. Braccini, M., Gardinazzi, Y., Roli, A., Villani, M.: Sensory-motor loop adaptation in boolean network robots. Sensors **24**(11), 3393 (2024)
15. Maes, P., Meyer, J.A., Mataric, M.J., Pollack, J., Wilson, S.W.: A developmental model for the evolution of complete autonomous agents. In: From Animals to Animats 4: Proceedings of the Fourth International Conference on Simulation of Adaptive Behavior. MIT Press, Cambridge (1996). ISBN 978-0-262-63178-5
16. Alexander, J.M.: Random boolean networks and evolutionary game theory. Philos. Sci. **70**, 1289–1304 (2003)
17. Bialek, W., et al.: Social interactions dominate speed control in poising natural flocks near criticality. Proc. Natl. Acad. Sci. U.S.A. **111**, 7212–7217 (2014)
18. Daniels, B.C., Krakauer, D.C., Flack, J.C.: Control of finite critical behaviour in a small-scale social system. Nat. Commun. **8**, 14301 (2017)
19. Kauffman, S.A.: At Home in the Universe: The Search for Laws of Self-Organization and Complexity. Oxford University Press, Oxford (1995)
20. Bornholdt, S., Kauffman, S.: Ensembles, dynamics, and cell types: revisiting the statistical mechanics perspective on cellular regulation. J. Theor. Biol. **467**, 15–22 (2019)
21. Shmulevich, I., Kauffman, S.A., Aldana, M.: Eukaryotic cells are dynamically ordered or critical but not chaotic. Proc. Natl. Acad. Sci. U.S.A. **102**, 13439–13444 (2005)
22. Villani, M., La Rocca, L., Kauffman, S.A., Serra, R.: Dynamical criticality in gene regulatory networks. Complexity **2018**, e5980636 (2018)
23. Villani, M., D'Addese, G., Kauffman, S.A., Serra, R.: Attractor-specific and common expression values in random boolean network models (with a preliminary look at single-cell data). Entropy **24**(3), 311 (2022)
24. Villani, M., Balugani, M., Serra, R.: The properties of pseudo-attractors in random boolean networks. In: Villani, M., Cagnoni, S., Serra, R. (eds.) Artificial Life and Evolutionary Computation. WIVACE 2023. Communications in Computer and Information Science, vol. 1977. Springer, Cham (2024)
25. Aldana, M., Coppersmith, S., Kadanoff, L.P.: Boolean dynamics with random couplings. In: Kaplan, E., Marsden, J.E., Sreenivasan, K.R. (eds.) Perspectives and Problems in Nolinear Science: A Celebratory Volume in Honor of Lawrence Sirovich, pp. 23–89. Springer, New York (2003)
26. Han, X., et al.: Construction of a human cell landscape at single-cell level. Nature **581**, 303–309 (2020)
27. Wang, R., Zhang, P., et al.: Construction of a cross-species cell landscape at single-cell level. Nucleic Acids Res. **51**(2), 501–516 (2023)

Costs of Information Gathering Strategies and Their Impact on Animal Dispersal

Antoine Sion[1(✉)], Matteo Marcantonio[2], Stefano Masier[3], and Elio Tuci[1]

[1] Faculty of Computer Science, University of Namur, Rue Grandgagnage 21, 5000 Namur, Belgium
{antoine.sion,elio.tuci}@unamur.be
[2] Landlife Ecospatial Labs, Mezzocorona, Italy
[3] Università Statale di Milano - UNIMI, Via Festa del Perdono, 7, 20122 Milano, Italy

Abstract. Animal dispersal is among the most impacted behaviours from habitat loss and fragmentation. Animals disperse in search of better conditions to ultimately increase their fitness. However, the dispersal process induces costs that can be translated as a loss in fecundity or as an increase in predation risks. In this study, we use an evolutionary multi-agent model to quantify how costs related to information acquisition shape information gathering strategies used in animal dispersal. We separate costs of private information, i.e., related to the direct acquisition of information sensed in the physical environment, from costs related to social information, i.e., derived from the observation of conspecifics behaviour. We find that a low mortality risk associated with dispersal movements selects non-informed dispersal in low-variability environments. However, high-variability environments select an information gathering strategy based on the exploration of multiple patches of habitat. A higher mortality risk causes drastic drops in population abundance and agents gather information from only one patch to limit their movements. Overall, private information is predominantly used in high-variability environments while social and private information are used equally in low-variability environments.

Keywords: Dispersal · Costs · Information gathering · Evolutionary modelling

1 Introduction

The impact of human activities on animal species is global, profoundly modifying the living conditions of animals and fragmenting their habitats [17]. To mitigate the impact of these environmental changes, we must first understand how animals' behaviours adapt and evolve in response to their modified living conditions. One behaviour of interest which is crucial for a majority of species is dispersal. Animals disperse in search of better resources for reproduction due to multiple factors such as varying environmental conditions, inbreeding avoidance

T. Carletti et al. (Eds.): WIVACE 2024, CCIS 2532, pp. 148–159, 2025.
https://doi.org/10.1007/978-3-031-93631-9_12

or resource competition [2]. Thus, dispersal is multi-causal and results in the movement of animals from and to different patches of habitat. Multiple costs such as energy expenditure or risk of predation occur during the dispersal process. Gathering information is especially costly, since animals invest time and energy into information acquisition.

We are particularly interested in the study of costs of information acquisition and their effects on evolving dispersal strategies. Costs of information acquisition can translate themselves in an expense of time and energy during information gathering, because of movement and investment in the development of sensory organs dedicated to the gain of information [15]. Two other types of cost are the increased mortality risk due to individuals exposing themselves during the prospecting phase, as well as the damage done to the individual over time. Without direct repercussion on the fecundity, those costs can nonetheless impact fitness. Experimental research quantifying the impact of dispersal costs is scarce due to the difficulty of accounting for multiple information sources as well as environmental variability. Evolutionary models of dispersal provide an interesting alternative to this problem because of the implementation of different rules related to the behavioural aspects of animals at the individual level, simultaneously allowing the observation of population-level dynamics. However, the majority of modelling studies on dispersal rely on a fixed dispersal rate or density-dependent information for agents to make a dispersal decision [1,6,9,12]. In our two previous works [13,14], we filled this gap by extending and enhancing a dispersal model [4] with genetic factors enabling or disabling the use of private or social information and by including separate information acquisition costs. This allowed us to study different types of environment with varying habitat qualities as well as multiple costs situations.

Previously, the perceptual range of our agents was fixed: agents were either equipped with a limited range meaning they were only able to acquire information from their natal patch, or with an extended range meaning they could sample additional information from a randomly chosen adjacent patch. In this paper, we aim to put the focus on the information gathering strategy that will be used at the population-level. Prospecting different patches of habitat is a key behaviour involved in informed dispersal and could help species adapt to environmental changes [10]. Costs linked to the exploration of the environment are substantial but difficult to evaluate through empirical experiments [2]. Works modelling dispersal often rely on the implementation of specific genes influencing prospecting strategies in their design [3,11]. Similarly, we add a supplementary gene in our modelling framework that conditions the number of patches that an agent can sample in its lifetime before taking a dispersal decision. Our agents can thus potentially acquire information from multiple adjacent patches, not only just one, before taking a dispersal decision. We consider the reduction of fecundity due to costs related to private or social information acquisition, with these costs increasing when gathering more information. We also include mortality risks occurring when moving from one patch of habitat to another.

2 Material and Methods

Our model simulates 40,000 agents (male/female ratio of 1:1) in a squared environment composed of 20×20 patches. The carrying capacity (habitat quality) of each patch varies at each generation. The number of offsprings in generation t depends on the number of female agents from generation $t-1$ present on a patch, as well as its quality. Agents can gather private (patch quality) or social (number of conspecifics) information on their natal patch and adjacent patches. Binary genes enabling the use of social and private information are s_1 and s_2 respectively. Real-valued genes are used to determine the contribution of each information type in the dispersal probability, a_1 corresponds to private information in the natal patch, b_1 to social information in the natal patch, a_2 to private information in the adjacent patch and b_2 to social information in the adjacent patch. Additionally, we introduce a gene corresponding to the number of adjacent patches where the agent gathers information from ($s_3 = [1, 8]$ for the eight nearest neighbours patches). During information gathering, the agent chooses s_3 times a random adjacent patch on which it acquires information.

Based on the acquired information, female agents take a dispersal decision once in their lifetime to move to an adjacent patch before reproduction. Agents inherit the genes from their mothers and generations are non-overlapping, with genes mutating between each generation (model states variables and their possible values are shown in Table 1). The steps in our model are the following:

Table 1. State variables used in the model

Parameter	Values/bounds
Probability of dying while dispersing/gathering information (μ)	$\{0.0; 0.01; 0.1\}$
Gene enabling the use of social information (s_1)	$\{0; 1\}$
Gene enabling the use of private information (s_2)	$\{0; 1\}$
Gene for the number of visited patches (s_3)	$[1, 8]$
Gene weighting social information in the natal patch (b_1)	$[-0.2, 0.2]$
Gene weighting social information in an adjacent patch (b_2)	$[-0.2, 0.2]$
Gene weighting private information in the natal patch (a_1)	$[-0.2, 0.2]$
Gene weighting private information in an adjacent patch (a_2)	$[-0.2, 0.2]$
Lower bound for the reproductive success of a female (r_{min})	$[0.75, 1.0]$
Upper bound for the reproductive success of a female (r_{max})	1.0
Relative cost for acquiring social information (c_1)	$[0.0, 1.0]$
Relative cost for acquiring private information (c_2)	$[0.0, 1.0]$
Probability of mutation of an allele (P_{mut})	0.00005
Initial quality of a patch (K_0)	100
Minimum quality of a patch (K_{min})	5
Standard deviation of the quality of a patch (σ)	$\{1; 10; 100\}$
Upper limit for the local population on a patch (N_{max})	500

Reproduction: new agents are born in each patch following this equation:

$$N_{next_i} = min(2N_{f_i}exp(r(1 - \frac{2N_{f_i}}{K_i})), N_{max}) \tag{1}$$

with N_{f_i} the number of females on the patch in the last generation, $r = 1$ a fixed population growth rate parameter, K_i the quality of the patch and N_{max} an upper limit for the number of new agents on the patch. New agents inherit the genes from their mother, selected from the local pool of females randomly with replacement in a weighted distribution. Mothers each have a reproductive weight that varies depending on the different costs used when acquiring information:

$$w = r_{max} + (c_1 s_1 + c_2 s_2)\frac{s_3}{8}(r_{min} - r_{max}) \tag{2}$$

with r_{min} and r_{max} bounding the value of the weight. c_1 is the relative cost for acquiring social information and $c_2 = 1 - c_1$ the relative cost for acquiring private information. A female using both sources of information ($s_1 = 1$ and $s_2 = 1$) and the highest number of patches ($s_3 = 8$) has the lowest possible weight (r_{min}). We multiply the total cost of acquisition obtained by the number of patches visited by a female gathering information (divided by the maximum number of visited patches). Our rationale is that an individual pays a total cost for acquiring information with the development of sensory organs for private and/or social information and a direct loss of energy due to movement, sensing and storing the information. While gathering information from an adjacent patch, females have a probability μ of dying due to an increased predation risk. When the reproduction phase ends, all agents from the previous generation die.

Varying Quality: the quality of each patch varies following a normal distribution:

$$K_i(t) = max(K_0 + \phi, K_{min}) \tag{3}$$

with ϕ an independent normal variable with a mean of zero and a standard deviation σ. High values of σ corresponds to high-variability environments, while low values create low-variability environments with spatially homogeneous patch qualities.

Mating: in the presence of a male, females mate and will pass their genes in the reproduction phase. If no male is present on a patch, females stay unmated and will not reproduce.

Dispersal: females compute a probability of dispersal for each visited patch with this equation:

$$P = \frac{1}{1 + exp(-a_1 Q_{1\,s} - a_2 Q_{2\,s} - b_1 N_{1\,s} - b_2 N_{2\,s})} \tag{4}$$

with Q_{1s} the sensed quality of the natal patch, Q_{2s} the sensed quality of the adjacent patch, N_{1s} the sensed number of conspecifics on the natal patch and N_{2s} the sensed number of conspecifics on the adjacent patch. Information is enabled or not following:

$$N_{1\,s} = s_1 N_1 \quad \text{and} \quad N_{2\,s} = s_1 N_2 \tag{5}$$

$$Q_{1\,s} = s_2 Q_1 \quad \text{and} \quad Q_{2\,s} = s_2 Q_2 \tag{6}$$

Then, females select the patch with the highest dispersal probability and take a dispersal decision. They also have a probability μ of dying during the transfer from their natal patch to the adjacent one, as when gathering information.

We study different environmental conditions with $\sigma = \{1; 10; 100\}$ corresponding to low, moderate and high-variability environments. We also vary information acquisition costs with $r_{min} = \{0.75; 0.80; 0.85; 0.90; 0.95; 1.0\}$, as well as relative costs between social and private information with $c_1 = \{0.0; 0.25; 0.5; 0.75; 1.0\}$. Additionally, different mortality risks when moving between patches are tested with $\mu = \{0; 0.01; 0.1\}$ to investigate the consequences of a lower predation risk during transfer. Results presented in the next section are obtained after 20,000 generations and each data point is computed over 10 different iterations.

3 Results

Heatmaps showing our results for three different μ values are presented in Fig. 1, 2, 3, 4, 5 and 6. With zero probability of dying when moving between patches (see Fig. 1), the total population size stays very high for all environmental conditions, around 40,000 agents, effectively maximising fitness. In low-variability environments ($\sigma = 1$), agents never use any type of information, even when the total or relative cost is zero. Thus, the average number of patches visited ranges from 1 to 8 since zero cost is associated with information gathering (no mortality risk when moving between patches and no reduction in fecundity since agents do not acquire information). In moderate-variability environments ($\sigma = 10$), the same trend can be observed except when the total cost of information is free ($r_{min} = 1.0$) or for free private information ($c_1 = 1.0$). For those cases, private information is used by all agents (with some additional social information in limited cases), gathered from a singular adjacent patch. Finally, with the highest environmental stochasticity $\sigma = 100$, agents rely on private information to take their dispersal decision in all studied conditions, gathering information from two different adjacent patches. Some social information is also taken into account only when the total cost of information is free. Figure 2 shows median values of genes involved in computing the dispersal probability, quantitatively representing the dispersal strategy chosen by the agents. For cases where information is used ($s_1 > 0$ or $s_2 > 0$ in Fig. 1), a_1 takes a negative value around -0.04 while a_2 is positive near 0.20. Genes related to social information do not take a specific value since few agents use this type of information. Thus, high quality adjacent patches attract agents by rising their dispersal probability whereas high quality natal patches lower this probability.

Some key differences appear when slightly rising the mortality risk during transfer between patches (see Fig. 3). In high-variability environments ($\sigma = 100$), the situation stays the same with agents acquiring almost exclusively private

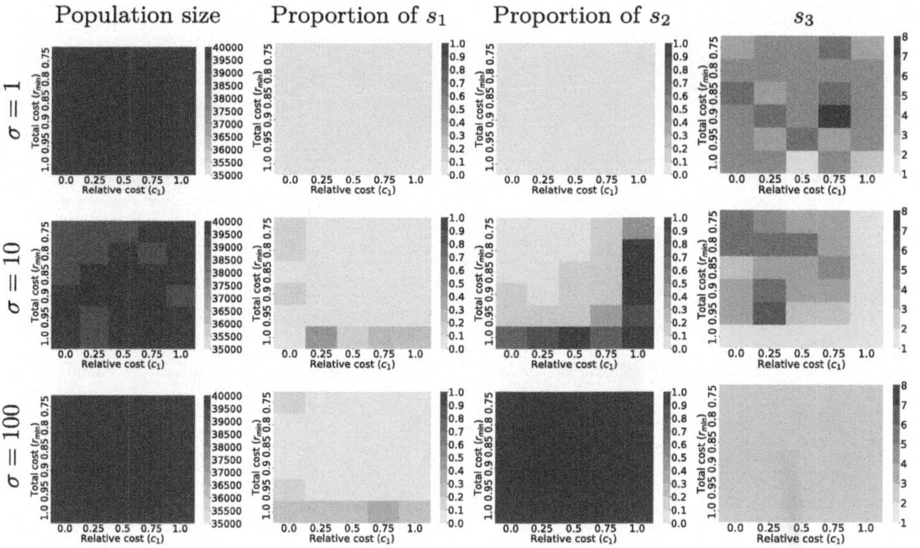

Fig. 1. Heatmaps showing median values found in the population after 20,000 generations for a mortality risk $\mu = 0$. Each column respectively shows the total population size, the proportion of females acquiring social information, the proportion of females acquiring private information and the average number of patches visited. Each row corresponds to a different level of environmental variation $\sigma = \{1; 10; 100\}$.

information on two distinct patches. With $\sigma = 10$, the quantity of agents using private information diminishes with the highest proportion being around half of the agents acquiring this type of information. As seen previously, social information is still used marginally when its acquisition is free. With very low environmental variations, no information is used at all. It can be noted that a slight rise in μ directly impacts the average number of patches visited, which is always equal to one when $\sigma = \{1; 10\}$. Dispersal strategies stay approximately the same as with $\mu = 0$ (see Fig. 4): a_1 takes a negative value and a_2 is close to 0.20. A small difference can be noticed when the cost of social information is free, with b_1 taking overall a positive value and b_2 value being negative. This translates a behaviour where agents will choose to disperse away from patches with high conspecific densities.

With the highest studied probability of dying (see Fig. 5), population size decreases dramatically, reaching a maximum of 37,000 agents in low-variability environments. For low and moderate environmental variations ($\sigma = \{1; 10\}$), agents rely on both information types when the total cost is free or with equal relative costs. In other cases, they prefer to use the cheapest information source. In the most difficult scenario ($\sigma = 100$), agents acquire private information in all costs situations with additional social information if the cost stays relatively low. Due to the high mortality risk, agents limit their movements between patches and gather information from a single adjacent patch. In terms of dispersal behaviours

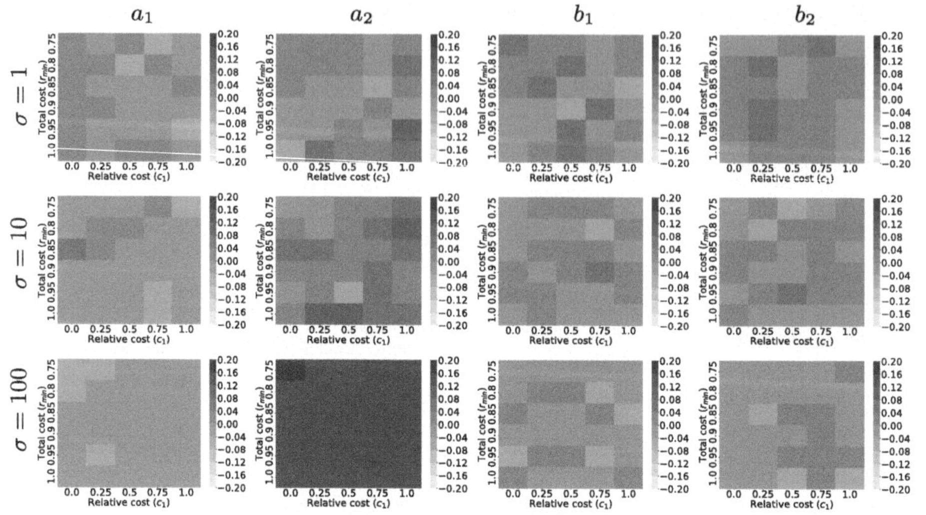

Fig. 2. Heatmaps showing the median values of real-valued genes found in the population after 20,000 generations for a mortality risk $\mu = 0$. Each column respectively shows a_1 (gene for private information in the natal patch), a_2 (gene for private information in the adjacent patch), b_1 (gene for social information in the natal patch), b_2 (gene for social information in the adjacent patch).

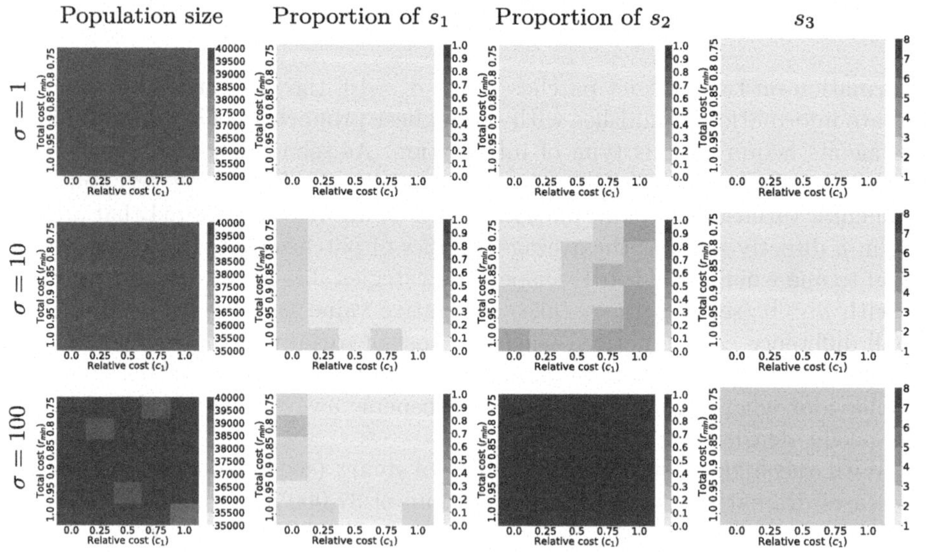

Fig. 3. Heatmaps showing median values found in the population after 20,000 generations for a mortality risk $\mu = 0.01$.

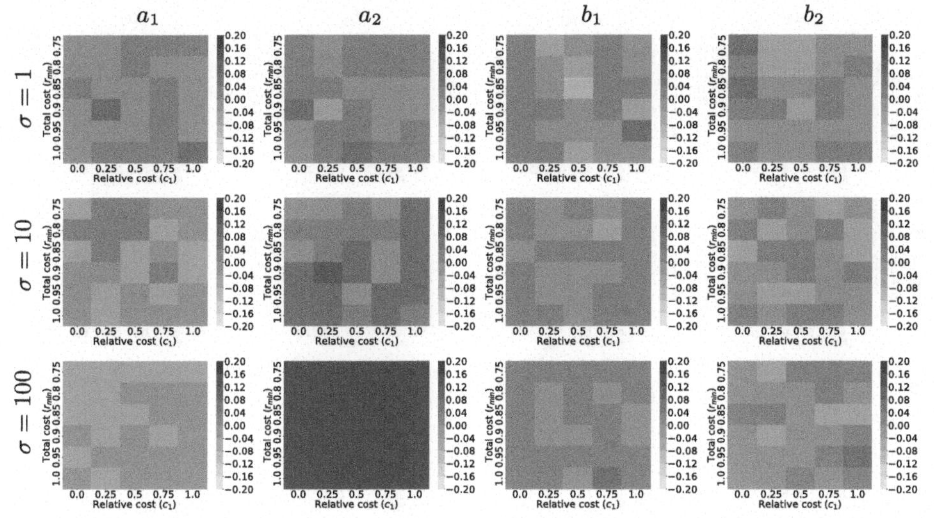

Fig. 4. Heatmaps showing the median values of real-valued genes found in the population after 20,000 generations for a mortality risk $\mu = 0.01$.

occurring when using information (see Fig. 6), a_1 is overall negative, a_2 positive, b_1 positive and b_2 negative. The magnitudes of the genes values rise when increasing the environmental variation to $\sigma = 100$, where private information carries more weight in the dispersal decision than social information.

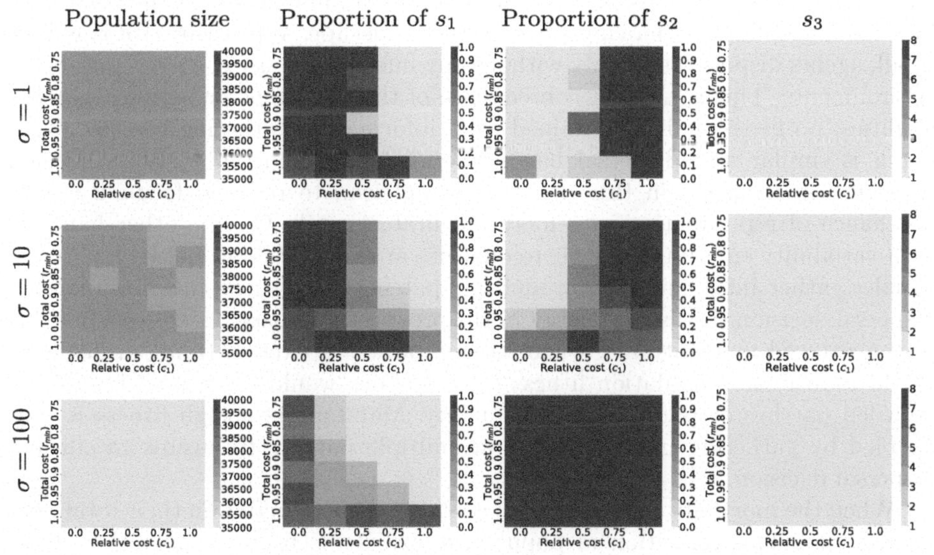

Fig. 5. Heatmaps showing median values found in the population after 20,000 generations for a mortality risk $\mu = 0.1$.

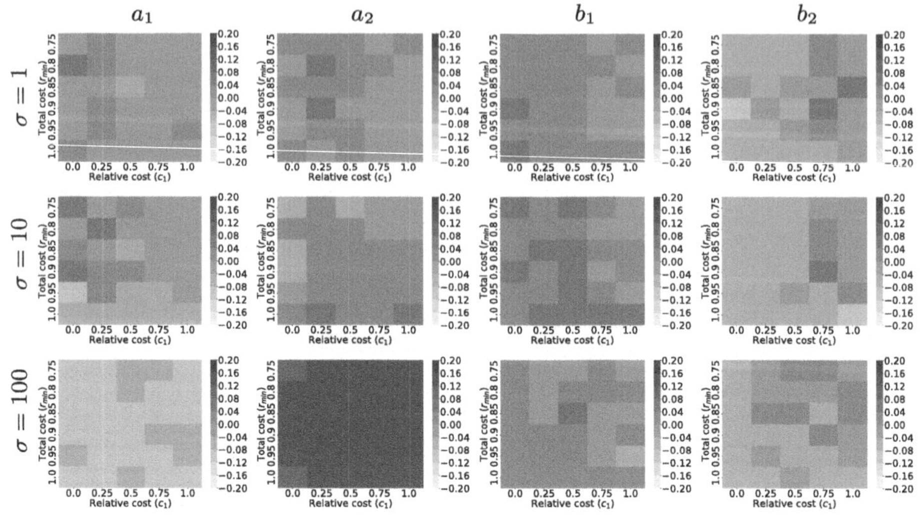

Fig. 6. Heatmaps showing the median values of real-valued genes found in the population after 20,000 generations for a mortality risk $\mu = 0.1$.

4 Discussion

In this study, we investigate how different dispersal behaviours evolve in gradually harsher environmental settings. Our results show that population fitness, represented by the total population size, is directly influenced by the risk taken by female agents when moving through the landscape. When this risk is null or small, agents disperse randomly without any information usage in low-variability environments. This is a direct consequence of the homogeneity between patches qualities: no fitness gain is obtained from information gathering because every patch is similar and dispersal has a low associated mortality risk. Choosing to gather information in this situation would lead to reduced fecundity, with no chance of reproduction in a more favourable patch. On the other hand, in high-variability environments where qualities are very different between patches, females gather information from multiple patches in order to take an informed dispersal decision. Investing energy in the prospecting phase is only possible due to the reduced mortality risk when moving between patches. Random dispersal would reduce the population fitness since females would often land blindly on crowded patches or patches of poor quality. Maintaining a high fitness is thus enabled by gathering information from multiple patches to ensure an efficient dispersal decision.

When the mortality risk is higher, agents struggle to maintain their total population size. This reduction in population size stems from the fact that information gathering is too costly. In low-variability environments, agents cannot disperse randomly as previously observed since they have a high chance of dying in the dispersal process. Thus, they rely on gathering information from only one adjacent patch to counter this negative effect and disperse only when they can reach a better patch with certitude. In high-variability environments, the same gathering strategy evolve but a significant loss in fitness is seen. Overall, the consequences of a higher risk while exploring the environment are double: agents cannot gather information from multiple patches since many of them would die in the process and the total population size shrinks due to deaths occurring during the dispersal phase.

The use of private information is predominant in high-variability environments for any probability of dying when moving between patches. Social information can be less reliable in those conditions because of the rapid changes in patch qualities: a high number of conspecifics indicate a previous good quality of the patch but this may not be true for the next generation [8]. However, in low-variability environments, agents rely on both information sources at equal acquisition costs, or on the cheapest one when these costs are asymmetric. Both sources are equally reliable in this scenario due to the homogeneous patch qualities, leading agents to minimise the fecundity costs related to information acquisition. It is interesting to note that relying on private information in high-variability environments was also the selected strategy in our previous studies with agents only able to sample information from one adjacent patch [13,14].

Finally, all evolved behaviours displayed positively density-dependent dispersal when using social information. Agents tend to leave high-density patches to disperse towards low-density ones. On the other hand, private information often carries more weight in the final dispersal decision, with agents being attracted to high quality patches.

5 Conclusion

Our study highlights that information gathering strategies heavily depend on the environmental conditions encountered by agents, namely the variation of environmental quality between patches of habitat and the overall mortality risk while moving through the landscape. The total cost of acquisition of information and the relative costs related to private and social information only influence the type of information use selected, but not the number of patches where information gathering takes place. Non-informed dispersal is predominant in low-variability environments but disappears when the mortality risk becomes too high. Gathering information from multiple patches is only useful in high-variability environments, provided that the mortality risk is small. When this risk becomes high, gathering a lot of information becomes too costly and agents limit their information sources.

Our information gathering strategy was kept simple in this work to focus on the effect of the multiple environmental parameters that bear on the total population fitness, with our agents choosing to take a dispersal decision only on the patch with the highest computed probability of dispersal in their visited neighbouring patches. Future works could explore more complex information gathering strategies, e.g. not limited to adjacent patches or with different methods used to compute the final dispersal decision made by our agents. Experimental work could also be conducted to validate our approach. We propose a possible experiment to evaluate the robustness of our model, using as model organism the spider mite *Tetranychus urticae*, a fast-reproducing yet complex organism with well-developed sensory organs and semi-social structure [16]. Their genome is also fully sequenced and annotated [5]. Following a standard protocol for experimental settings using this species (see for instance [7]), a network of patches (leaf cuts from different plants, see below) interconnected by a hostile matrix (Parafilm bridges surrounded by wet cotton) could be arranged. Mites would be free to explore the patches around the one they originated from and then decide whether to disperse or remain in their original patch. Costs of dispersal could be artificially manipulated, for example by removing random patches or individuals from the system. Patch quality could be manipulated by introducing low- or high-quality food types (e.g., corn vs bean leaves) or by varying the freshness of leaves cut from the same host plant, for instance via desiccation. Finally, individuals' densities could also be manipulated to maintain the population below the carrying capacity of the system and avoid population crash.

Individuals could be collected from the same stock population, placed into setups undergoing different treatments, and then followed for several generations (spider mites have a generation time of 10 days at 30 °C). After the predetermined number of generations, genome of the different populations could be sequenced looking for differences in allele proportion. These real-world findings could confute of confirm what we propose in our in silico model.

Acknowledgments. This work was supported by Service Public de Wallonie Recherche under grant n° 2010235 - ARIAC by DIGITALWALLONIA4.AI. Computational resources have been provided by the Consortium des Équipements de Calcul Intensif (CÉCI), funded by the Fonds de la Recherche Scientifique de Belgique (F.R.S.-FNRS) under Grant No. 2.5020.11 and by the Walloon Region.

References

1. Bach, L.: On the evolution of conditional dispersal under environmental and demographic stochasticity. Evol. Ecol. Res. **9**(4), 663–673 (2007)
2. Bonte, D., et al.: Costs of dispersal. Biol. Rev. **87**(2), 290–312 (2012). https://doi.org/10.1111/j.1469-185X.2011.00201.x
3. Delgado, M.M., Bartoń, K.A., Bonte, D., Travis, J.: Prospecting and dispersal: their eco-evolutionary dynamics and implications for population patterns. Proc. Roy. Soc. B Biol. Sci. **281**(1778), 20132851 (2014). https://doi.org/10.1098/rspb.2013.2851
4. Enfjäll, K., Leimar, O.: The evolution of dispersal - the importance of information about population density and habitat characteristics. Oikos **118**(2), 291–299 (2009). https://doi.org/10.1111/j.1600-0706.2008.16863.x
5. Grbić, M., et al.: The genome of Tetranychus urticae reveals herbivorous pest adaptations. Nature **479**, 487–92 (2011). https://doi.org/10.1038/nature10640

6. Ádám Kun, Scheuring, I.: The evolution of density-dependent dispersal in a noisy spatial population model. Oikos **115**(2), 308–320 (2006). https://doi.org/10.1111/j.2006.0030-1299.15061.x
7. Mortier, F., Masier, S., Bonte, D.: Genetically diverse populations spread faster in benign but not in challenging environments. Ecology **102**(6), e03345 (2021). https://doi.org/10.1002/ecy.3345
8. Nieberding, C.M., Marcantonio, M., Voda, R., Enriquez, T., Visser, B.: The evolutionary relevance of social learning and transmission in non-social arthropods with a focus on oviposition-related behaviors. Genes **12**(10), 1466 (2021). https://doi.org/10.3390/genes12101466
9. Poethke, H.J., Hovestadt, T.: Evolution of density-and patch-size-dependent dispersal rates. Proc. Roy. Soc. London Ser. B Biol. Sci. **269**(1491), 637–645 (2002). https://doi.org/10.1098/rspb.2001.1936
10. Ponchon, A.: Prospecting for informed dispersal: reappraisal of a widespread but overlooked ecological process. Preprints (2024). https://doi.org/10.20944/preprints202401.0428.v1
11. Ponchon, A., Scarpa, A., Bocedi, G., Palmer, S., Travis, J.: Prospecting and informed dispersal: understanding and predicting their joint eco-evolutionary dynamics. Ecol. Evol. **11** (2021). https://doi.org/10.1002/ece3.8215
12. Ruxton, G.D., Rohani, P.: Fitness-dependent dispersal in metapopulations and its consequences for persistence and synchrony. J. Anim. Ecol. **68**(3), 530–539 (1999). https://doi.org/10.1046/j.1365-2656.1999.00300.x
13. Sion, A., Marcantonio, M., Masier, S., Tuci, E.: On the evolution of dispersal strategies under the costs of acquisition of private and social information. Ecol. Model. **501**, 111004 (2025). https://doi.org/10.1016/j.ecolmodel.2024.111004
14. Sion, A., Marcantonio, M., Tuci, E.: Influence of the costs of acquisition of private and social information on animal dispersal. In: Brock, O., Krichmar, J. (eds.) From Animals to Animats 17, pp. 223–235. Springer, Cham (2025). https://doi.org/10.1007/978-3-031-71533-4_17
15. Travis, J., et al.: Modelling dispersal: an eco-evolutionary framework incorporating emigration, movement, settlement behaviour and the multiple costs involved. Methods Ecol. Evol. **3**(4), 628–641 (2012). https://doi.org/10.1111/j.2041-210X.2012.00193.x
16. Van Petegem, K., et al.: Kin competition accelerates experimental range expansion in an arthropod herbivore. Ecol. Lett. **21**(2), 225–234 (2018). https://doi.org/10.1111/ele.12887
17. Wong, B.B., Candolin, U.: Behavioral responses to changing environments. Behav. Ecol. **26**(3), 665–673 (2015). https://doi.org/10.1093/beheco/aru183

Cooperation in Evolved Modular Soft Robots

Paolo Baldini[1], Michele Braccini[1], Filippo Benvenuti[1], Eric Medvet[2](✉), Andrea Roli[1], and Francesco Rusin[2]

[1] Department of Computer Science and Engineering, University of Bologna, Campus of Cesena, Bologna, Italy
{p.baldini,m.braccini,filippo.benvenuti7,andrea.roli}@unibo.it
[2] Department of Engineering and Architecture, University of Trieste, Trieste, Italy
emedvet@units.it, francesco.rusin2@phd.units.it

Abstract. Self-assembly is the ability of natural beings to aggregate in a suitable structure in order to tackle complex tasks that could not be solved by single individuals. Robotics tries to replicate this ability to solve complex problems using artificial embodied physical agents. Nevertheless, the potential of self-assembly in robotics has still to be fully exploited. Here we propose self-assembly with voxel-based soft robots, a type of modular soft robots. The goal is to evolve the controller of the robots, inducing the formation of a single structure to move beyond a hole in the terrain that could not be crossed by individual robots. We perform experiments in simulation in increasingly complex arenas. The results show that evolved robots are capable of forming a connected chain and overall succeed in their mission. We also discuss limitations and improvements of the proposed approach.

Keywords: Self-assembly · Soft robotics · Modular robotics · Evolutionary robotics · Neural networks

1 Introduction

The ability to cooperate is a foundational capability for many biological species. When single individuals are not enough the group intervenes, cooperatively facing difficult situations. This allows tackling problems otherwise unbearable, or for which a single individual would spend a great amount of effort. Among different types of cooperation, here we focus on *self-assembly*, which in fact can be found at any scale in nature [18]. In general, self-assembly can be defined as the property of a group of separate entities to autonomously cooperate and come together to form patterns and structures. Many examples are available in nature, notably, the creation of columns and bridges in ant colonies to cross gaps [13].

Due to its importance in tackling difficult tasks, the ability to self-assemble started to be explored also in artificial collective systems. One of the first multi-robot systems able to self-assemble was proposed by O'Grady et al. [12] with the aim of moving over rough terrains. Tuci et al. [17] extended the work by controlling the robots with an artificial neural network (ANN). Additionally, they

T. Carletti et al. (Eds.): WIVACE 2024, CCIS 2532, pp. 160–172, 2025.
https://doi.org/10.1007/978-3-031-93631-9_13

tested the ability of collective transportation of heavy objects to a nest, focusing on the design of mechanisms for movement coordination. Recently, Recently, Swissler and Rubenstein [15] have proposed the creation of robots able to form non-lattice structures (*i.e.*, with omnidirectional connections) with the aim to climb on each other and self-assemble into bridges. A recent and detailed survey on the state of the art in robotic self-assembly was provided by Bray and Groß [2]. Here we explore the idea of self-assembly applied to voxel-based soft robots (VSRs) [5,7], also known as virtual soft robots [11]. VSRs differ from classical robots in that their body is made of a soft material, allowing for more degrees of freedom in movement [3]. This simplifies the problem of coordinated movement, as each VSR can deform to accompany others movements. Additionally, soft-robots reduce the risk of working with humans, minimizing the effect of crashes [1,6].

We propose evolving the controller of VSRs to obtain a self-assembling behavior in order to complete a task requiring collaboration. Specifically, we consider the case of robots that have to pass over a hole in the ground. The long-term vision is that of soft-robots able to cooperatively navigate complex environments, face a variety of tasks, and safely interact with humans.

2 Experimental Setting

We set up an experiment in which a group of VSRs has to cooperate to pass over a hole in the ground. All the VSRs are identical square-shaped voxel, controlled by a copy of the same controller. The core of the controller is an ANN generated by an evolutionary algorithm (EA). We investigate the performance of controllers produced employing three different fitness functions (FFs). To avoid bias in the results, we run the EA on different arenas characterized by the size of the hole and its distance from the VSRs. Additionally, we investigate how the number of VSRs composing the group affects the success ratio in the experiment.

Each VSR is equipped with 18 sensors, which allow it to perceive the environment and other VSRs [16]: a rotation-angle sensor, a area-ratio sensor that senses the ratio between the current area of the voxel and the rest area (of $1\,\text{m} \times 1\,\text{m}$); a contact sensor which returns 1 if there is at least one contact and -1 otherwise; a proximity sensor, on a single side, sensing the distance to the closest object up to a range $d = 5\,\text{m}$ along a direction normal to the side; two speed sensors in the two orthogonal directions with respect to the current orientation of the voxel; four attachment sensors, one for each side, sensing 1 if the side is attached to another VSR and -1 otherwise; four side-compression sensors, one for each side, sensing the ratio between the current side length and the rest length (of $1\,\text{m}$); four directional-communication sensors, one for each side, each reading communication signals (*i.e.*, scalar numbers) sent by other VSRs within a given range. Additionally, each VSR is endowed with an endogenous signal periodically oscillating and acting as a clock. All sensor output values are normalized in $[-1, 1]$. We choose this configuration for the sensors of the VSRs based on previous experience and studies [14].

Each VSR acts by means of 12 actuators. Four springs, one per side, enable motion by changing their rest value. Four attach/detach actuators, one per side, control aggregation with other VSRs. When an actuation value exceeds a positive threshold τ for a given side and another VSR is close enough in that direction, the two VSRs attract each other (eventually physically attaching). Conversely, when the value exceeds $-\tau$ and the side is attached to another VSR, it detaches from it. Four communication systems, one per side, allow sending scalar signals along a direction normal to the side.

The sensory and the endogenous signals are fed to an ANN controller that produces the control and communication signals. Therefore, the ANN must have 19 input nodes and 12 output nodes. We set up a fully connected feed-forward ANN with 3 hidden layers with 16, 12, and 12 neurons respectively. Including the bias, this results in a total of 836 weights, with tanh as activation function in the neurons. We apply the ANN not at every simulation step, *i.e.*, at a frequency of 60 Hz, but less frequently, at 10 Hz: this way, we avoid vibrating behaviors that would not work in real life [9]. We optimize the ANN weights using an EA, namely covariance matrix adaptation evolution strategy (CMA-ES) [4]. We use the same ANN in every VSR during a simulation. In CMA-ES, we set the population size to 24 and run it for 417 iterations[1].

At the beginning of each simulation, we place the VSRs at fixed positions on the ground, on the left of the hole (see Fig. 1a). The evaluation is based on a FF which computes a fitness value based on the horizontal x-position reached by the VSRs with respect to the start of the arena (the x-axis is oriented left to right). We employ three different FFs, namely: (a) Average group x-position $\mathcal{AVG}$; (b) Maximum VSR x-position $\mathcal{MAX}$; (c) Minimum VSR x-position $\mathcal{MIN}$. $\mathcal{AVG}$ computes the average of the group x-position (see Eq. (1)). $\mathcal{MAX}$ considers the right border x-position of the rightmost VSR of the group (see Eq. (2)). $\mathcal{MIN}$ considers the left border x-position of the leftmost VSR of the group (see Eq. (3)). The result is that $\mathcal{AVG}$ requires most of the group to move right, or a subgroup to move much farther; $\mathcal{MAX}$ just needs one VSR performing well; $\mathcal{MIN}$ requires the entire group to perform well. As the hole might trap the VSRs, a fitness computed with any of the FFs with a value greater than the rightmost boundary of the hole denotes a situation in which some robots (if not all) successfully traversed it. This can happen only if the VSRs successfully self-assemble or, more broadly, if they cooperate. Indeed, the hole is larger than a single VSR and deeper than the height of the piled group, requiring cooperation to pass it. Note that the VSRs before the hole are not guaranteed to know how to self-assemble. They may indeed wander randomly without aggregating.

[1] This value depends on the maximum number of evaluations that we allow: $10'000$.

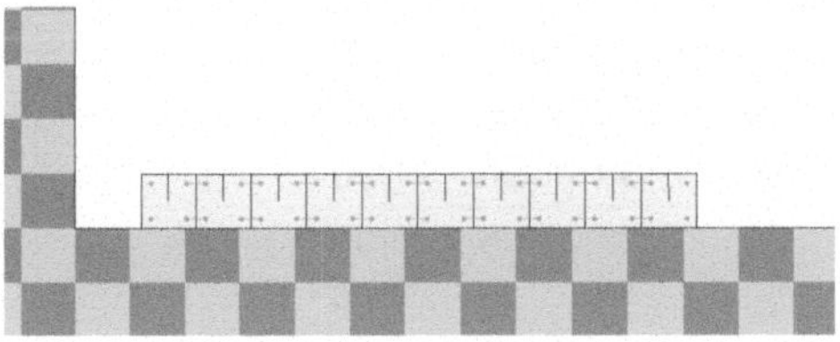
(a)

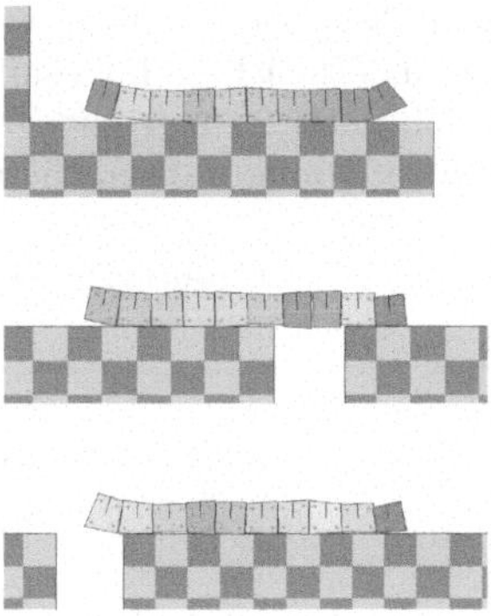
(b)

Fig. 1. In (a), 10 VSRs at the beginning of the simulation; the hole is 20 m distant from the position of the leftmost VSR. In (b), the VSRs group manages to overcome the hole of type w_2 (*i.e.*, a hole larger than 2 VSRs).

The formulas of the FFs are the following:

$$f_{\mathcal{AVG}}(\ ) = \frac{1}{n}\sum_{i=1}^{n} x_{c,i} \tag{1}$$

$$f_{\mathcal{MAX}}(\ ) = \max(\{x_{r,i} | i \in [1, n]\}) \tag{2}$$

$$f_{\mathcal{MIN}}(\ ) = \min(\{x_{l,i} | i \in [1, n]\}) \tag{3}$$

where:

- n is the number of VSRs composing the group;
- i is the index of the VSR in the group;
- x_c, x_l, x_r are respectively the center x position of the VSR, and the left and right border x-position of the VSR at the end of the evaluation.

We perform the experiments in six different arenas characterized by a flat ground with a hole. Half of the arenas present a hole at 12 m distance from the starting VSRs position, while in the other half the hole is at 20 m. The hole can assume three different widths: $w_1 = l + \varepsilon$, $w_2 = 2l + \varepsilon$, and $w_3 = 3l + \varepsilon$, with $l = 1$ m being the voxel side length (see Fig. 1). In other words, the first arena contains a hole slightly larger than one VSR, the second contains a hole slightly larger than two VSRs, and the third contains a hole slightly larger than three VSRs. The goal is to verify the ability to self-assemble in conditions of increasing complexity. We test groups of different sizes: 4, 6, 8, and 10 VSRs. The simulation time of each group of VSRs is limited to 30 s regardless of the arena used. In this time the group is required to pass over the hole and move as far as possible to the right. We perform 30 runs (or replicas) of CMA-ES for each arena. We use JGEA [10] for running the EA and 2D-VSR-Sim [8] for running the simulations.

Table 1. Number of replicas with a performance sufficient to be considered a "success", with each of the FFs. The results refer to runs in arenas with hole of size $1+\varepsilon$ at distance 12 m (left) and 20 m (right) from the first VSR. Since we performed 30 replicas per experiment configuration, the maximum value obtainable is 30.

Group size	Distance 12 m			Distance 20 m		
	$\mathcal{AVG}$	$\mathcal{MAX}$	$\mathcal{MIN}$	$\mathcal{AVG}$	$\mathcal{MAX}$	$\mathcal{MIN}$
4	29	27	30	30	20	29
6	30	30	30	30	26	30
8	30	30	30	29	28	30
10	30	28	29	30	30	29

3 Results

To correctly analyze the results, we first need to reason about the meaning of the fitness computed with each FF. $\mathcal{AVG}$ considers the average x-position of the group of VSRs. This means that outliers have a small influence on the computed fitness value. A high value represents a group that overall performs well. Conversely, a low value indicates a group that overall performs poorly. Therefore, $\mathcal{AVG}$ incentivizes the VSRs to cooperate, as the average x-position of the group matters more than the individual one. Nevertheless, analyzing the collective behavior only by looking at this fitness value possibly hides some peculiar behaviors. A group that succeeds pushing a single or few individuals beyond the hole would still have a relatively low fitness, hiding this partial success. Differently, $\mathcal{MAX}$ considers the furthest x-position reached by a VSR. This means that there is not any explicit request to maximize the number VSRs beyond the hole, requiring instead just one VSR to pass it. It follows that the fitness value computed with $\mathcal{MAX}$ does not allow understanding how the group as a whole performed, again hiding some information. Finally, $\mathcal{MIN}$ only considers the leftmost VSR, which is the nearest to the start of the arena. The idea is to incentivize cooperation, as the fitness of the group depends on the position of the worst VSR. This again hides some information, as we are not aware of the fitness of any subgroup. Overall, regardless of the FF employed, we can use the right side of the hole as a coarse grain threshold for a first discrimination of the results.

The results show that the performance (*i.e.*, the fitness of the best individual at the end of the evolution) in arenas with hole of size $1+\varepsilon$ is mostly greater than the threshold, regardless of the FF employed (see Table 1). This implies that the whole group succeeds in passing the hole for $\mathcal{MIN}$, most of it pass the hole for $\mathcal{AVG}$, and at least one VSR passes the hole for $\mathcal{MAX}$. Nevertheless, the performance decreases drastically when considering the arenas with hole $2+\varepsilon$ (see Table 2). Here, just few groups obtain a performance greater than the threshold, meaning that most of the VSRs do not pass the hole. Specifically, when considering $\mathcal{MAX}$, every failure implies that no VSR succeeded in passing

Table 2. Number of replicas with a performance sufficient to be considered a "success", with each of the FFs. The results refer to runs in arenas with hole of size $2+\varepsilon$ at distance 12 m (left) and 20 m (right) from the first VSR. Since we performed 30 replicas per experiment configuration, the maximum value obtainable is 30.

Group size	Distance 12 m			Distance 20 m		
	$\mathcal{AVG}$	$\mathcal{MAX}$	$\mathcal{MIN}$	$\mathcal{AVG}$	$\mathcal{MAX}$	$\mathcal{MIN}$
4	0	0	0	0	2	0
6	0	2	1	4	6	1
8	2	5	2	5	8	1
10	5	1	0	6	13	2

the hole. The performance decrease is even more evident in arenas with hole of size $3+\varepsilon$. In this situation, we do not register any success. The problem is likely due to the impossibility of the group to create a stable structure that can resist the gravitational force. This happens when many VSRs start to lose touch with the ground. Because of this lack of success, we decide to not show any result for arenas with hole of size $3+\varepsilon$. Overall, $\mathcal{MIN}$ seems the best for simple arenas, obtaining the same or better success rate than the other FFs while ensuring cooperation. $\mathcal{MAX}$ allows obtaining more (probably partial) successes in the hardest arenas, but performs worse than the other FFs in simpler ones. Finally, $\mathcal{AVG}$ seems to perform as well as $\mathcal{MIN}$ in the simple arenas, but performs better than it in the hardests.

Until now our analysis just considered a binary result: beyond the hole or not. However, this view is somehow limiting, as it does not describe how well the groups performs after passing the hole (*i.e.*, how far they reach). Figure 2 shows the distribution of the performance for each FFs, group size, and hole distance, in arenas with hole of size $1+\varepsilon$. As visible, the maximum performances are more or less the same regardless of the FF employed. Nevertheless, the overall distribution changes according to the measure we use. Specifically, $\mathcal{AVG}$ and $\mathcal{MIN}$ present very compact distributions near the maximum. This indicates that the performance is stable across different groups. Nevertheless, this is not the case for $\mathcal{MAX}$. The distributions obtained employing this FF are indeed much wider, indicating that the results are less stable. When considering arenas with holes of size $2+\varepsilon$, we see instead that $\mathcal{MIN}$ produces solutions that perform very poorly (see Fig. 3). Differently, $\mathcal{AVG}$ and $\mathcal{MAX}$ still succeed in passing the hole, but the difference in distributions between the two decreases.

As anticipated at the beginning of this section, the performance does not give us full information on the ability to pass the hole. Therefore, we consider the final x-position of all the VSRs of the group to understand how many passed it. We normalize the results dividing the number of VSRs beyond the hole by their total number in the experiment configuration. Figure 4 shows the results in arenas with hole of size $1+\varepsilon$. Since $\mathcal{MIN}$ requires the VSRs to cross altogether, the results tend to be polarized: they succeed, or they do not. Solutions which

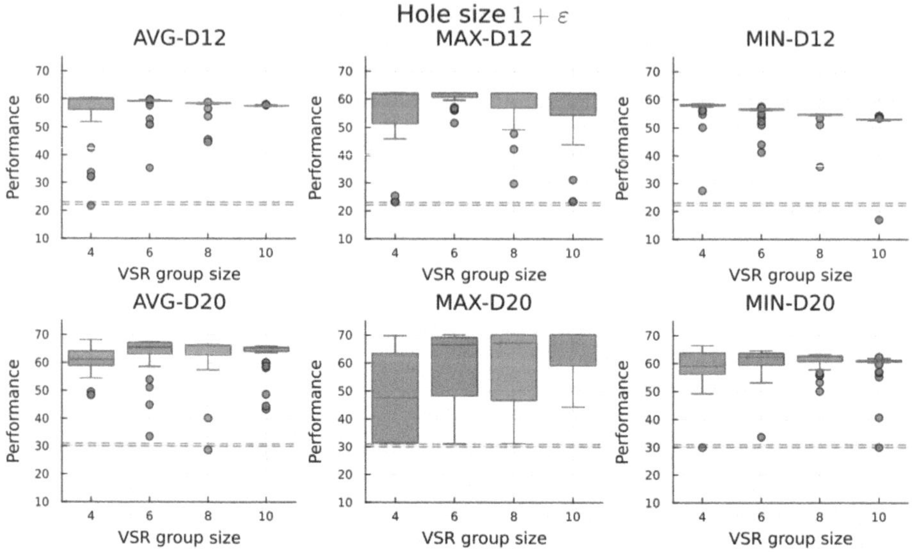

Fig. 2. Performance distribution in the arenas with hole size $1+\varepsilon$. Each distribution, one per experiment configuration, consists of 30 values. The configurations of the experiment differ by the FF used, the distance of the hole from the starting x-position of the VSRs, and the group size. The plots on the first, second, and third column respectively represent experiments run with FFs $\mathcal{AVG}$, $\mathcal{MAX}$, and $\mathcal{MIN}$. The plots on the first and second row respectively represent experiments with the hole at distance 12 m and 20 m from the x-position of the first VSR. The red band in the plots represents the hole.

allow just some VSRs to pass are completely ignored during evolution, and thus do not often appear in the final results. $\mathcal{AVG}$ presents similar results, but being less strict it does not discard solutions which leave some VSRs behind. This results in some of the runs having just part of the group beyond the hole. For $\mathcal{MAX}$ the ratio of VSRs left behind increases significantly, especially in the arenas with the hole at 20 m distance. Nevertheless, most of the results still show groups larger than one VSR passing the hole, indicating that the evolution still somehow selects cooperation. When considering arenas with hole of size $2+\varepsilon$, the pass-ratio of $\mathcal{MIN}$ drops almost always to zero (see Fig. 5). This is again due to the polarization induced by this FF. Even the passing ratios of $\mathcal{AVG}$ and $\mathcal{MAX}$ decrease, becoming more similar to each other. These results confirm the previous considerations on the effectiveness of the different FFs, indicating $\mathcal{MIN}$ as the best for simple arenas, $\mathcal{MAX}$ as the best for hardest ones, and $\mathcal{AVG}$ as an effective trade-off between the other two.

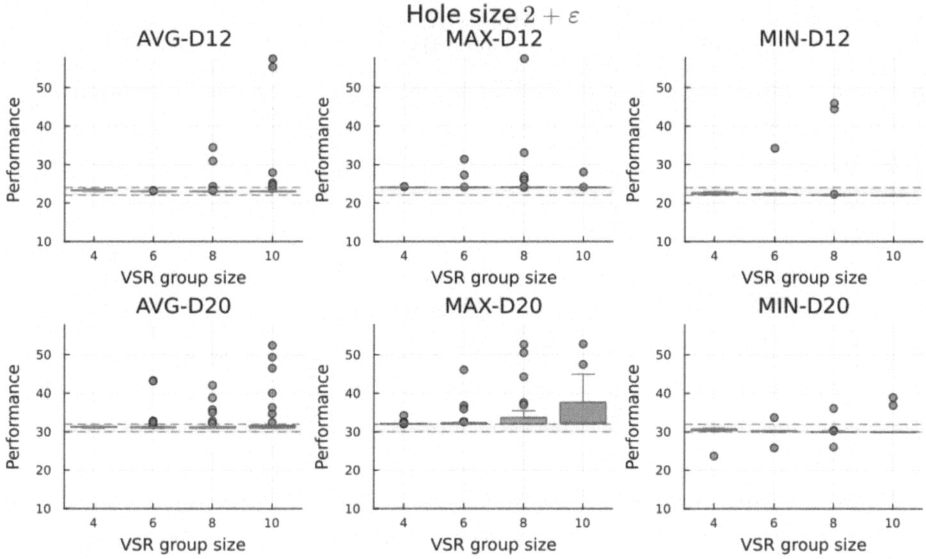

Fig. 3. Performance distribution in the arenas with hole size $2+\varepsilon$. Each distribution, one per experiment configuration, consists of 30 values. The configurations of the experiment differ by the FF used, the distance of the hole from the starting x-position of the VSRs, and the group size. The plots on the first, second, and third column respectively represent experiments run with FFs $\mathcal{AVG}$, $\mathcal{MAX}$, and $\mathcal{MIN}$. The plots on the first and second row respectively represent experiments with the hole at distance 12 m and 20 m from the x-position of the first VSR. The red band in the plots represents the hole. (Color figure online)

Another important aspect to investigate is the impact of the number of VSRs on the performance and passing ratio. Starting from the former, we recognize that the size of the group has a different impact according to the considered FF (see Figs. 2 and 3). Increasing the VSRs in $\mathcal{AVG}$ seems to increase the performance of the group. In arenas with hole of size $1+\varepsilon$, a difference is clear mostly for the smaller groups of size 4. In arenas with hole of size $2+\varepsilon$, the difference seems more evident across all the groups. For $\mathcal{MAX}$, the increase in performance is evident in arenas with both the hole sizes. Specifically, increasing the size of the group seems to affect the mean and spread of the distributions significantly. For $\mathcal{MIN}$ we witness instead a different trend. In the arenas with the smaller hole size, the performance seems to decrease at the increase in the group size. Differently, results obtained in arenas with hole of size $2+\varepsilon$ do not show any noticeable difference across different group sizes. These peculiar trends are due to the fact that having more VSRs requires a higher number of individuals passing the hole. This obviously complicates the task, as with $\mathcal{MIN}$ a single error affects the fitness of the whole group.

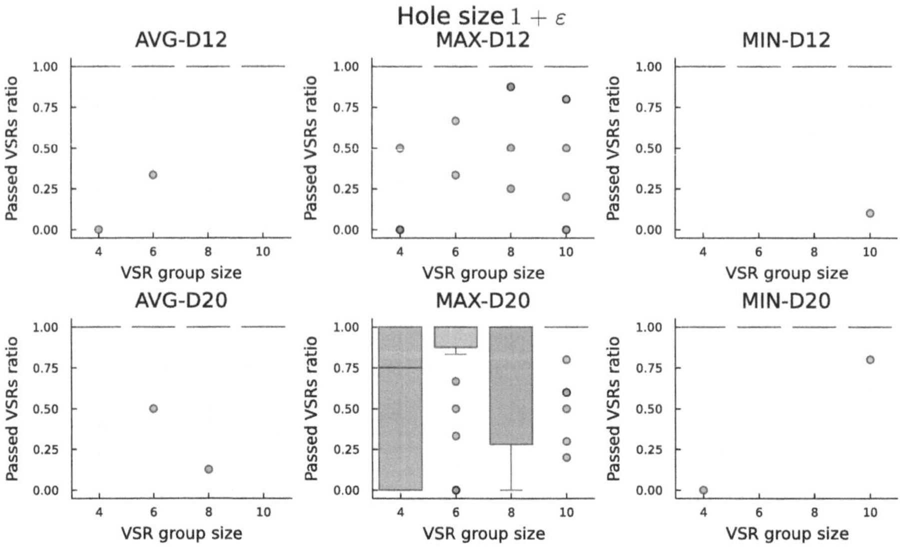

Fig. 4. Ratio distribution of VSRs which have passed the hole of size $1+\varepsilon$. For each of the 30 replicas of experiment configurations, we calculate the number of VSRs beyond the hole and divide it by the group size. A value of 1 means that all the VSRs passed the hole. A value of 0 means that no VRS passed the hole. Each distribution, one per experiment configuration, consists of 30 values. The configurations of the experiment differ by the FF used, the distance of the hole from the starting x-position of the VSRs, and the group size. The plots on the first, second, and third column respectively represent experiments run with FFs $\mathcal{AVG}$, $\mathcal{MAX}$, and $\mathcal{MIN}$. The plots on the first and second row respectively represent experiments with the hole at distance 12 m and 20 m from the x-position of the first VSR.

We witness similar results when considering the passing ratio of a group. For $\mathcal{AVG}$ the metric increases slightly when increasing the group size in the arenas with hole of size $2+\varepsilon$ at distance 20 m. For $\mathcal{MAX}$ we recognize an increment in arenas with both the hole sizes. This seems to confirm that $\mathcal{MAX}$ is more sensible to the group size than the other two FFs. For $\mathcal{MIN}$ we do not witness any particular difference.

Overall, both the results on the pure performance and the passing ratio show that increasing the group size leads to an improvement. The degree of the improvement strongly depends on the FF employed.

4 Discussion

In this work we demonstrate the feasibility of self-assembly in a group of VSRs to pass over a hole in the ground. We achieve this by evolving the VSR controller by means of an EA. Three different FFs guide the evolutionary process, namely: $\mathcal{AVG}$, $\mathcal{MAX}$, $\mathcal{MIN}$. We find that the performance of the group depends on

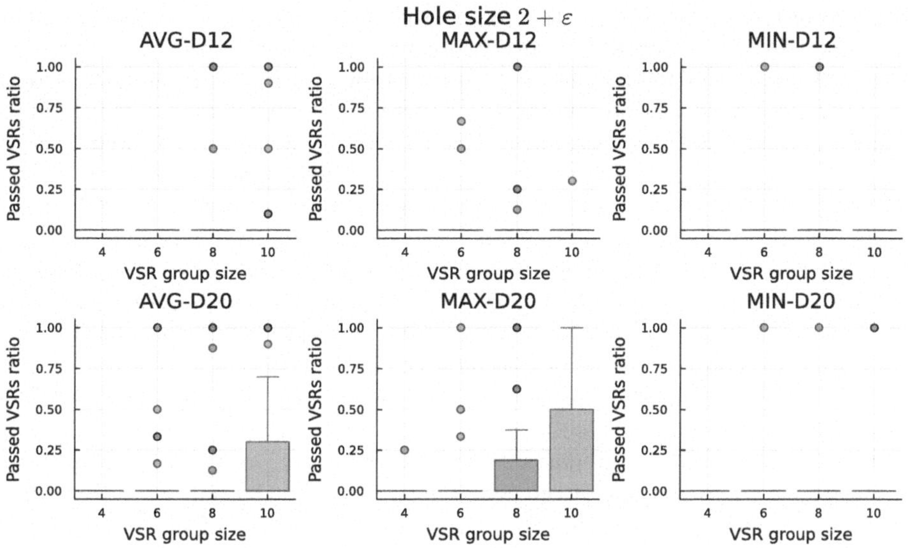

Fig. 5. Ratio distribution of VSRs which have passed the hole of size $2+\varepsilon$. For each of the 30 replicas of experiment configurations, we calculate the number of VSRs beyond the hole and divide it by the group size. A value of 1 means that all the VSRs passed the hole. A value of 0 means that no VRS passed the hole. Each distribution, one per experiment configuration, consists of 30 values. The configurations of the experiment differ by the FF used, the distance of the hole from the starting x-position of the VSRs, and the group size. The plots on the first, second, and third column respectively represent experiments run with FFs $\mathcal{AVG}$, $\mathcal{MAX}$, and $\mathcal{MIN}$. The plots on the first and second row respectively represent experiments with the hole at distance 12 m and 20 m from the x-position of the first VSR.

the FF employed. Specifically, $\mathcal{MIN}$ performs well when the arena is simple, but it is very sensible to failures. As this metric considers the x-position of the leftmost individual, a single VSR falling in the hole is enough to hinder the fitness of all the group. This evaluation approach is especially strict, as it ignores the difference between an entire group falling in the hole and just some of its members. The result is that partial successes that could be further improved by the EA are discarded, and every epoch of the evolution results in a sort of random search. It follows that $\mathcal{MIN}$ is enough to solve easy problems, but becomes ineffective as soon as the complexity of the task increases and refinements of the solutions become necessary. $\mathcal{MAX}$ follows a diametrically opposite approach by considering the x-position of the rightmost individual. This feedback is more easily exploited by the EA during the evolutionary search. Indeed, a controller succeeding in passing at least one VSR is forwarded and used in the generation of the next population. Nevertheless, this FF still presents some issues, as maximizing the passing ratio only happens if it gives advantage beyond the hole. For instance, having a larger group on the right side of the hole can enable a more effective motion, increasing the fitness and thus pushing for an increased passing ratio. Overall, we notice that many controllers produced with $\mathcal{MAX}$ leave a significant amount of VSRs behind. This may indicate that increasing the group size beyond the hole does not necessarily improve the motion, or that this further refinement is too complex to achieve. Finally, $\mathcal{AVG}$ represents a combination of the two previous FFs, trying to provide a more fine-grained evaluation of the controllers. This considers indeed the average x-position of all the VSR, overcoming the problem of considering only the best or the worst individual. The expected outcome is the maximization of the size and speed of the group beyond the hole, as it would have evolutionary advantage against single individuals and slow groups. Overall, we identify a successful trend induced by the use of $\mathcal{AVG}$. This is however comparable to that of $\mathcal{MAX}$ in arenas with hole of size $2+\varepsilon$, rising the question on whether the finer evaluation of $\mathcal{AVG}$ is really beneficial. We believe that despite similar results, $\mathcal{AVG}$ is still worth using as it provides better and most stable results in arenas with hole of size $1+\varepsilon$, and can potentially be more effective when the size of the group grows largely. Overall, we believe that the best FFs are $\mathcal{AVG}$ and $\mathcal{MAX}$, depending on nuances of our goal. $\mathcal{MIN}$ seems instead not effective enough, being too sensible to single failures. Finally, as the FFs induce different search behaviors, they might be combined, each to be used in different phases of the search.

Another factor that possibly affects the success of the group of VSRs is their initial distance from the hole. Indeed, self-assembling requires some space to operate. In this work we test two different distances in order to assess if they lead to any change in the results. Specifically, we test with holes 12 m and 20 m distant from the starting position of the VSRs group. We find some differences in the results. Increasing the distance of the hole overall improves the performance. We believe this is due to the fact that the group of VSRs has more space to create a stable structure, synchronize the movement, and acquire momentum to pass the hole. Nevertheless, we find an exception when using $\mathcal{MAX}$ in arenas with hole of size $1+\varepsilon$. In this scenario, the performance obtained decreases with the increase of the distance. The motivation behind this different behavior is currently unclear.

We also investigate if the size of the VSRs group affects the results. We identify a positive correlation between the group size and its effectiveness in the task. This means that increasing the number of individuals in the group seems to generally improve the performance. The motivation is probably that a larger group is more stable, allowing to withstand the gravitational force dragging part of it into the hole. Additionally, a larger group can possibly employ more advanced motion strategies, moving further beyond the hole. This general considerations do not apply however to $\mathcal{MIN}$, which display negligible and sometimes even decreasing performance with the increase in size. We believe this is due to the fact that increasing the number of individuals increases the probability of fails, affecting the worst-based fitness calculation of $\mathcal{MIN}$.

One interesting aspect that we observed in some preliminary results, but that was not replicated in the final ones, is that soft-robotics enables strategies that are not possible with stiff robots. More specifically, we saw the group of VSRs finding a solution for the arena with the hole of size $3+\varepsilon$. The VSRs form a chain which heads directly towards the bottom of the hole. By maintaining the chain rigid, the tail gains centripetal acceleration (as it happens in a catapult). When the tail of the chain perceives this sudden acceleration, it detaches, being thrown on the other side of the hole. This peculiar strategy is in fact only possible if we consider using soft-robots, which can withstand being toss on the ground. This is obviously not the focus of the work, and it was neither replicated in the final results, nevertheless, we consider it interesting and a clear example of what soft-robotics enables.

5 Conclusion

In this work we assess if it is possible to evolve the controller of VSRs to self-assemble and solve a hole-passing task. To verify the scalability of the approach, we experiment with multiple arenas of increasing complexity and with groups of different size. Additionally, we investigate how different FFs affect the performance of the group.

Our results show that it is possible to generate a controller that induces self-assembly and allows passing the hole. The performance varies according to the complexity of the arena, ranging from almost 100% of success in the simplest arena to 0% in the most complex one. Specifically, we note that increasing the distance between the hole and the group of VSRs generally increases the performance. We attribute this effect to the possibility of the group to form more stable and effective structures. We find that the performance of the group also depends on the FF employed. Specifically, evaluating the group according to its worst individual (*i.e.*, $\mathcal{MIN}$) generates underperforming groups in complex arenas. Conversely, using the best individual (*i.e.*, $\mathcal{MAX}$) produces good results, but introduces some fluctuations. The best FF overall seems to be $\mathcal{AVG}$, which produces more stable, but still good, results. Finally, we identify scalability in the approach, with the group performance usually increasing with its size. This does not happen however when employing the $\mathcal{MIN}$ FF. In this case, the performance seems even negatively affected by the increase in group size. This is probably due to the fact that increasing the number of individuals increases the probability of fails, affecting the worst-based fitness calculation of $\mathcal{MIN}$.

We intend to perform additional experiments to verify if a pre-training or a scaffolding may lead to better results in the case of large holes. Indeed, the EA starts the evolution in each arena without any previous knowledge. This requires rediscovering fundamental behaviors every time. We expect that evolving VSRs able to solve simpler arenas will speed up the process and increase the percentage of success.

References

1. Arnold, T., Scheutz, M.: The tactile ethics of soft robotics: designing wisely for human-robot interaction. Soft Rob. **4**(2), 81–87 (2017)

2. Bray, E., Groß, R.: Recent developments in self-assembling multi-robot systems. Curr. Robot. Rep. **4**(4), 101–116 (2023)
3. Cheney, N., Bongard, J., Lipson, H.: Evolving soft robots in tight spaces. In: Proceedings of the 2015 Annual Conference on Genetic and Evolutionary Computation, pp. 935–942 (2015)
4. Hansen, N., Ostermeier, A.: Completely derandomized self-adaptation in evolution strategies. Evol. Comput. **9**(2), 159–195 (2001)
5. Hiller, J., Lipson, H.: Automatic design and manufacture of soft robots. IEEE Trans. Rob. **28**(2), 457–466 (2011)
6. Jørgensen, J., Bojesen, K.B., Jochum, E.: Is a soft robot more "natural"? Exploring the perception of soft robotics in human–robot interaction. Int. J. Soc. Robot. **14**(1), 95–113 (2022)
7. Legrand, J., Terryn, S., Roels, E., Vanderborght, B.: Reconfigurable, multimaterial, voxel-based soft robots. IEEE Robot. Autom. Lett. **8**(3), 1255–1262 (2023)
8. Medvet, E., Bartoli, A., Lorenzo, A.D., Seriani, S.: 2d-VSR-sim: a simulation tool for the optimization of 2-D voxel-based soft robots. SoftwareX **12**, 100573 (2020)
9. Medvet, E., Bartoli, A., Pigozzi, F., Rochelli, M.: Biodiversity in evolved voxel-based soft robots. In: Proceedings of the Genetic and Evolutionary Computation Conference. ACM (2021)
10. Medvet, E., Nadizar, G., Manzoni, L.: JGEA: a modular java framework for experimenting with evolutionary computation. In: Proceedings of the Genetic and Evolutionary Computation Conference Companion, pp. 2009–2018 (2022)
11. Mertan, A., Cheney, N.: Investigating premature convergence in co-optimization of morphology and control in evolved virtual soft robots. In: European Conference on Genetic Programming (Part of EvoStar). Springer (2024)
12. O'Grady, R., Groß, R., Mondada, F., Bonani, M., Dorigo, M.: Self-assembly on demand in a group of physical autonomous mobile robots navigating rough terrain. In: Advances in Artificial Life, pp. 272–281. Springer, Heidelberg (2005)
13. Reid, C., Lutz, M., Powell, S., Kao, A., Couzin, I., Garnier, S.: Army ants dynamically adjust living bridges in response to a cost-benefit trade-off. Proc. Natl. Acad. Sci. **112**(49), 15113–15118 (2015)
14. Rusin, F., Medvet, E.: How perception, actuation, and communication impact the emergence of collective intelligence in simulated modular robots. Artif. Life 1–18 (2024)
15. Swissler, P., Rubenstein, M.: Fireant3D: a 3D self-climbing robot towards non-latticed robotic self-assembly. In: 2020 IEEE/RSJ International Conference on Intelligent Robots and Systems (IROS), pp. 3340–3347 (2020)
16. Talamini, J., Medvet, E., Bartoli, A., Lorenzo, A.D.: Evolutionary synthesis of sensing controllers for voxel-based soft robots. In: The 2019 Conference on Artificial Life. MIT Press (2019)
17. Tuci, E., Groß, R., Trianni, V., Mondada, F., Bonani, M., Dorigo, M.: Cooperation through self-assembly in multi-robot systems. ACM Trans. Auton. Adapt. Syst. **1**(2), 115–150 (2006)
18. Whitesides, G., Grzybowski, B.: Self-assembly at all scales. Science **295**(5564), 2418–2421 (2002)

Energy-Aware Neural Architecture Search: Leveraging Genetic Algorithms for Balancing Performance and Consumption

Antoine Gratia[1(✉)], Paul Temple[2], Pierre-Yves Schobbens[1], and Gilles Perrouin[1]

[1] PReCISE/NaDI, Computer Science Faculty, University of Namur, Namur, Belgium
antoine.gratia@unamur.be
[2] University of Rennes, CNRS, Inria, IRISA, Rennes, France

Abstract. In recent years, the field of deep learning has been propelled forward by increasingly complex and resource-intensive neural network models. Despite their impressive performance, their training consumes much energy, incurring significant environmental impacts. Even more so, when neural architecture search (NAS) methods train hundreds of thousands of models to benchmark optimization. We offer to create greener benchmarks using genetic algorithms. We leverage our previous CNNGen approach producing random CNN topologies based on our context-free grammar. We then use the NSGA-II genetic algorithm to create topologies balancing performance and energy consumption. We rely on machine-learning-based predictors that estimate candidates' performance and energy consumption to avoid training during generations, saving significant computational costs. This paper reports on our experiments and discusses further developments.

Keywords: Genetic Algorithms · Neural Architecture Search · Green AI

1 Introduction

Deep learning has made remarkable strides, driving diverse applications such as image and speech recognition [6,9], natural language processing [16], and autonomous vehicles. These advancements are largely attributed to the increasing size and complexity of models [2,17]. Yet, we can wonder to what extent we need to make the models more complex if the gain in performance is marginal. Conversely, can we shrink the deep learning models down while maintaining performance? Identifying the optimal architecture for a given task is a challenging and time-consuming endeavor. To address this, Neural Architecture Search (NAS) [8] formulates an iterative optimization problem to design neural network models that achieve the best performance. Datasets have been developed

T. Carletti et al. (Eds.): WIVACE 2024, CCIS 2532, pp. 173–185, 2025.
https://doi.org/10.1007/978-3-031-93631-9_14

to benchmark generated models and validate NAS optimization approaches. NASBench-101 [19] is probably the most popular and can generate thousands of unique architectures by systematically combining connections and network functions within predefined cells. However, exploring the space of neural architectures is resource-intensive and has a significant carbon footprint. For instance, training LLaMA [16] involves thousands of high-end GPUs, resulting in an estimated carbon footprint of 173 tCO2eq (*i.e.,* tons of CO2 equivalent). To put this into perspective, training a single Transformer model for NLP tasks can emit as much carbon as the lifetime emissions of five cars [15]. Despite this, current NAS techniques do not account for resource consumption and carbon footprint in their algorithms and datasets, focusing solely on performance (accuracy). Moreover, we argue that NAS benchmarks, such as NASBench-101, lack architectural diversity. These benchmarks are based on an initial architecture that already predicts well (above 80% accuracy [19]), allowing only limited variations within cells. Consequently, most generated architectures exhibit training accuracies that exceed 99% on CIFAR-10 when trained for 108 epochs. Additionally, reproducing results is challenging due to the unavailability of essential artifacts, such as fully trainable model code. Our previous work, CNNGen [5], a convolutional neural network (CNN) generator based on a domain-specific language (DSL) and a context-free grammar, was designed to construct CNN models from scratch and provide the expected artifacts (*e.g.*, models, parameters, performances and energy consumption, and code). In addition, we have proposed several predictors for energy consumption and performance. They rely on the description of the models to predict the expected accuracy and/or energy consumption without training the models. Thus, we can explore many models for a limited cost. In this work, we propose using genetic algorithms (GA) to generate new CNN models that balance energy consumption and performance. The GA iteratively evolves a population of CNN models, selecting and recombining them based on a fitness function that considers both energy efficiency and accuracy. Note that we rely again on the estimators to evaluate the performance and energy consumption of new population members. This approach not only fosters greater architectural diversity, but also promotes the development of more sustainable CNN exploration and optimization methods. Concretely, we experimented with the off-the-shelf NSGA-II algorithm. Our results show that our workflow can optimize a topology after five hundred generations, with exhibiting certain CNNs with a slightly better performance and significant decrease in energy consumption. We also discuss the role of predictors and the complexity of predicting performance and accuracy in such a complex CNN topological space.

The rest of the paper is as follows. Section 2 presents the main concepts of neural architecture search and our previous work, which allows us to produce diversified topologies. Section 3 describes the genetic algorithm we experimented with and how it operates with our CNN models as well as our accuracy and energy consumption predictors, while Sect. 4 reports on experiments. Section 5 offers a critical reflection on predictor performance, while Sect. 6 wraps up the paper.

2 Background

Neural Architecture Search (NAS) focuses on automating the design of neural network models, focusing on search space, search strategy, and performance estimation. Search strategies include random search, Bayesian optimization, evolutionary methods, and reinforcement learning [11,14,20]. Performance estimation techniques, such as learning curve extrapolation, surrogate models or network morphisms, aim to reduce computational costs [1,4,8,12,13]. AutoML, exemplified by Auto-Keras [7,8], simplifies the usage of machine learning models for individuals with limited expertise. While Auto-Keras starts with an existing architecture and modifies it through morphing, it requires a good starting point and explores a small neighborhood around the current architecture. State-of-the-art NAS techniques [10] also require a reference architecture to start the search. This architectural choice is left to the experts.

Multi-Objective Genetic Algorithms (MOGAs) are optimization techniques that evolve a population of candidate solutions towards optimization strategies such as Pareto-optimal fronts, addressing problems with multiple conflicting objectives. Inspired by natural selection, MOGAs use crossover, mutation, and selection to explore the solution space. They aim to find a diverse set of trade-off solutions rather than a single optimal solution, making them particularly useful in complex scenarios in which objectives cannot be easily aggregated. Techniques like non-dominated sorting and crowding distance ensure convergence towards and maintenance of diverse solutions across optimization strategies. MOGAs have been applied effectively in various fields such as engineering design, economics, and machine learning. As argued previously, we want to combine MOGAs with NAS to generate new predicting models that would both perform with high accuracy and with a reduced energy consumption compared to current popular models.

Domain-Specific Languages (DSLs) are high-level software languages capturing the essence of a particular domain via specialized abstractions [18]. A DSL typically provides a concise, expressive syntax that enables developers to easily and accurately model complex systems and capture domain knowledge with constraints, reducing the risk of errors or inconsistencies. Then, dedicated generators derive artifacts (code, documentation) automatically from these models. DSLs exist for numerous domains (*e.g.*, software engineering, biology, or robotics) and at various abstraction levels (from programming GPUs to language modeling).

CNNGen [5] is a novel dataset-agnostic approach to generate energy-aware benchmarks for CNNs. CNNGen captures the CNN space using context-free grammar and specifies them with a DSL. CNNGen starts with a description of the topology of the CNN model to be created, which is called a description sentence. The topology refers to the succession of layers that will be stacked (but without any specification of the parameters). As we said before, CNNGen comes with domain-specific rules regarding the succession of layers preventing the generation of unrealistic models. Once the topology is generated, a second phase assigns coherent parameters to every single layer. Along with the description sentence, CNNGen generates various artifacts: an image representation (in

PNG format) of the model associated with the sentence, a TensorFlow runnable code, and a report containing the performance and energy consumption of the trained model associated with the description sentence. By generating random description sentences that meet the generation rules, we can create our dataset of CNN models similar to NAS benchmarks but, unlike NAS approaches, it does not necessitate finding a well-performing CNN model first. In the end, if we want to search for a well-performing CNN model, we simply need to guide the exploration of the gigantic CNN topology with a tool implementing an optimization strategy. In this work, we used MOGAs to maximize the model's accuracy and minimize its energy consumption. The problem is that MOGAs need to know the accuracy and energy consumption associated with each model to decide whether they should be kept for future generation. Instead of training each model, we rely on predictors.

Predictors [5] have also been proposed along CNNGen to avoid fully training every single generated CNN model. As stated before, CNNGen reports both the accuracy and the energy consumption of generated models, thus, we proposed two distinct kinds of predictors. The first predictor assesses the accuracy by analyzing the Python code generated by CNNGen, which was designed to run specific CNN models based on their description sentences. This predictor concentrated on the various architectural layers of the CNN and extracted their respective hyperparameters. The primary objective was to identify and learn correlations between the CNN layers and their hyperparameters, as well as the overall performance of the architecture.

The second predictor is a Regressor Decision Tree which analyzes four manually defined features to predict the energy consumption of CNN models. These features are the number of layers, the number of epochs used for training, the floating-point operations per second (FLOPs), and the total number of parameters in the model. In this paper, we considered a random forest for better energy prediction.

Similarly to ours, Mo and Iacca use a surrogate machine learning model to estimate the performance of models to avoid training and accelerate the evolution of the genetic algorithm [13]. Our approach extends this idea to a multi-objective setting. The same authors exploited learning curve estimation still requires (minimal) training of the models to estimate their performance [12], while our work adopts a radical no-training approach during search.

3 Sustainable NAS Framework

Our approach relies on CNNGen to create an initial population of neural networks and uses NSGA-II and our improved predictors to evolve them.

3.1 NSGA-II

NSGA-II (Non-dominated Sorting Genetic Algorithm II) is a popular multi-objective optimization genetic algorithm designed to solve complex problems with multiple conflicting objectives [3]. The algorithm finds a set of optimal solutions, known as the Pareto front, where no solution can be improved regarding one of the objectives without degrading another. NSGA-II starts by generating an initial population of potential solutions and evaluates their performance on each objective. These solutions are sorted based on dominance: a solution is said to dominate another if it is better in at least one objective and is not worse in any others. The algorithm then uses a ranking system to classify solutions into different non-dominated fronts, with the first front being the best, etc. NSGA-II employs a crowded comparison operator to ensure diversity amongst the solutions. This operator ranks solutions within the same front based on their proximity to other solutions, preferring diversity instead of peaking all the solutions in the same neighborhood. The process involves selection, crossover, and mutation operations to create a new generation of solutions. With each generation, NSGA-II explores the solution space, aiming to find a well-distributed set of high-quality solutions along the Pareto front. The algorithm continues iterating through generations until a stopping criterion is met, such as a maximum number of generations. Cuurently, our CNN generator approximatively generates 1% of architectures that do not compile. Yet, because of the mutations and crossover, and despite specific precautions, the number of errors, and therefore the number of architectures that do not compile, increases up to 10%. To avoid keeping these architectures in the pool of solutions and further help NSGA-II, we give these architectures the highest possible float value. The final output is a diverse set of solutions representing the trade-offs between the different objectives. Decision-makers simply have to choose any solution among these.

3.2 Improved Predictors

We utilized the predictors proposed by CNNGen [5] but observed that their predictions were not sufficiently accurate. To improve reliability, we modified these predictors by creating a new dataset consisting of 40,000 architectures generated with CNNGen. We also enhanced the preprocessing of input data, optimising transformations and performing a comprehensive grid search to identify better hyperparameters for more accurate performance predictions (i.e. transformer model). For the energy predictor, we applied feature engineering techniques to extract new features that could enhance prediction accuracy. Our analysis identified several key features as particularly relevant, including the number of layers, FLOPs, parameters, convolutional layers, and fully connected layers, as well as the presence or absence of global average pooling, max pooling, and flattening operations. Additionally, we experimented with various machine learning techniques, including decision tree-based methods, support vector regression, and random forests. After evaluating their performance, we concluded that random forest yielded the most accurate predictions, making it the most suitable approach for improving energy estimation in our framework. In the end, regarding accuracy prediction, we have used a transformer model on the Python code associated with a CNN architecture (as done previously in CNNGen) while for the prediction of the energy consumption, we have used a random forests composed of 50 trees that use the aforementioned retrieved features from the feature engineering study.

4 Experimentations

4.1 Experimental Protocol

Whenever a new CNN model is used, we rely on CNNGen to produce a random description sentence used to generate the associated model. For instance, CNNGen is called to build our initial population that will evolve through genetic algorithms. To assess fitness, we employ performance and energy consumption predictors to avoid the need for training architectures. These predictors evaluate both the performance (accuracy) of architectures and energy consumption. Both elements are combined in the fitness function to form a multi-objective optimization problem to guide the search process. But, these predictors need to be trained first. Again, we randomly generated 40,000 models from CNNGen. We trained them on CIFAR-10 datasets for 20 epochs. Then, our 40,000 models are divided into 80% for training the predictors, and the remaining 20% are used for evaluating the predictors. Among the 80% dedicated to training, 20% of them are used for validation. As stated before, we only consider NSGA-II. We describe their parameterization and present the results hereafter.

4.2 NSGA-II

NSGA-II generates a Pareto front of non-dominated solutions regarding the different objectives. Thus, in our case, the resulting set of CNN models presents different trade-offs between accuracy and energy consumption, leaving the final choice to the end-user. To conduct this experiment, we used the version of NSGA-II provided with JMetal, a Java library, in its version 6.2.2[1]. We set the size of the initial population and the offspring sizes to 10. We ran the genetic optimization for 500 generations. We have also set the cross-over and mutation rates at 60% and 10% respectively.

4.3 Predictors Parameterisation

As said previously, we rely on a Transformer model based on Python code associated with CNN models to predict the performance (*i.e.,* the accuracy) of models generated via CNNGen. This predictor is the same as the one used in our previous work [5]. Regarding the predictor for energy consumption, we have used a random forest composed of 50 decision trees. These trees are limited to a maxi-

[1] https://github.com/MOEAFramework/JMetal-Plugin/releases/tag/v6.2.2.

mum depth of 10 and use a mean square error criterion for output. The possible splits use the features listed in Sect. 3.2, defined after using a feature engineering analysis.

4.4 Results

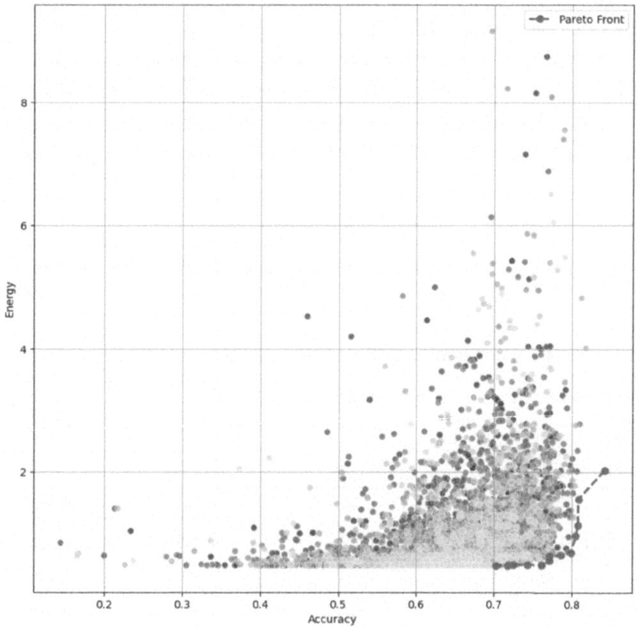

Fig. 1. Results of the NSGA-II optimization process for 500 generations. The X-axis represents the predicted accuracy for all the architectures. The Y-axis represents the predicted value of energy (CO^2 equivalent in grams). The red curve (at the bottom-right) reports the Pareto front. (Color figure online)

Figure 1 presents the results of one NSGA-II optimization run. This scatter plot presents the trade-off between accuracy and energy consumption for various CNN architectures generated using CNNGen. Each point represents an individual architecture, with accuracy plotted on the x-axis and energy consumption on the y-axis. The red curve highlights the Pareto front, which consists of architectures that achieve the best possible trade-off between these two competing objectives. From the distribution of points, we observe that architectures with lower accuracy tend to have lower energy consumption, forming a dense cluster at the bottom of the plot. As accuracy increases, energy consumption generally increases, suggesting that more accurate models require higher computational resources. However, the spread of points indicates significant variation in energy consumption for architectures with similar accuracy, implying that some models are more energy-efficient than others at the same performance level. The Pareto front follows a characteristic upward curve, where architectures along this frontier represent the optimal balance between accuracy and energy. Any point not on the Pareto front is a suboptimal solution, meaning that another architecture exists with either a higher accuracy and the same energy consumption, or a lower energy consumption and the same accuracy, or being better at both objectives at the same time. The shape of the front suggests that beyond a certain accuracy threshold, increasing performance comes with a disproportionately higher energy cost. This observation highlights the challenge of designing CNN architectures that maximize accuracy while remaining energy-efficient. Overall, this analysis emphasizes the importance of multi-objective optimization in neural architecture search. While higher accuracy is desirable, the energy cost must be carefully considered, especially for applications where computational efficiency is crucial.

Table 1. Generation 0 - CO^2 equivalent in grams. NaN: The architecture could not compile.

Name	Pred. Accuracy	True Accuracy	Pred. Energy	True Energy
Architecture 0	0,74	0,69	1,1	1,06
Architecture 1	0,65	0,63	0,63	0,73
Architecture 2	NaN	NaN	NaN	NaN
Architecture 3	0,57	0,56	1,03	0,96
Architecture 4	0,44	0,59	0,54	0,58
Architecture 5	0,69	0,48	2,69	3,03
Architecture 6	0,58	0,66	1,19	1,33
Architecture 7	0,73	0,63	1,10	1,16
Architecture 8	0,71	0,60	3,07	2,95
Architecture 9	0,72	0,67	0,64	0,62

Table 2. Generation 250 - CO^2 equivalent in grams. NaN: The architecture could not compile.

Name	Pred. Accuracy	True Accuracy	Pred. Energy	True Energy
Architecture 0	0.70	0.66	0.56	0.57
Architecture 1	0.49	0.41	0.48	0.39
Architecture 2	0.67	0.56	0.86	1.12
Architecture 3	0.77	0.71	0.64	0.71
Architecture 4	0.61	0.63	0.96	0.98
Architecture 5	0.46	0.53	0.66	0.82
Architecture 6	0.66	0.55	0.56	0.57
Architecture 7	0.64	0.52	0.72	0.76
Architecture 8	0.75	0.73	1.69	1.89
Architecture 9	0.68	0.67	1.44	1.51

To complete this general analysis, Tables 1, 2 and 3 present individual performance and consumption of our architectures at generation 0, 250, and 500.

We can observe that, on average, the true accuracy of generated models slightly decreases (0.61 for generation 0 vs 0.58 for generation 500), representing a 5% decrease. However, energy consumption decreases more significantly (1.38 for generation 0 vs 0.84 for generation 500), representing a decrease of 39%. If we look at individuals, we find interesting compromises. For example, architecture 7 in generation 500 achieves almost the same performance as architecture 8 in generation 250 for half of the energy consumed. This demonstrates the capability of our approach to balance accuracy and energy consumption.

Table 3. Generation 500 - CO^2 equivalent in grams. NaN: The architecture could not compile.

Name	Pred. Accuracy	True Accuracy	Pred. Energy	True Energy
Architecture 0	NaN	NaN	NaN	NaN
Architecture 1	0.52	0.50	0.53	0.50
Architecture 2	0.61	0.59	0.48	0.39
Architecture 3	0.72	0.49	1.08	1.09
Architecture 4	0.74	0.49	0.98	0.99
Architecture 5	0.62	0.65	1.05	1.09
Architecture 6	NaN	NaN	NaN	NaN
Architecture 7	0.73	0.72	0.91	0.91
Architecture 8	0.68	0.62	0.82	0.95
Architecture 9	NaN	NaN	NaN	NaN

5 Discussion and Threats to Validity

5.1 Discussion

In the previous section, we presented our results for our sustainable NAS workflow. In particular, we integrated performance and energy predictors into the workflow to reduce computational costs. While we conducted standard procedures to train and evaluate these predictors, the genetic algorithm may focus on generating more and more architectures that were outside the data distribution of our initial training set. Since the overall optimization goal is to focus on only a tiny subset of the possible CNN architectures that CNNGen can produce, it is not surprising that, over time, the data distribution drifts toward new architectures that were difficult to grasp with random generation. Thus, we assume that the more generations NSGA-II produces, the less accurate our predictors will be if they remain as-is. Yet, this can be improved in different ways. For instance, using more training data to get better predictors. This may require tuning the CNNGen generation of description sentences to look for architectures with higher accuracies or lower energy consumption. But, training more CNN models to train our predictors is not a panacea: *i)* it requires more resources which negatively impacts sustainability, *ii)* cross-over and mutations will still contribute to out-of-distribution topologies. A more promising idea would be to retrain our estimators regularly after a specific number of generations. For instance, for every 200 NSGA-II generations, we launch a retraining procedure on our predictors to follow the evolution of the generated CNN architectures. This way, the predictors will follow the drift in the data distribution which may control the difference between the predicted and actual values of both accuracy and energy consumption. Of course, other strategies could be used, such as active learning or continuous retraining.

The analysis of the tables across different generations reveals that the predictors used for estimating both accuracy and energy consumption are not entirely reliable. In Generation 0 1, while some predictions are relatively close to the true values, discrepancies are noticeable, such as in Architecture 5, where the predicted precision (0.69) is significantly higher than the true accuracy (0.48). Additionally, some architectures, like Architecture 4, exhibit an inversion where the predicted accuracy is lower than the actual value, suggesting inconsistencies in the predictor's estimations. This may lower the confidence of using the proposed architectures by final users. However, we can observe that the predictions remain, mainly, in a 5% error envelope, the estimations on the energy consumption seem to be less stable (ranging from 2% to 15%), showing once again that further effort should be made in considering energy consumption as one of the goal of optimizing CNN (and more generally deep learning) architectures. As the generations progress, the presence of NaN values in Generation 500 (see Table 3) highlights another critical issue: the increasing occurrence of architectures that cannot be compiled. This could be a consequence of the mutation and crossover mechanisms of NSGA-II.

While Tables 1 and 2 show only 1 architecture that could not be compiled (over 20 architectures), Table 3 reports 4 of such architectures. Our cross-over and mutation operators strive to limit topology-breaking changes. Yet the effect of randomness cannot be completely avoided, notably regarding the determination of some hyperparameters.

Besides, the tables suggest that the energy estimations from the predictors also suffer from inaccuracies. In several cases, the predicted energy consumption does not align well with the true values, such as in Generation 250 (see Table 2), where Architecture 2 was predicted to have an energy consumption of 0.86 but in reality consumed 1.12. This discrepancy indicates that the energy predictor, but it also applies to the accuracy predictor, requires further refinement to ensure reliability in guiding the search process. Another concerning trend is that as the search progresses, the performance of certain architectures deteriorates instead of improving. In Table 3, some architectures (*e.g.,*, Architecture 3 and Architecture 4) exhibit a drop in true accuracy despite relatively high predictions. This further reinforces the risk that mutations and crossovers are creating architectures that lie outside the training distribution of the predictor, leading to increasingly unreliable predictions.

5.2 Threats to Validity

5.2.1 Internal Validity

Threats to internal validity concern the implementation and experimental protocol. Regarding NSGA-II we used JMetal, a popular GA framework. While our optimization problem is complex and require exploring a large exploration space, we dedicated a substantial budget for GA to search for optimal solutions so that they can converge. Yet, maybe increasing even more this budget may lead to better results.

5.2.2 External Validity

Our discussion illustrates an external threat to validity. Despite our efforts, our results may not (and are not) apply to every possible CNN topology one can create. We also exhibited different behavior of NSGA-II. Other types of GA may yield different results.

6 Conclusion

Deep learning models are more and more used but they are expensive to train, especially in terms of energy consumption. While one may think that more complex models are the best as they may exhibit higher performances (*i.e.*, accuracy), this is not always the case. Depending on the context in which they will operate, seeking smaller models that use less energy (and thus are of reduced size) but perform similarly is also an interesting option. This also echoes the growing global concern of reducing carbon emissions. The results of our study highlight both the strengths and limitations of our CNN architecture search approach, particularly in balancing accuracy and energy efficiency. The optimization process using NSGA-II has demonstrated the ability to explore a diverse set of architectures, leading to promising trade-offs between performance and computational cost. However, the analysis of our predictors has revealed critical areas for improvement. One of the primary challenges observed is the inaccuracy of the predictors for both accuracy and energy consumption. Across multiple generations, discrepancies between predicted and actual values suggest that the predictors struggle to generalize beyond their training distribution. Furthermore, the presence of architectures in later generations that cannot be compiled indicates that increasing the search time (by increasing the number of generations) can be counter-productive despite reaching for new CNN architectures that seem to meet our optimisation criteria. We assume that applying multiple times the NSGA-II mutation and crossover operators may lead to inconsistencies in the description sentences of our CNN architectures since NSGA-II is not aware of constraints that are encoded into CNNGen. Thus, increasing the number of random changes on a description sentence causes more changes to lead to an architecture that cannot be compiled. Additionally, the observed performance plateau and occasional drops in accuracy suggest that the current search process may be generating suboptimal architectures due to prediction errors. In some cases, predicted highly accurate models turned out to perform worse than expected, emphasizing the need for more robust predictive mechanisms. Similarly, energy estimations have shown inconsistencies, which could mislead the optimization process when selecting architectures based on energy efficiency. To address these limitations, several improvements are necessary. Enhancing predictors accuracy through more extensive and diverse training datasets, refined feature engineering, and advanced machine learning techniques (such as ensemble methods or uncertainty-aware models) could help mitigate out-of-distribution errors. Additionally, incorporating periodic retraining of predictors or integrating constraints on mutation and crossover could reduce the likelihood of generating architectures that are poorly estimated. Despite these challenges, the current results demonstrate the feasibility of optimizing CNN architectures with a trade-off between performance and energy efficiency. By refining our predictors and improving the search process, we can further enhance the reliability of our method, ultimately contributing to the development of more sustainable and efficient deep learning models.

Acknowledgement. This work was supported by Service Public de Wallonie Recherche under grant n° 2010235 – ARIAC by DIGITALWALLONIA4.AI. Gilles Perrouin is an FNRS Research Associate.

References

1. Cai, H., Chen, T., Zhang, W., Yu, Y., Wang, J.: Efficient architecture search by network transformation. In: McIlraith, S.A., Weinberger, K.Q. (eds.) Proceedings of the Thirty-Second AAAI Conference on Artificial Intelligence, (AAAI-18), the 30th Innovative Applications of Artificial Intelligence (IAAI-18), and the 8th AAAI Symposium on Educational Advances in Artificial Intelligence (EAAI-18), New Orleans, Louisiana, USA, 2–7 February 2018, pp. 2787–2794. AAAI Press (2018). https://www.aaai.org/ocs/index.php/AAAI/AAAI18/paper/view/16755
2. Canziani, A., Paszke, A., Culurciello, E.: An analysis of deep neural network models for practical applications. arXiv preprint arXiv:1605.07678 (2016)
3. Deb, K., Pratap, A., Agarwal, S., Meyarivan, T.: A fast and elitist multiobjective genetic algorithm: NSGA-II. IEEE Trans. Evol. Comput. **6**(2), 182–197 (2002). https://doi.org/10.1109/4235.996017
4. Domhan, T., Springenberg, J.T., Hutter, F.: Speeding up automatic hyperparameter optimization of deep neural networks by extrapolation of learning curves. In: Proceedings of the 24th International Conference on Artificial Intelligence, IJCAI 2015, pp. 3460–3468. AAAI Press (2015)
5. Gratia, A., Liu, H., Satoh, S., Temple, P., Schobbens, P.Y., Perrouin, G.: CNNGen: a generator and a dataset for energy-aware neural architecture search. In: 32nd European Symposium on Artificial Neural Networks, Computational Intelligence and Machine Learning (2024)

6. Hinton, G., et al.: Deep neural networks for acoustic modeling in speech recognition: the shared views of four research groups. IEEE Signal Process. Mag. **29**(6), 82–97 (2012). https://doi.org/10.1109/MSP.2012.2205597
7. Jin, H., Chollet, F., Song, Q., Hu, X.: AutoKeras: an AutoML library for deep learning. J. Mach. Learn. Res. **24**(6), 1–6 (2023). http://jmlr.org/papers/v24/20-1355.html
8. Jin, H., Song, Q., Hu, X.: Auto-keras: an efficient neural architecture search system. In: Proceedings of the 25th ACM SIGKDD International Conference on Knowledge Discovery & Data Mining, pp. 1946–1956. ACM (2019)
9. Lecun, Y., Bottou, L., Bengio, Y., Haffner, P.: Gradient-based learning applied to document recognition (1998)
10. Liu, H., Simonyan, K., Yang, Y.: DARTS: differentiable architecture search (2019)
11. Miikkulainen, R., et al.: Evolving deep neural networks. In: Artificial Intelligence in the Age of Neural Networks and Brain Computing, pp. 293–312. Elsevier (2019)
12. Mo, H., Iacca, G.: Accelerating evolutionary neural architecture search for remaining useful life prediction. In: International Conference on Bioinspired Optimization Methods and Their Applications, pp. 15–30. Springer (2022)
13. Mo, H., Iacca, G.: Evolutionary neural architecture search on transformers for RUL prediction. Mater. Manuf. Process. **38**(15), 1881–1898 (2023)
14. Real, E., Aggarwal, A., Huang, Y., Le, Q.V.: Regularized evolution for image classifier architecture search. In: Proceedings of the AAAI Conference on Artificial Intelligence, vol. 33, pp. 4780–4789 (2019)
15. Strubell, E., Ganesh, A., McCallum, A.: Energy and policy considerations for deep learning in NLP. In: Proceedings of the 57th Annual Meeting of the Association for Computational Linguistics, pp. 3645–3650. Association for Computational Linguistics, Florence (2019). https://doi.org/10.18653/v1/P19-1355, https://aclanthology.org/P19-1355
16. Touvron, H., et al.: LLaMA: open and efficient foundation language models (2023)
17. Villalobos, P., Sevilla, J., Besiroglu, T., Heim, L., Ho, A., Hobbhahn, M.: Machine learning model sizes and the parameter gap. arXiv preprint arXiv:2207.02852 (2022)
18. Wąsowski, A., Berger, T.: Domain-Specific Languages: Effective Modeling, Automation, and Reuse. Springer (2023)
19. Ying, C., Klein, A., Real, E., Christiansen, E., Murphy, K., Hutter, F.: NAS-bench-101: towards reproducible neural architecture search (2019)
20. Zoph, B., Le, Q.V.: Neural architecture search with reinforcement learning. In: International Conference on Learning Representations (ICLR) (2017). https://arxiv.org/abs/1611.01578

Explainable Multi-label Classification for Predictive Maintenance

Antoine Hubermont[1,2](✉), Aymeric Vellinger[1], and Elio Tuci[1]

[1] University of Namur, 5000 Namur, Belgium
antoine.hubermont@unamur.be
[2] Telespazio Belgium, 6600 Bastogne, Belgium

Abstract. This paper extends previous work in predictive maintenance that implement a self-adaptive evolutionary strategy ($SA - ES$) for feature selection in multi-label anomaly detection tasks. We incorporate Shapley Additive exPlanations (SHAP), an explainable artificial intelligence (XAI) method, to evaluate the relevance and classification performance of features selected by the $SA - ES$, and compare their effectiveness against those with the highest absolute Shapley values. A comparative analysis is performed using three multi-label classifiers on a public predictive maintenance dataset, with evaluation conducted through $5 - fold$ cross-validation. Our findings validate the efficiency of the $SA - ES$ in reducing feature dimensions and minimizing Hamming Loss. Additionally, we present insightful visualizations and interpretations for multi-label anomaly classification, facilitating the application of predictive maintenance in real-world industrial scenarios.

Keywords: Machine Learning · Predictive Maintenance · multi-label classification · XAI

1 Introduction

Maintenance is a critical focus in the industrial sector to ensure service continuity and system reliability. With the rapid advancements in the artificial intelligence field, the implementation of optimised maintenance strategies has become increasingly accessible [9]. Predictive maintenance strategy (hereafter PdM) takes place between the maintenance based on scheduled operations (preventive) and the repair actions once a problem occurred (corrective). PdM prevents the unnecessary replacement of operational components and avoids service disruptions through effective anomaly classification. PdM is primarily concerned with two approaches identified by authors in [22]: prognostics and diagnostics. The prognostic approach estimates the remaining useful life (RUL) of components. The diagnostic approach relies on the detection and prediction of anomalies, which could indicate potential failures or deviations from normal operation. Typically, the diagnostic approach involves single-label classification for anomaly detection, under the assumption that detected anomalies are mutually exclusive.

T. Carletti et al. (Eds.): WIVACE 2024, CCIS 2532, pp. 186–197, 2025.
https://doi.org/10.1007/978-3-031-93631-9_15

However, this assumption is not systematic as stated in [7] where the authors compare the single-label and multi-label approaches for anomaly classification in a *PdM* dataset, the *Microsoft Azure Dataset* [2].

Regardless of the type of approach (i.e., single or multi-label), *PdM* applications have to be reliable and trustworthy, to effectively cut costs of spare parts replacement operations and avoid negligence in executives decision-making [18]. Methods in the field of eXplainable Artificial Intelligence (*XAI*) has been developed to ease the interpretability of *PdM* algorithms and ensure an appropriate integration into existing maintenance strategies. Among them, "Shapley Additive exPlanations" (*SHAP*) [10] is a model agnostic method based on the game theory to illustrate the importance of the features that led to a prediction. The *SHAP* framework is based on Shapley values that compute the expected marginal contribution of a player in a cooperative game. In the *PdM* context, a player is a feature, and the cooperative game is an anomaly predicted with a classifier. Several works investigated explainability with *SHAP* in single-label classification problems for various *PdM* applications such as water pumping systems [21], fault detection in hard drive systems [5], and fault detection on bearings with LSTM [12]. In the context of XAI on multi-label classification problems, authors in [4] applied *SHAP* on three well-known multi-label datasets namely, *yeast*, *water quality* and *foodtruck* datasets available on the *Multi-Label Classification Dataset Repository* [1]. Regarding multi-label problems on predictive maintenance, authors in [14] propose a solution to the lack of explainability of Random Forest with a LionForest on multi fault diagnosis on the *AI4I* dataset [13]. This dataset has a low number of features and LionForest issues a set of rules to explain predictions. Regarding explainability in *PdM* multi-label classification problems, no work has been carried out in the field with *SHAP*.

In this paper, we aim to extend our previous work [8], where we successfully reduced the number of features required for multi-label anomaly detection using a self-adaptive evolutionary strategy as a feature selection algorithm. We apply the XAI method on the *Azure PdM* dataset to provide insights into how features contribute in predictions and to verify the consistency of our self-adaptive algorithm with feature importance derived from SHAP, an expensive but near-optimal feature selection tool based on game theory [11]. Additionally, we conduct comparisons using three multi-label classifiers and analyse the differences in *SHAP* values across them. Furthermore, we compare the performance of classifiers trained on features selected by the self-adaptive evolutionary strategy with those trained on the top-ranked features based on absolute Shapley values

Importantly, we are not suggesting to rely entirely on *SHAP* as a feature selection tool as extensively discussed in [6] since our aim is to place our optimal algorithm into perspective with game theory approach. Finally, we will discuss the contributions of *SHAP* to enhance the comprehension of anomaly classification in multi-label tasks for trustworthy predictive maintenance strategies.

2 Methodology

In the following section, we describe the methods used to study the feature importance of three state of the art multi-label classifiers on the *Azure PdM* dataset.

2.1 Classifiers

Multi-label problems require dedicated approaches, namely problem transformation and algorithm adaptation as stated in [20]. Problem transformation changes the nature of the task by dividing a multi-label problem into a set of single-label problems. In contrast, algorithm adaptation keeps the multi-label nature of the problem by changing the structure of classification algorithm to handle more than one label. Problem transformation methods have drawbacks such as information loss, poor consideration of label dependency and a low generalisation capability. To support this study, we select three multi-label classifiers from our previous work [8] which belongs to the ensemble approach as defined in [3] which mitigate the disadvantages by combining and aggregating problem transformation methods with voting and bagging mechanisms.

The selected classifiers are Binary Relevance (BR) [16,20], Classifier Chains (CC) [16], and Pruned Set (PR) [15]. BR is the most straightforward approach, treating each label as a separate binary classification problem, with the final prediction being the aggregation of all the binary classifications. CC also frames the task as a set of binary classifiers, but structures them in a chain, where each classifier benefits from the predictions of the previous ones in the chain. For both BR and CC, the $J48$ decision tree is used as the base binary classifier, following recommendations in [3]. In the case of PR, it removes the least frequent labels from the training set and reintroduces them in association with more frequent labels. The algorithm implementations comes from the $scikit-multilearn$ library [19] and its $MEKA$ wrapper [17].

2.2 Dataset and Preprocessing

Table 1. Dataset meta characteristics

Characteristics	*Azure*
Instances	291300
Features	27
Labels	4
Cardinality	0.02
Density	0.005

The classifiers as defined in Sect. 2.1 are trained on one multi-label dataset in the *PdM* field, the Microsoft Azure Predictive Maintenance (hereafter *Azure*). *Azure* is a synthetic dataset issued from the collection "Predictive Maintenance Modelling Guide" in the Microsoft Azure suite [2]. The dataset contains the records of 100 synthetic machines, each monitored by four sensors that capture data on vibration, rotation, pressure, and voltage. These sensors record data every hour over the span of one year. The dataset includes troubleshoot records that provide information on recoverable errors and non-recoverable failures, which can affect 4 generic components on the machines. The failures are the classification targets and they are defined according to the component they affect namely, C1, C2, C3, and C4. Hourly monitoring introduces noise into the recorded data. To mitigate this noise, we apply feature engineering by incorporating past information into the dataset through time-shifted features, resulting in the creation of lagged features as described in previous studies. The time used to create the lagged features, known as the lagging window, is set to 3h and 24h in this study. For each sensor, two lagged features are created using a moving average (avg) and a moving standard deviation (std). Consequently, the total number of lagged features added to the Azure dataset is 16 (4 sensors $\times$ 2 lagging windows $\times$ 2 for avg and std). Additional information included in the dataset comprises the time since the last maintenance operation on each component (in hours), the number and type of errors that occurred in the last n hours (where n is defined by the lagging window), and the machine service age. To further reduce the effect of random data variation, we have changed the recording frequency by selecting features every 3 h instead of every hour.

The Table 1 summarises the characteristics of the *Azure* dataset. Regarding the multi-label aspect, cardinality refers to the average number of labels per instance; density refers to the average number of label per instance weighted by the number of unique labels. The dataset has a very small proportion of anomalies to be classified, as indicated by the low cardinality and density values. This constraint highlights the importance of reliable detection, further emphasizing the need to explain the impact of features on the classifier's decisions.

2.3 Feature Selection

The Self-Adaptive Evolutionary Strategy ($SA-ES$) as presented in [8], which is part of the family of $(\mu+\lambda)-ES$ algorithms where the parents (μ)-the selected individuals from the current population that contribute their genetic information to produce new individuals-and the offspring (λ)-the new individuals generated through combination and mutation of the parents' genetic information-are considered for selection into the next generation, which is one iteration of the evolutionary process.

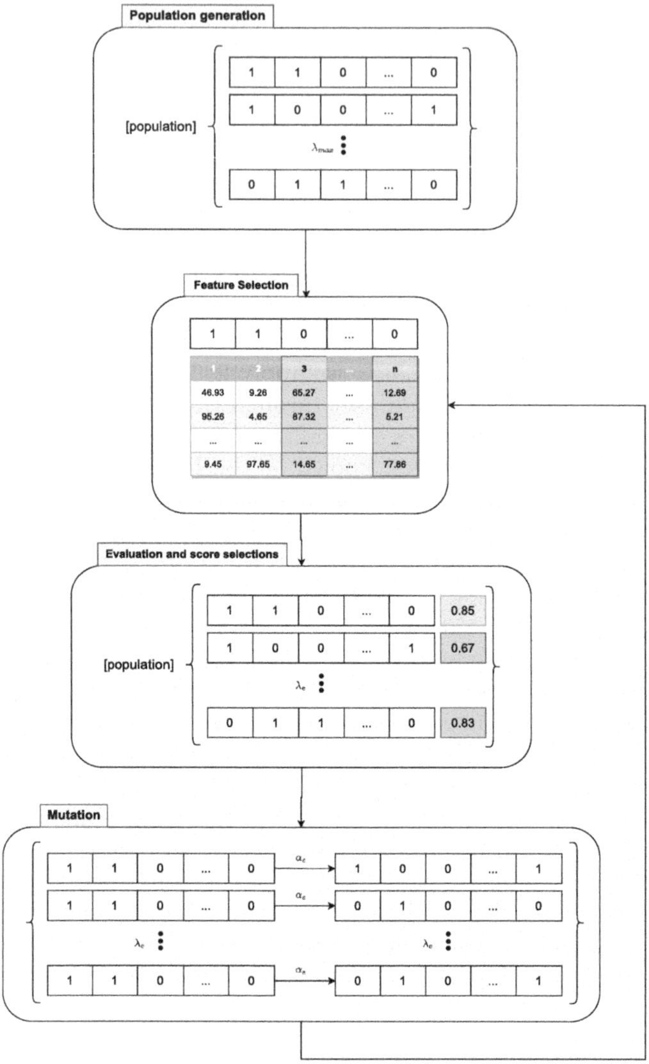

Fig. 1. Evolutionary workflow.

For feature selection, each individual is a binary array representing dataset features, with '1' indicating a selected feature and '0' indicating an unselected one. In each generation, we use bit-mask crossover to combine bits from two parent solutions based on a random binary mask. Then, we apply bit-flip mutation, flipping selected bits. The mutation probability is updated each generation based on the previous generation's Hamming loss fitness function performance. A summary workflow diagram of the algorithm is described in Fig. 1.

2.4 SHAP

Shapley Additive exPlanations primary purpose is to explain the output of machine learning models by quantifying the contribution of each feature to a particular prediction or the uncertainty generated by this feature in the prediction. The underlying logic behind this quantification is the measure of the individual contribution of a feature to a prediction. This measure can be determined by testing all feature subsets with and without the feature and then average over the number of all the possible subsets. The Shapley value of a feature is the weighted average of its marginal contributions across subsets as defined in the original paper [10].

SHAP framework proposes multiple tools to compute the Shapley values such as $TreeExplainer$ which is fast and designed for tree based classifier and $KernelExplainer$ which is slower but model agnostic. Even if the base classifiers selected in this study are decision trees, their multi-label implementations are not supported in $TreeExplainer$ and we have therefore selected the $KernelExplainer$ to run the visualisations.

2.5 Evaluation

To evaluate fairly the performance between the classifiers trained on a set of features selected by $SA - ES$ and the set of the most important features, we trained every classifiers with their default parameters and we used a $5 - fold$ cross validation to minimise bias. Finally, we base the classifier evaluation on the Hamming Loss. This metric is widely used and has the advantage of reporting the misclassification rate among all the label where 0 is the perfect score.

3 Results

Our findings indicate that the type of classifier impacts feature importance, as measured by absolute Shapley values, when trained on the $Azure$ dataset. The complete feature importance distribution per class for the BR classifier is shown in Fig. 2. From this visualization, the features can be categorized into two groups based on their ranking. The first group consists of the most influential features on anomalies, particularly those between the *comp3* and *comp4* features. The second group includes features with the least impact on the classes, represented by the lower-ranked features, starting from $vibrationstd\_24h$.

It is noteworthy that the features in the first category have the greatest influence on anomalies across all classes. For instance, *error5* stands out as the most influential feature, exhibiting the highest absolute Shapley values. Similarly, comp3 plays a significant role in classifying class 3 anomalies, has moderate importance for anomalies 1 and 4, and has a minor impact on anomalies 2.

Thus, it is possible to determine which features most strongly influenced the BR classifier in classifying a specific anomaly by identifying those with the highest importance. For example, class 1 is primarily influenced by the features $voltmean\_24h$ and $error1$, class 2 by $rotatemean\_24h$, class 3 by $pressuremean\_24h$ and $error4$, and class 4 by $vibrationmean_24h$ and $error5$. The feature importance analysis for the BR classifier reveals strong correlations between certain features and specific anomalies, where some features have an almost exclusive influence. For instance, $error4$ and $pressuremean\_24h$ predominantly influence the classification of anomaly 3, while $rotatemean\_24h$ and $error2$ mainly affect the classification of anomaly 2.

For the feature importance rankings of the CC and PR classifiers, the features can be divided into two groups, similar to the BR classifier, based on their ranking. The highest-ranked features for the CC classifier are shown in Fig. 3, and for the PR classifier in Fig. 4. Notably, the top-ranked features are consistent across all the considered classifiers, highlighting the stability of their contributions to classification. As a result, the lowest-ranked features for CC and PR are not displayed in this study, as they closely similar with those presented in Fig. 2. However, within the group of most influential features, the ranking order varies depending on the classifier. The top three features for CC and PR are similar, with environmental features such as $pressure$ and $vibration$, and error based features being the most important with $error5$. In contrast, the BR classifier places more importance on component-related features, such as $comp3$ and $comp2$.

Analysing feature importance from a label perspective across the three classifiers shows that both BR and CC tend to focus classification on a single type of label. For instance, the feature $vibrationmean\_24h$ predominantly influences the classification of class 4 anomalies, while $rotatemean\_24h$ is mostly relevant for class 2 anomalies. In contrast, PR classification is more balanced across labels, with $vibrationmean\_24h$ impacting both class 4 and class 2 anomalies. It is important to note that certain features consistently hold significant importance for the same labels across different classifier types, although their influence may vary depending on the classifier, as seen with $rotatemean\_24h$. Additionally, some features, like $comp4$, have mixed importance, influencing the classification of multiple anomalies across classifiers.

To validate the consistency of our SA-ES method, we compared the Hamming loss across BR, CC, and PR classifiers trained on a set of features selected by the $SA-ES$ and another set with the top-ranked features based on the absolute Shapley values. The $5-fold$ cross-validation results illustrated in Fig. 5 shows that classifiers trained on the $SA-ES$ selected features consistently outperform those trained on top-ranked Shapley features. The best results were achieved by the CC classifier using the $SA-ES$ selected features. This performance gap between of the $SA-ES$ over $SHAP$-based feature selection can be attributed to

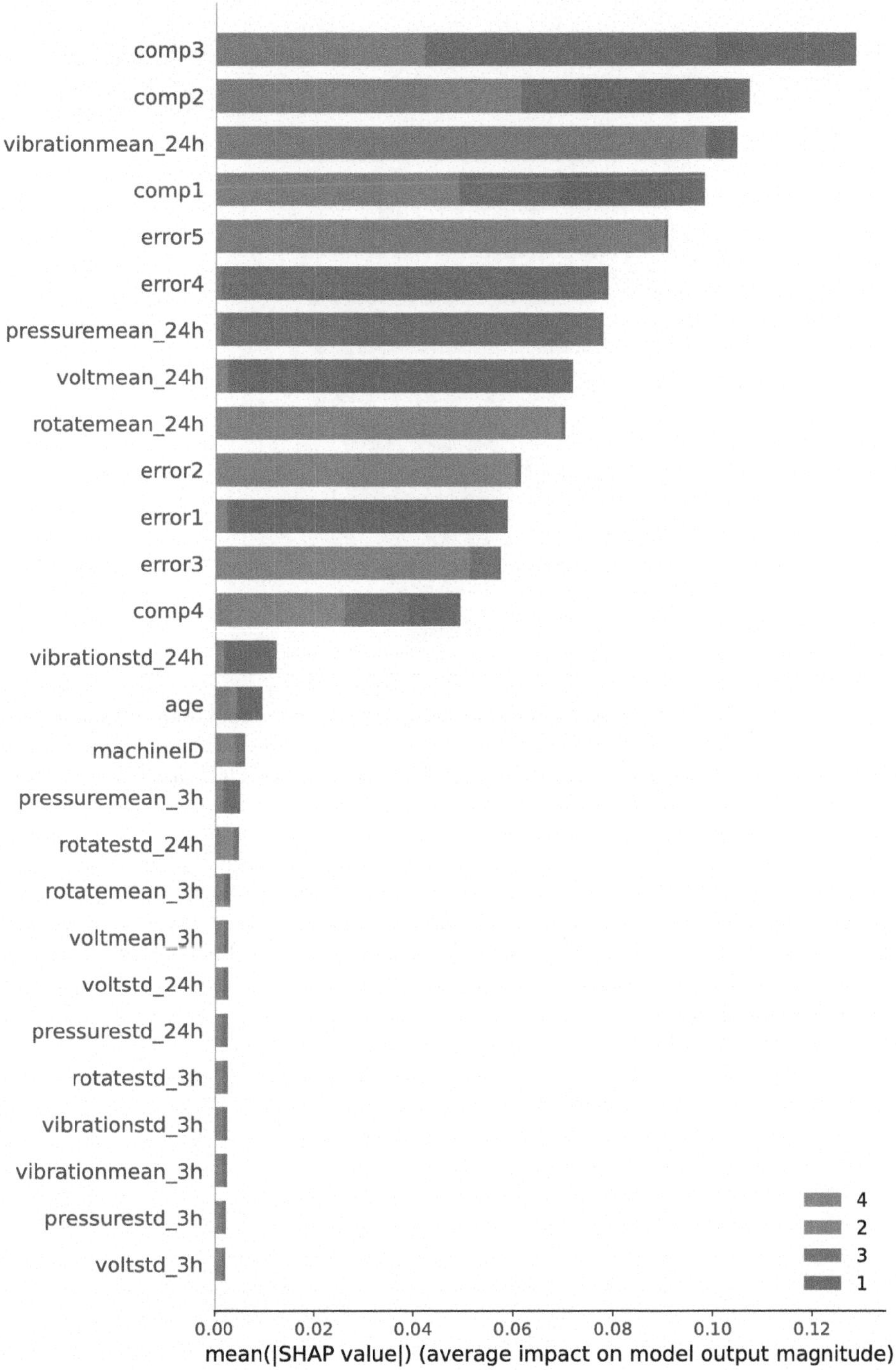

Fig. 2. Complete feature importance distribution by class for *BR* classifier trained on the baseline *Azure* dataset.

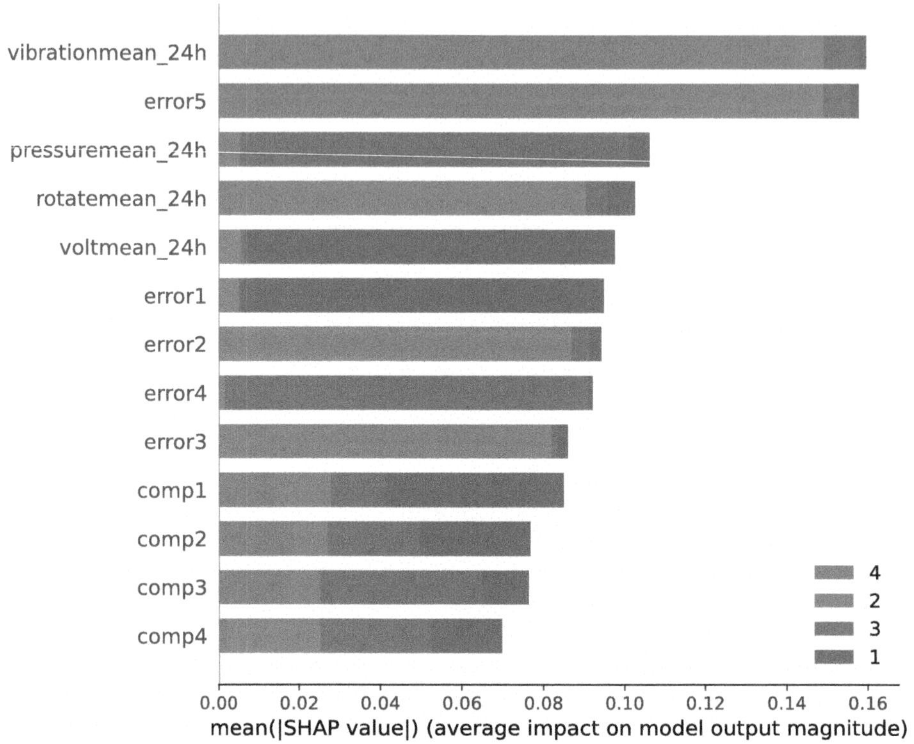

Fig. 3. Complete feature importance distribution by class for *CC* classifier trained on the baseline *Azure* dataset.

the number of selected features: *SHAP* uses only 13 features, while the $SA-ES$ selects 18 for *CC* and *BR*, and 14 for *PR*. Despite that that fewer features are selected based on the Shapley values ranking, these results confirm that this ranking method is not optimal for feature selection, aligning with findings from previous researches [6].

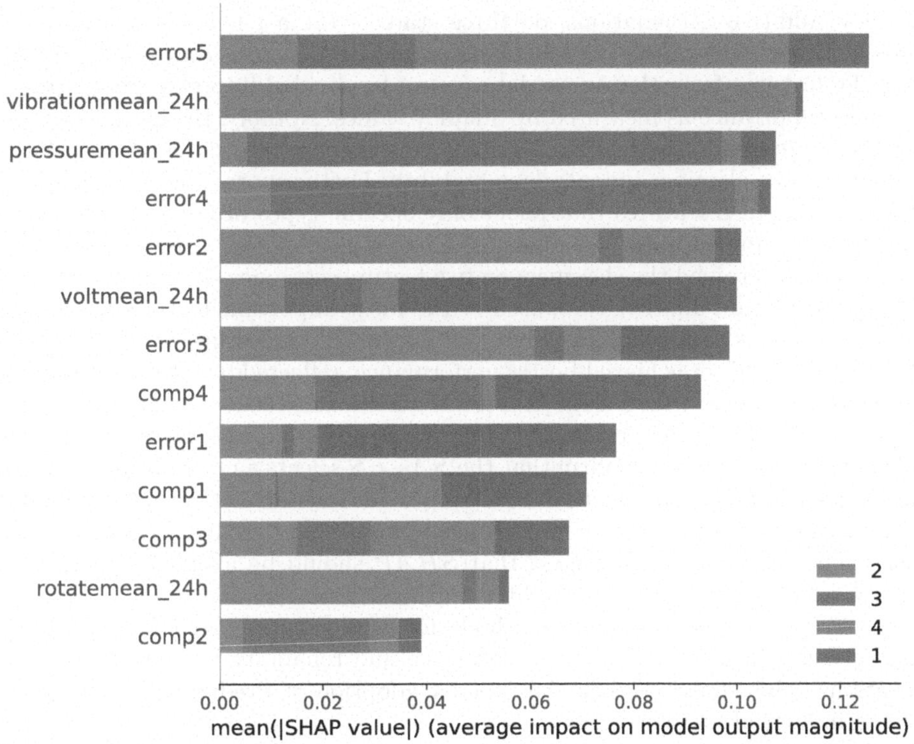

Fig. 4. Complete feature importance distribution by class for *PR* classifier trained on the baseline *Azure* dataset.

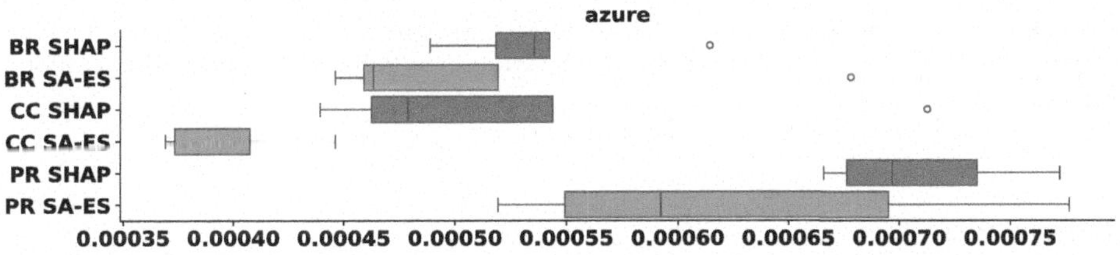

Fig. 5. Hamming Loss evaluation scores for *Azure* dataset between classifiers trained on the top ranked features in the sense of *SHAP* and classifiers trained on a set of features selected by the $SA - ES$.

4 Conclusions

In this study, we successfully extend a previous work [8] with consistent feature selection with the self-adaptive evolutionary strategy (SA-ES) as well as providing insightful visualizations to enhance understanding of how features are utilized in a multi-label anomaly classification task using the *Azure* predictive maintenance dataset. We applied a well know XAI technique based on the game theory,

Shapley Additive exPlanations, on three state of the art multi-label classifiers namely, Binary Relevance, Classifier Chains and Pruned Set. Our results indicate that the features from the Azure dataset can be divided into two groups based on their importance as measured by Shapley values. Notably, the group containing the most important features remains consistent across different classifiers, demonstrating classifier-independent stability. Furthermore, the visualizations clearly distinguish features that affect only specific types of failures from those with broader, mixed influence, offering valuable insights for decision-makers.

Finally, we trained the classifiers on two feature sets: one selected by the $SA-ES$ algorithm and another composed of the most important features according to $SHAP$ values. The results validate our approach, as classifiers trained on the $SA-ES$ selected set achieved better performance, reflected by a lower Hamming Loss. We emphasize that $SHAP$ is not intended for feature selection, despite its clear feature importance ranking. While it may be tempting to use SHAP for this purpose, our results confirm that the $SA-ES$ selects a more optimal feature subset for classification and that $SHAP$ provides, as an XAI tool, meaningful visualizations to study feature importance in multi-label predictive maintenance tasks. These findings also suggest that $SHAP$ should be avoided as a feature selection method, as noted in the literature.

Overall, our study presents methods for implementing XAI in multi-label anomaly classification, enhancing confidence and reliability through clear visualizations. This contributes to the broader adoption of Predictive Maintenance in real industrial systems.

Acknowledgments. The authors acknowledge the SPW-recherche (Walloon Region) and Telespazio Belgium throughout the research funding WIN4DOC under Grant No. 2110246. Computational resources have been provided by the Consortium des Équipements de Calcul Intensif (CÉCI), funded by the Fonds de la Recherche Scientifique de Belgique (F.R.S.-FNRS) under Grant No. 2.5020.11 and by the Walloon Region.

References

1. Multi-label classification dataset repository - knowledge discovery and intelligent systems - KDIS - university of córdoba. http://www.uco.es/kdis/mllresources/
2. Predictive Maintenance Modelling Guide Data Sets. https://gallery.azure.ai/Experiment/Predictive-Maintenance-Modelling-Guide-Data-Sets-1
3. Bogatinovski, J., Todorovski, L., Džeroski, S., Kocev, D.: Comprehensive comparative study of multi-label classification methods. Expert Syst. Appl. **203**, 117215 (2022). https://doi.org/10.1016/j.eswa.2022.117215
4. Chen, S.: Interpretation of multi-label classification models using shapley values (2021). https://doi.org/10.48550/arXiv.2104.10505
5. Ferraro, A., Galli, A., Moscato, V., Sperlì, G.: Evaluating eXplainable artificial intelligence tools for hard disk drive predictive maintenance. Artif. Intell. Rev. **56**(7), 7279–7314 (2023). https://doi.org/10.1007/s10462-022-10354-7

6. Fryer, D., Strümke, I., Nguyen, H.: Shapley values for feature selection: the good, the bad, and the axioms. IEEE Access **9**, 144352–144360 (2021). https://doi.org/10.1109/ACCESS.2021.3119110
7. Hubermont, A., Tuci, E., De Quattro, N.: Simultaneous failures classification in a predictive maintenance case. In: ESANN 2023 Proceedings, pp. 537–542. Ciaco - i6doc.com (2023). https://doi.org/10.14428/esann/2023.ES2023-129
8. Hubermont, A., Vellinger, A., Antonic, N., Tuci, E.: Feature selection for multi-label classification in predictive maintenance. In: Quintián, H., et al. (eds.) The 19th International Conference on Soft Computing Models in Industrial and Environmental Applications, SOCO 2024, pp. 52–61. Springer, Cham (2025)
9. Kanawaday, A., Sane, A.: Machine learning for predictive maintenance of industrial machines using IoT sensor data. In: 2017 8th IEEE International Conference on Software Engineering and Service Science (ICSESS), pp. 87–90 (2017). https://doi.org/10.1109/ICSESS.2017.8342870. iSSN 2327-0594
10. Lundberg, S., Lee, S.I.: A unified approach to interpreting model predictions (2017)
11. Man, X., Chan, E.: The best way to select features? arXiv preprint arXiv:2005.12483 (2020)
12. Mansouri, T., Vadera, S.: Explainable fault prediction using learning fuzzy cognitive maps. Expert. Syst. **40**(8), e13316 (2023). https://doi.org/10.1111/exsy.13316
13. Matzka, S.: Explainable artificial intelligence for predictive maintenance applications. In: 2020 Third International Conference on Artificial Intelligence for Industries (AI4I), pp. 69–74. Institute of Electrical and Electronics Engineers (2020). https://doi.org/10.1109/AI4I49448.2020.00023
14. Mylonas, N., Mollas, I., Bassiliades, N., Tsoumakas, G.: Local multi-label explanations for random forest. In: Machine Learning and Principles and Practice of Knowledge Discovery in Databases, pp. 369–384. Springer (2023). https://doi.org/10.1007/978-3-031-23618-1_25
15. Read, J., Pfahringer, B., Holmes, G.: Multi-label classification using ensembles of pruned sets. In: 2008 Eighth IEEE International Conference on Data Mining, pp. 995–1000 (2008). https://doi.org/10.1109/ICDM.2008.74
16. Read, J., Pfahringer, B., Holmes, G., Frank, E.: Classifier chains for multi-label classification. Mach. Learn. **85**(3), 333–359 (2011). https://doi.org/10.1007/s10994-011-5256-5
17. Read, J., Reutemann, P., Pfahringer, B., Holmes, G.: MEKA: a multi-label/multi-target extension to Weka. J. Mach. Learn. Res. **17**(21), 1–5 (2016). http://jmlr.org/papers/v17/12-164.html
18. Sasikala, B., Sachan, S.: Exploring the Frontiers of Artificial Intelligence and Machine Learning Technologies. San International Scientific Publications (2024). https://doi.org/10.59646/efaimltC3/133
19. Szymański, P., Kajdanowicz, T.: A scikit-based Python environment for performing multi-label classification. ArXiv e-prints (2017)
20. Tsoumakas, G., Katakis, I.: Multi-label classification: an overview. IJDWM **3**, 1–13 (2007)
21. Upasane, S.J., Hagras, H., Anisi, M.H., Savill, S., Taylor, I., Manousakis, K.: A type-2 fuzzy-based explainable AI system for predictive maintenance within the water pumping industry. IEEE Trans. Artif. Intell. **5**(2), 490–504 (2024). https://doi.org/10.1109/TAI.2023.3279808
22. Wen, Y., Rahman, M., Xu, H., Tseng, B.: Recent advances and trends of predictive maintenance from data-driven machine prognostics perspective. Measurement **187**, 110276 (2021). https://doi.org/10.1016/j.measurement.2021.110276

Cross-Entropy Regularization with Mutual Information in Training CNNs

Ch. Zaugg[1,3](✉), R. Ingold[3], D. Trendafilov[2], and A. Fischer[3,4]

[1] ZHAW, ACSS, Technikumstrasse 9, 8400 Winterthur, Switzerland
christoph.zaugg@zhaw.ch
[2] Namur Institute for Complex Systems, University of Namur, Namur, Belgium
dari-borisov.trendafilov@unamur.be
[3] Department of Informatics, DIVA, Fribourg University, Boulevard de Pérolles 90, 1700 Fribourg, Switzerland
{christoph.zaugg,rolf.ingold,andreas.fischer}@unifr.ch
[4] iCoSys, HES-SO, Boulevard de Pérolles 80, 1700 Fribourg, Switzerland

Abstract. We examine the learning behavior of a shallow and a deep convolutional neural network performing classification tasks on subsets of two databases. Our investigation focuses on the label, the input, and the prediction layer, and we compute the mutual information between these layers epoch-wise using Rényi's matrix-based entropy functional. We evaluate the data processing inequality to interpret the learning behavior in a consistent information-theoretic framework. Our primary goals are to 1) clarify the relation between the two training objectives of minimizing the cross-entropy and maximizing the mutual information between the label and the prediction layer, 2) gradually switch from the first to the second training objective, and 3) interpret the impact of the latter transition. One of the main contributions is the proposed novel method for regularizing the cross-entropy objective and assessing the neural network's learning activity.

Keywords: Neural networks · Rényi's entropy functional · Data processing inequality · InfoMax

1 Introduction

Recent advances in deep learning provide evidence that with enormous datasets and computing power, arbitrary big models could offer unprecedented performance. However, this learning process has the well-known downside of over-fitting. More recent advances aim to eliminate this bias and optimize learning using sparse datasets and a shorter training process, also known as one-shot learning [10,16].

This paper presents an initial study exploiting generic information-theoretic functionals to optimize convolutional neural networks (CNNs) and identify model saturation over time. The aim is to speed up the training process and mitigate over-fitting.

T. Carletti et al. (Eds.): WIVACE 2024, CCIS 2532, pp. 198–207, 2025.
https://doi.org/10.1007/978-3-031-93631-9_16

In training two types of CNNs in classification tasks on two databases, we focus on the label layer Y, the input layer X, and the prediction layer $\hat{Y}$ and epoch-wise compute the mutual information (MI) values $I(Y;X)$, and $I(Y;\hat{Y})$. We compare both values to assess the data processing inequality (DPI):

$$I(Y;X) \geq I(Y;\hat{Y}). \quad (1)$$

The DPI is a necessary condition in an information-theoretic framework and serves to cross-check the approximated MI values. In [19], the authors investigate chains of DPIs to optimize the learning behavior of a Multilayer Perceptron (MLP).

We clarify the relation between two training objectives, i.e., minimizing cross-entropy (CE) and maximizing $I(Y;\hat{Y})$. Our loss function – a mixture of CE and $I(Y;\hat{Y})$ – depends dynamically on the training epoch, which brings the benefit of simultaneously minimizing CE and maximizing $I(Y;\hat{Y})$. The resulting insight could provide an automated trigger for terminating the training and avoid potential over-fitting.

We base our reasoning on two pillars: an approximation method for computing MI values between high-dimensional continuous random variables and a specific parameter optimization procedure.

In Sect. 2, we embed our work into related work, and in Sect. 3, we describe the details of the methodology. In Sect. 4, we present the setting of our experiments, and Sect. 5 contains the results of the two training objectives. In Sect. 6, we formulate our conclusions and provide an outlook to future work.

2 Related Work

The information bottleneck (IB) principle in [13], or the dimpled manifold model (DMM) in [11] are paradigms to explain the learning behavior of deep neural networks (DNNs). The first stems from information theory and attributes learning to the optimal trade-off between compression and predictability. The second takes a geometrical perspective. The dimension of an image manifold in a classification task tends to be much lower than the dimension of the embedding space. After a few training epochs, the decision boundary is close to the image manifold, and additional training only causes tiny dimples on the decision boundary to bring incorrectly classified data points to the correct side.

Our study is founded on the framework of information theory. It elucidates the learning behavior of a CNN performing a classification task in an early training phase after transient behavior caused by random initialization has subsided.

The authors of [6] estimate and maximize $I(X;\hat{Y})$ mainly for unsupervised learning tasks, i.e., their training objective is Linsker's InfoMax principle stated in [7]. They base their Deep InfoMax (DIM) on the Mutual Information Neural Estimator (MINE) introduced in [1] that provides lower bounds for estimating MI values, and the work in [8] tightens these bounds. In [14], the authors maximize $I(Y;\hat{Y})$ by estimating a probability density function (PDF) using a kernel density and propose $I(Y;\hat{Y})$ as a regularization term for hinge, squared, logistic and exponential loss functions. In our study, we investigate how maximizing $I(Y;\hat{Y})$ relates to minimizing cross-entropy.

In [4], the authors present Rényi's matrix-based entropy functional of order α and its convergence to the MI as the order α tends to one. The method only relies on estimating the kernel width of the radial basis function (RBF) and does not require a PDF. The authors of [18] estimate the kernel width based on Silverman's rule of thumb [12] and use Rényi's matrix-based entropy functional to approximate MI values between various layers of DNNs up to 50′000 training epochs. In [15], the authors optimize the kernel width by maximizing the alignment between matrices as proposed in [2]. They train the DNNs up to 5′000 training epochs and elaborate on the compliance of the DPIs.

Our study, investigating the early training phase of two types of CNNs, aims at identifying a learning saturation criterion that could prevent over-fitting. The significant fluctuation in the MI values poses a major challenge during that phase. Instead of following MINE's lower bounds, we resorted to Rényi's matrix-based entropy functional, which is known to converge. The consistency of the DPIs serves as a validation guideline.

3 Methodology

3.1 Approximation

We apply Rényi's matrix-based entropy functional of order α [18] to the layers Y, X, and $\hat{Y}$. Considering a batch a of size N in one of these layers, we compute the kernel matrix from the row vectors a_i as follows:

$$K_{ij} = \exp\left(-\frac{\|a_i - a_j\|^2}{s^2}\right). \tag{2}$$

Normalizing the kernel matrix K ensures the data processing inequality (Eq. 1):

$$A_{ij} = \frac{K_{ij}}{\sqrt{A_{ii}} \cdot \sqrt{A_{jj}}}. \tag{3}$$

We compute the contribution of the batch to the entropy $H_\alpha(A)$ using $\alpha = 1.01$:

$$H_\alpha(A) = \frac{1}{1-\alpha} log_2\left(tr(A^\alpha)\right). \tag{4}$$

In Eq. 5 and 6, the matrix B corresponds to another layer and is obtained by applying Eq. 2 and 3 to the batch b. We approximate the joint entropy $H_\alpha(A, B)$ by inserting the trace normalized Hadamard product of A and B:

$$H_\alpha(A, B) = H_\alpha\left(\frac{A \circ B}{tr(A \circ B)}\right) \tag{5}$$

and compute $I_\alpha(A;B)$ as follows:

$$I_\alpha(A;B) = H_\alpha(A) + H_\alpha(B) - H_\alpha(A,B). \tag{6}$$

Averaging over the batches approximates the MI between the layers considered, e.g., Y and X, or Y and $\hat{Y}$.

3.2 Optimal Kernel Width

Computing the kernel matrix in Eq. 2 requires estimating the kernel width s of the RBF. The $(i,j)^{th}$ entry of the label matrix takes the value one if the ith and the jth labels in a batch are equal, and zero elsewhere. The label matrix captures the clustering of the labels in a batch.

To make the clustering in another layer resemble as much as possible, Cristianini [2] proposes the kernel width s^* that maximizes the cosine, i.e., the alignment between the label and the kernel matrix associated with the other layer. Figure 1 illustrates the choice of the optimal kernel width s^* in our experiments.

In all of our experiments, we maximize the cosine between the label layer Y and kernel matrix of the prediction layer $\hat{Y}$. While a CNN learns, the alignment increases during training, which makes the optimal kernel width s^* depend on the training epoch. On the other hand, the value of s^* sets the scale in Eq. 2 for computing the MI values. With our tools, we can interpret the learning behavior in a consistent information-theoretic framework only during the epochs in which s^* stays within a typical value over a particular observation horizon, e.g., 100 training epochs in Fig. 1(b). Our experiments reveal that the DPI (cf. Eq. 1) is valid during the observation horizon at the beginning of the training. Section 4

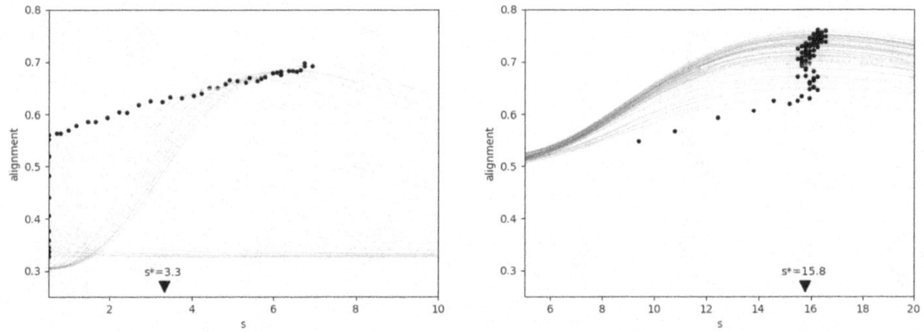

(a) Shallow CNN, MNIST, 50 epochs (b) ResNet18, Fashion-MNIST, 100 epochs

Fig. 1. Choice of optimal kernel width: We partition the intervals on the abscissa into 100 equidistant kernel width candidates. Each training epoch is represented with a grey curve whose ordinate value is the batch mean of the alignments, i.e., the cosine values between the label matrix and the kernel matrix of the prediction layer evaluated at the candidate kernel width. The black dots denote the maxima on the grey curves. The optimal parameter s^* is the average of the black dots' abscissa values.

describes in more detail both types of CNNs (shallow and ResNet18) and the types of databases (MNIST and Fashion-MNIST) we experimented with.

4 Experiments

4.1 Data Sets

We experimented with labeled images of the MNIST [3] and the Fashion-MNIST [17] databases. The examples consist of ten categories, with an image resolution (1, 28, 28). In order to keep the computational complexity low in this initial study, we extracted rather small balanced subsets of 5'000 training and 1'000 validation and test examples from these databases. Figure 2 presents random samples of 16 labeled images.

4.2 CNN Architectures

We study two CNNs architectures, a shallow one and ResNet18 [9]. The shallow CNN contains four hidden layers. The first two layers perform a convolution followed by a max pooling, and the last two are fully connected linear layers. A rectified linear (ReLU) activation function terminates each hidden layer. The prediction layer, that assigns the label with the highest score to an image, is fully connected. The shallow CNN has approximately 44K network parameters.

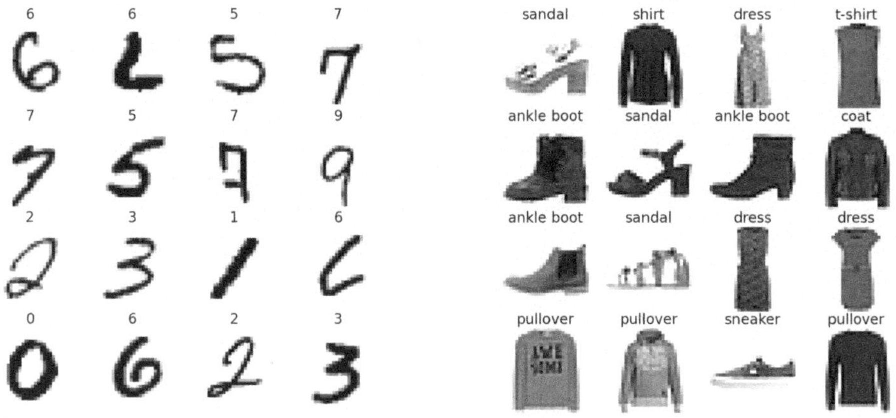

(a) MNIST (b) Fashion-MNIST

Fig. 2. Random samples of the two databases MNIST and Fashion-MNIST, used in our experiment, with resolution of (1, 28, 28).

To contrast the shallow CNN, we also experimented with one type of ResNet. The authors in [5] partition the layer sequence of a CNN into blocks. Instead of training the network parameters of the entire layer sequence, they train the parameters of the residual functions assigned to each block. We studied ResNet18 [9] with roughly 200 times as many network parameters as the shallow CNN. In the experiments with ResNet, we use images with resolution (3, 224, 224).

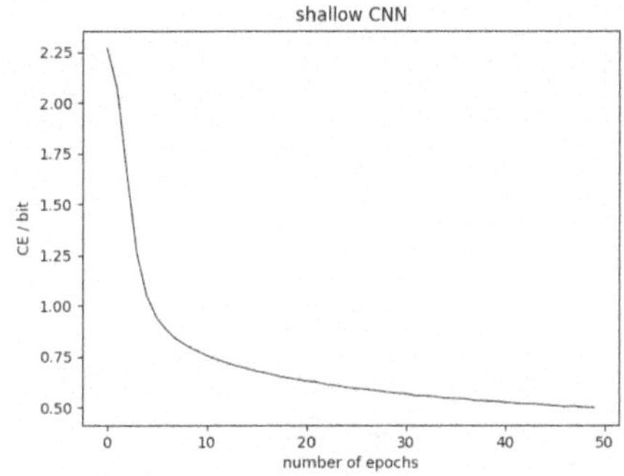

(a) Minimization of CE

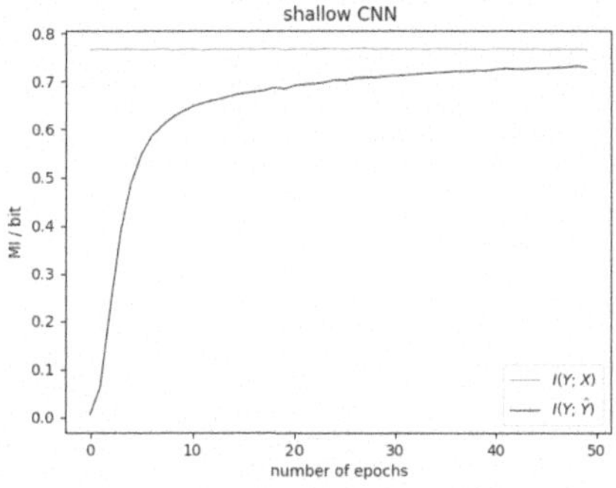

(b) Maximization of $I(Y;\hat{Y})$

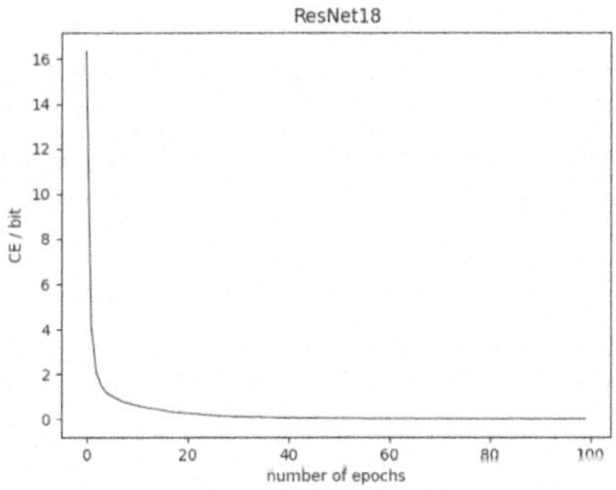

(c) Minimization of CE

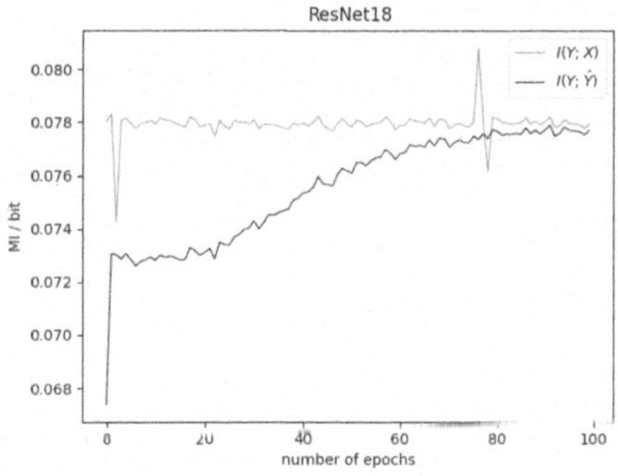

(d) Maximization of $I(Y;\hat{Y})$

Fig. 3. Minimizing CE implies maximizing $I(Y;\hat{Y})$ on the Fashion-MNIST data. In (a), we minimize the CE of the shallow CNN and observe in (b) that $I(Y;\hat{Y})$ increases. In (c), we minimize the CE of ResNet18 and observe in (d) that $I(Y;\hat{Y})$ increases. The results on the MNIST data for both types of networks are similar.

5 Results

5.1 Approximate Equivalence

Our results (see Fig. 3) on the Fashion-MNIST data reveal that minimizing cross-entropy (CE) maximizes the mutual information $I(Y;\hat{Y})$ between the label Y

and the prediction $\hat{Y}$ layers. This trend holds for the shallow CNN trained over 50 epochs (cf. Fig. 3ab) and for ResNet18 trained over 100 epochs (cf. Fig. 3cd). The grey curves in Fig. 3bd denote $I(Y;X)$. We observed that the data processing inequality (Eq. 1) is violated once (cf. Fig. 3d), possibly due to a numeric instability. The experimental results for the MNIST data are similar.

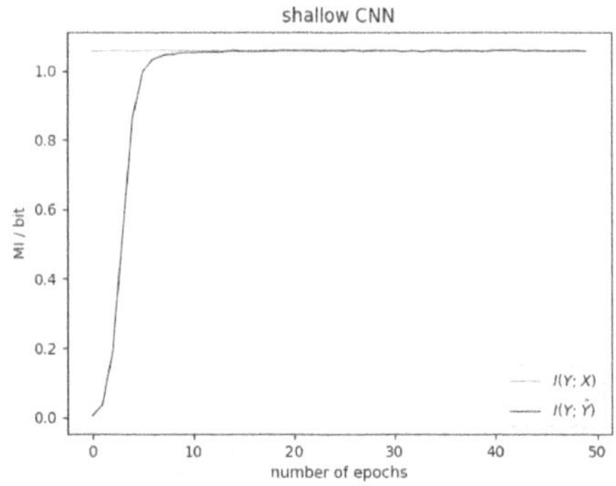

(a) Maximization of $I(Y;\hat{Y})$

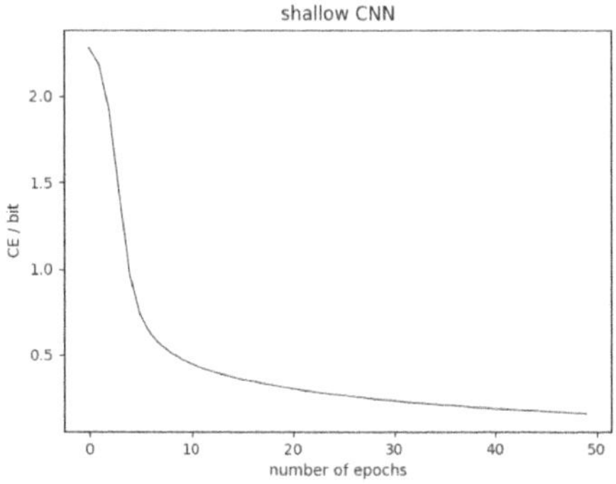

(b) Minimization of CE

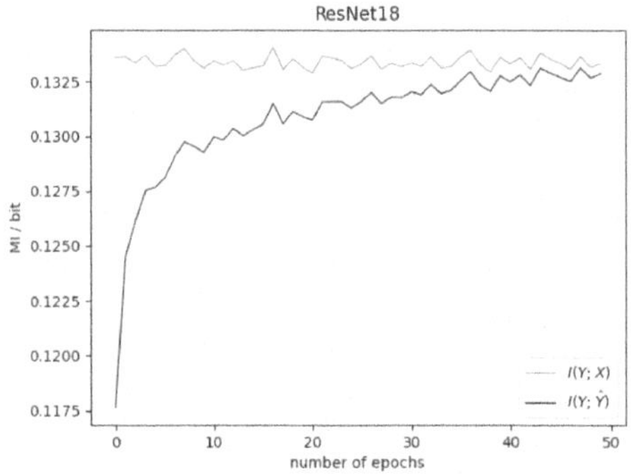

(c) Maximization of $I(Y;\hat{Y})$

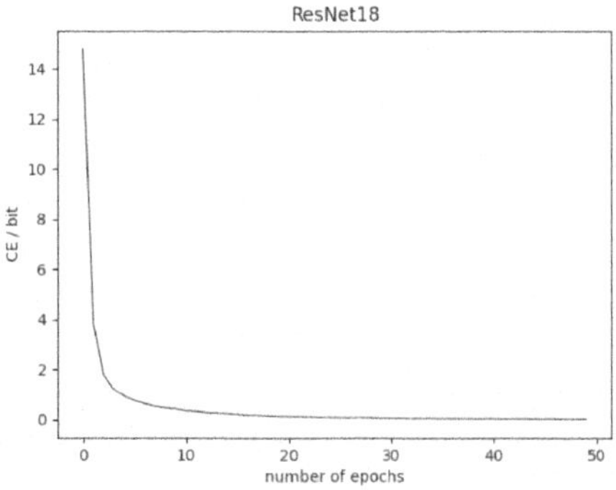

(d) Minimization of CE

Fig. 4. Maximizing $I(Y;\hat{Y})$ minimizes CE via Eq. 7 on the MNIST data: In (a), we maximize $I(Y;\hat{Y})$ of the shallow CNN and observe in (b) that the CE decreases. In (c), we maximize $I(Y;\hat{Y})$ of ResNet18 and observe in (d) that the CE decreases. The results on the Fashion-MNIST data for both types of networks are similar.

The maximization of $I(Y;\hat{Y})$ raises the problem of a vanishing gradient for both CNNs. To avoid that we use CE as an ignitor, i.e., we introduce a transition factor f that depends on the training *epoch* such that the training objective L at $epoch = 0$ coincides with CE and gradually allows $I(Y;\hat{Y})$ to gain influence as the number of training epochs grows:

$$L = f(epoch) \cdot CE - (1 - f(epoch)) \cdot I(Y;\hat{Y}). \tag{7}$$

Figure 4 presents the results from minimizing the loss L (Eq. 7) with $f(epoch) = (1 + epoch)^{-0.5}$ on the MNIST training data. We observed that any transition factor f depending on the training epoch that lets L coincide with the cross-entropy objective at the beginning of the training and then gently allows $I(Y;\hat{Y})$ to gain influence yields a non-vanishing gradient of $I(Y;\hat{Y})$, even though the gradient at $epoch = 0$ vanishes as Fig. 6ab reveal. Hence, we can approximately maximize $I(Y;\hat{Y})$ and minimize CE in parallel. Figure 4ab reflects this for the shallow CNN and Fig. 4cd for ResNet18, respectively.

5.2 Cross-Entropy Regularization

We interpret Eq. 7 as a novel method to regularize cross-entropy with the mutual information between the label and the prediction layer. In Fig. 5, we illustrate the effect of training ResNet18 on MNIST. If we minimize CE, the test loss remains above the training loss (see Fig. 5(a)). If we minimize L the test and the training loss almost coincide (see Fig. 5(b)).

5.3 Saturation of $I(Y;\hat{Y})$

Our experiments demonstrate that $I(Y;\hat{Y})$ increases with ongoing training. In an information-theoretic interpretation, the network reduces the uncertainty in the prediction layer as much as possible, while observing the label layer. The saturation of $I(Y;\hat{Y})$ is a natural criterion to asses the learning activity. We can approximately compute this quantity and its gradient. Figure 6 presents the ℓ^2 norm of the gradient $I(Y;\hat{Y})$ to characterize the learning of shallow CNN on the MNIST data. In Fig. 4(a), we see that $I(Y;\hat{Y})$ saturates after just a few training epochs. The bump in Fig. 6(a) corroborates the observation. ResNet18

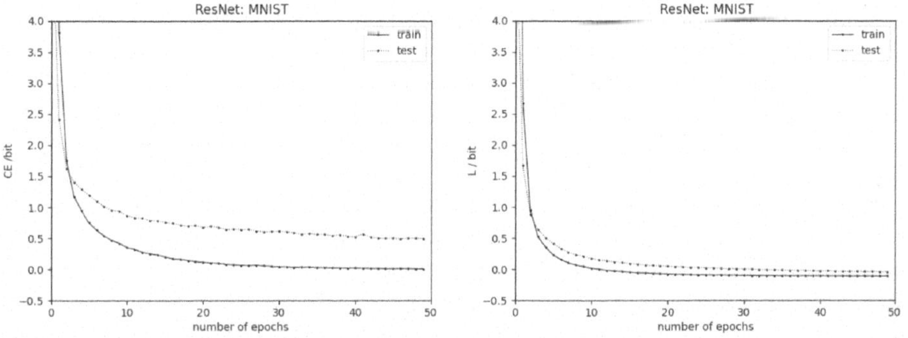

(a) Minimization of CE (b) Minimization of L

Fig. 5. Regularization property on the MNIST data. In (a), we minimize CE and observe a persistent gap between training and test loss. In (b), we minimize L according to Eq. 7 and observe how the gap closes. The results on the Fashion-MNIST data for both types of networks are similar.

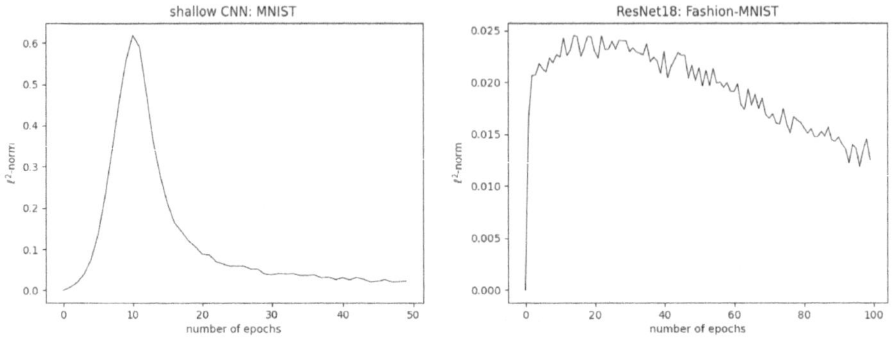

(a) I(Y;Ŷ) (b) I(Y;Ŷ)

Fig. 6. The gradient of $I(Y;\hat{Y})$ in training CNNs with L according to Eq. 7. In (a), the gradient of $I(Y;\hat{Y})$ quickly vanishes for the shallow CNN trained on the MNIST data. The result is consistent with a fast saturation of $I(Y;\hat{Y})$ in Fig. 4(a), revealing that learning has come to an end. In (b), the gradient of $I(Y;\hat{Y})$ slowly decreases for ResNet18 trained on the Fashion-MNIST data and keeps updating $I(Y;\hat{Y})$, revealing that learning goes on.

on Fashion-MNIST shows a more extended learning activity in Fig. 4(c), and Fig. 6(b) shows its potential for additional learning.

6 Conclusion and Future Work

Using Rényi's matrix-based entropy functional to approximate $I(Y;\hat{Y})$ causes its gradient to vanish. We overcome this difficulty by leveraging cross-entropy as an ignitor, as we propose in Eq. 7. Our results confirm the interpretation of the second term in Eq. 7 as regularization of cross-entropy. We have a similar result for a different loss function as the authors of [14]. The key difference is that our regularization depends dynamically on the training epoch. If we minimize the loss according to formula Eq. 7, we simultaneously reduce the cross-entropy and increase $I(Y;\hat{Y})$. Maximizing $I(Y;\hat{Y})$ only minimizes the difference between the label and prediction distributions and ignores the quality of the representations in the layers.

With our method, we can observe CNNs' learning behavior at the beginning of the training phase in a consistent information theoretical framework. Using the mutual information between the label and the prediction layer is a novel cross-entropy regularization method that could mitigate over-fitting. Future work will investigate the proposed regularization's impact on the generalization property with larger data sets and clarify the role of the ad-hoc defined transition factor in (Eq. 7).

Acknowledgements. This project has received funding from the European Union's Horizon 2020 research and innovation programme under the Marie Skłodowska-Curie grant agreement No 101034383.

References

1. Belghazi, M.I., et al.: MINE: mutual information neural estimation. arXiv preprint arXiv:1801.04062 (2018)
2. Cristianini, N., Shawe-Taylor, J., Elisseeff, A., Kandola, J.: On kernel-target alignment. In: Advances in Neural Information Processing Systems, vol. 14 (2001)
3. Deng, L.: The MNIST database of handwritten digit images for machine learning research. IEEE Signal Process. Mag. **29**(6), 141–142 (2012)
4. Giraldo, L., Rao, M., Principe, J.C.: Measures of entropy from data using infinitely divisible kernels. IEEE Trans. Inf. Theory **61**(1), 535–548 (2014)
5. He, K., Zhang, X., Ren, S., Sun, J.: Deep residual learning for image recognition. In: Proceedings of the IEEE Conference on Computer Vision and Pattern Recognition, pp. 770–778 (2016)
6. Hjelm, R.D., et al.: Learning deep representations by mutual information estimation and maximization. arXiv preprint arXiv:1808.06670 (2018)
7. Linsker, R.: Deriving receptive fields using an optimal encoding criterion. In: Advances in Neural Information Processing Systems, vol. 5 (1992)
8. Mroueh, Y., Melnyk, I., Dognin, P., Ross, J., Sercu, T.: Improved mutual information estimation. In: Proceedings of the AAAI Conference on Artificial Intelligence, vol. 35, pp. 9009–9017 (2021)
9. PyTorch: Resnet18 model (2024). https://pytorch.org/vision/2.0/models/generated/torchvision.models.resnet18.html. Accessed 31 Aug 2024
10. Santoro, A., Bartunov, S., Botvinick, M., Wierstra, D., Lillicrap, T.: Meta-learning with memory-augmented neural networks. In: International Conference on Machine Learning, pp. 1842–1850. PMLR (2016)
11. Shamir, A., Melamed, O., BenShmuel, O.: The dimpled manifold model of adversarial examples in machine learning. arXiv preprint arXiv:2106.10151 (2021)
12. Silverman, B.W.: Density Estimation for Statistics and Data Analysis. Routledge (2018)
13. Tishby, N., Zaslavsky, N.: Deep learning and the information bottleneck principle. In: 2015 IEEE Information Theory Workshop (ITW), pp. 1–5. IEEE (2015)
14. Wang, J., Wang, Y., Zhao, S., Gao, X.: Maximum mutual information regularized classification. Eng. Appl. Artif. Intell. **37**, 1–8 (2015)
15. Wickstrøm, K., Løkse, S., Kampffmeyer, M., Yu, S., Principe, J., Jenssen, R.: Information plane analysis of deep neural networks via matrix-based Renyi's entropy and tensor kernels. arXiv preprint arXiv:1909.11396 (2019)
16. Woodward, M., Finn, C.: Active one-shot learning. arXiv preprint arXiv:1702.06559 (2017)
17. Xiao, H., Rasul, K., Vollgraf, R.: Fashion-MNIST: a novel image dataset for benchmarking machine learning algorithms. arXiv preprint arXiv:1708.07747 (2017)
18. Yu, S., Principe, J.C.: Understanding autoencoders with information theoretic concepts. Neural Netw. **117**, 104–123 (2019)
19. Zaugg, C., Ingold, R., Füchslin, R., Fischer, A.: How to turn a leaky learner into a sealed one. In: Artificial Life and Evolutionary Computation: 17th Italian Workshop, WIVACE 2023, Venice, Italy, 6–8 September 2023, Revised Selected Papers, p. 29. Springer (2024)

Implementation of the Response to Synchronization in E-Puck2 Robots

Cinzia Tomaselli[1], Gonzalo Marcelo Ramírez-Ávila[2,4,5], Lucia Valentina Gambuzza[1], Mattia Frasca[1], Elio Tuci[3,4], and Timoteo Carletti[2,4(✉)]

[1] Department of Electrical, Electronic and Computer Science Engineering, University of Catania, Catania, Italy

[2] Department of Mathematics, University of Namur, Namur, Belgium
timoteo.carletti@unamur.be

[3] Department of Computer Science, University of Namur, Namur, Belgium

[4] Namur Institute for Complex Systems, naXys, University of Namur, Namur, Belgium

[5] Instituto de Investigaciones Físicas, and Planetario Max Schreier, Universidad Mayor de San Andrés, La Paz, Bolivia

Abstract. The response to synchronization is a phenomenon observed in several firefly species, where male ensembles synchronize their rhythmic flashes by triggering a response from females which ends the courtship process. In this work, we present a robotic implementation of this phenomenon by using a team of static e-puck2 robots that integrate oscillatory dynamics to mimic the flashing rhythm of the fireflies. To this end, robots communicate with each other via infrared (IR) and follow a distributed control law. They are divided into two groups: one representing the male population with bursting dynamics and the other representing females with non-bursting behavior. Our experimental results demonstrate that response to synchronization is robust with respect to the presence of realistic features such as obstacles and information loss. These factors play a significant role in refining the original model and enhancing its applicability in real-world scenarios.

Keywords: Coupled oscillators · bursty oscillators · E-puck robots · Synchronization

1 Introduction

Synchronization is the phenomenon according to which coupled (similar) self-sustained oscillators self-organize their rhythms to exhibit an (almost) unison rhythm without the need of any external control [28]. It is a widespread emerging phenomenon resulting from the interaction among individual oscillators [3,4]. It is found in the synchronous communication in several animals emitting sound as frogs [1], katydids [25] and cicadas [11], or communicating by other mechanical signals as in the case of crabs [2], just to mention a few. One of the most

T. Carletti et al. (Eds.): WIVACE 2024, CCIS 2532, pp. 208–222, 2025.
https://doi.org/10.1007/978-3-031-93631-9_17

interesting examples of synchronization in animals is the phenomenon observed in several firefly species [6]. This has inspired numerous studies focused on modeling fundamental aspects of flashing behavior [20,22,26] or designing electronic circuits that mimic such behavior [12,17]. Although simultaneous firefly flashing is a paradigmatic example of synchronization [5,8,32,34] and is related to mating and courtship [5,10,18], most of the research focused on male synchronization as the main factor of courtship, and few works considered the role of the female [9,24,29]. In some works, flash emission from females has been studied as predatory behavior to chase males of other species [19,38]. Other studies have explored the female response to controlled flash sequences that mimic the flashing pattern of conspecific males [7,21,23], revealing that females are sensitive to male signals, i.e., the female response is triggered or enhanced by male synchronization. This phenomenon is known as the *response to synchronization* [31].

This phenomenon is present in several firefly species in which the male and female rhythmic features might be quite different [14,33], among these species there is the genus *Photinus carolinus* [13]. Firefly courtship in several species consists thus of male synchronization followed by the female response, whose flashing pattern might exhibit simultaneous flashing or some other ordered flashing patterns. It is noteworthy to point out that groups of females cannot give rise to ordered flashing patterns without the presence of males.

The response to synchronization is described by the mathematical model discussed in [30] which considers dissimilar oscillators, one type of which is not capable of synchronizing without the presence of the second group. The purpose of this work is the experimental validation of this model using a team of e-puck2 robots. The robotic implementation, indeed, allows for testing the model's robustness against real-world factors such as message loss and the presence of obstacles that may affect neighborhood detection.

Our approach aligns with previous studies that implemented mathematical models of complex systems using robots, such as kilobots for naming game [37], or Elisa-3 robots for exploring collective dynamics through decentralized control [35,36]. In this context, synchronization has been explored under various scenarios. For instance, [16] explores a swarm of synchronized robots emitting pulses during navigation, with robots near a target increasing their oscillation frequency to induce the others to follow them. Another example is [27] which investigates synchronization in mobile pulse-coupled oscillators inspired by fireflies behavior, revealing that synchronization behaviors depend on agent speed and interaction parameters, leading to slow, fast, or inhibited synchronization regimes. Additionally, [35] considers a system of randomly moving robots, each integrating a Rössler oscillator and coupled with nearby units to reach synchronization.

For the purpose of this work, each robot will be endowed with an internal dynamic describing a non-smooth oscillator "producing a flash" during the discharging stage that enables communication with other robots and consequently mimics the flashing behavior of a *Photinus carolinus* firefly male or female. We show that interacting "male" robots are able to synchronize while interacting

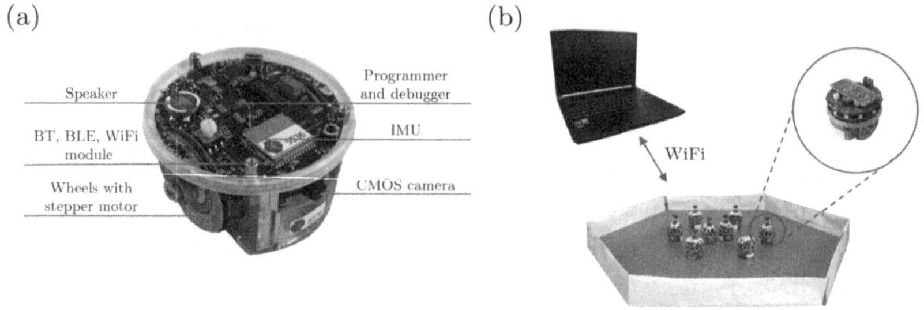

Fig. 1. (a) The *e-puck2* robot. (b) The experimental setup.

"female" robots cannot; however, when the "females" are mingled with the "male" robots, the former can respond to the synchronized signal received by the males and produce regular temporal patterns. Then, the experimental results are compared and complemented with the analysis of the non-smooth differential equations model implemented to describe the interaction of simulated populations of "male" and "female" robots whose internal dynamics mimic the firefly pulsating pattern. As a first preliminary investigation of the experimental setup and the role of the model parameters, we decided to keep the robots at fixed positions. The more general framework of moving robots will be presented in a forthcoming companion paper.

A relevant result of our work is that realistic conditions do not impede the response to synchronization phenomenon, proving the robustness of the model. In addition, the findings from our work might endow the model with essential aspects for more realistic situations. The latter can be useful for practical applications and for improving simulation features.

The paper is organized as follows. In Sect. 2, we describe the basic characteristics of the robots and the experimental setup. The model, its features, and its implementation in e-puck robots are detailed in Sects. 3 and 4. The experimental results and their interpretation are given in Sect. 5 and finally, we give the conclusions and perspectives of the work.

2 Description of the Robots and Experimental Setup

The experiments have been carried out using *e-puck2* robots equipped with the *Pi-puck* extension (see https://www.gctronic.com/e-puck2.php). Each of these robots has a cylindrical body characterized by a diameter of $d = 7\,[cm]$, a height of $h = 5.5$ [cm], and a weight of $w = 150$ [g]. It can move on a flat surface using a differential drive system with two wheels powered by stepper motors. Each robot is equipped with several sensors, shown in Fig. 1(a). For the experiment described in this paper, we only use the range and bearing board (more details at https://www.gctronic.com/doc/index.php?title=Others_Extensions#Range_and_bearing). This is a circular board featuring 12 IR emitters and 12 IR receivers uniformly distributed along its circumference, providing a decentralized communication system based on infrared technology with frequency modulation. This allows the robots to send messages of two bytes at a frequency of approximately 40 [Hz] and to determine the range and bearing of the transmitter at distances of up to about $d_M = 1$ [m], which can be tuned by adjusting the transmission power of the infrared transmitters.

The experimental setup, shown in Fig. 1(b), consists of N robots placed in an arena of hexagonal shape, with each side measuring $l = 60$ [cm]. The setup also includes a local computer that communicates with the robots via WiFi to initiate experiments and collect data from them at the end of each experiment.

3 The Model

In [30], both male and female fireflies are modeled as oscillators that alternate between an active stage, during which they oscillate, and a resting stage, during which they remain inactive and do not respond to external inputs. Males are represented by bursting oscillators (BOs), which exhibit a series of rapid oscillations (bursts of spikes) for each active stage, while females are modeled as non-bursting oscillators (NBOs), showing a single oscillation per active stage. Each oscillation consists of a charge and discharge stage whose duration is influenced by interactions with other units. However, the resting stage has a constant duration as it is unaffected by external interactions.

The dynamics of each isolated oscillator i is characterized by an internal variable $v_i(t)$, which increases until $V_u = \frac{2}{3}V_M$, decreases until $V_l = \frac{1}{3}V_M$, or remains constant at V_l during the charging, discharging, or resting stages, respectively, where V_M is a constant value. This variable is complemented by the binary function $y_i(t; n_i)$ that determines the type of oscillator by n_i (number of spikes per burst) and its stage: active when $y_i(t; n_i) = 1$, resting when $y_i(t; n_i) = 0$. Specifically, the dynamics is described by the following equations:

$$\dot{v}_i(t) = \left[\frac{\ln 2}{T_{c,i}} (V_M - v_i(t))\, \epsilon_i(t) - \frac{\ln 2}{T_{d,i}} v_i(t)\, (1 - \epsilon_i(t))\right] y_i(t; n_i)$$

$$y_i(t; n_i) = \sum_{k=0}^{M_i(t)} [H(t - kT_{p,i} - \Delta\Phi_i) - H(t - (k+1)T_{p,i} + T_{s,i} - \Delta\Phi_i)] \tag{1}$$

where $T_{p,i}$ is the total duration of the phrase (complete cycle), $M_i(t)$ is an integer obtained from the integer division of t by $T_{p,i}$ (in other words, the number of considered phrases), H is the Heaviside step function, and $\Delta\Phi_i$ is a phase delay that plays the role of an initial condition. The parameters $T_{c,i}$, $T_{d,i}$, and $T_{s,i}$ indicate the duration of the charging, discharging, and resting stages and determine the value of the phrase as follows: $T_{p,i} = T_{s,i} + n_i(T_{c,i} + T_{d,i})$. These parameters, illustrated in Fig. 2, assume different values depending on the type of oscillator. For BOs, the durations of the phrase, as well as the resting, charging, and discharging stages, are denoted as $T_{p,m}$, $T_{s,m}$, $T_{c,m}$, and $T_{d,m}$, respectively. In the case of NBOs, these parameters are instead given by $T_{p,f}$, $T_{s,f}$, $T_{c,f}$, and $T_{d,f}$. However, the primary distinction between the two types of oscillator lies in the value of $n_i \in \mathbb{N}$, which is $n_m > 1$ for BOs and 1 for NBOs. Consequently, given

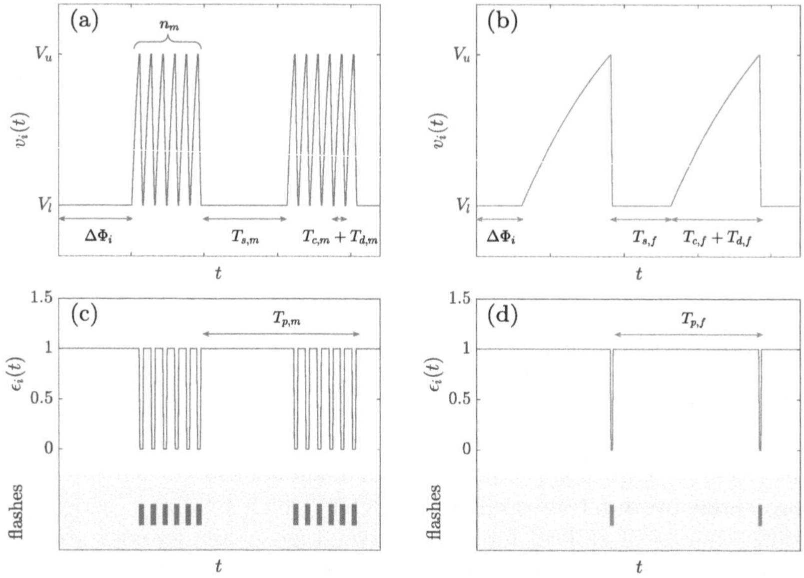

Fig. 2. Dynamics of a BO (a) and the corresponding evolution of $\epsilon_i(t)$ (c). Dynamics of a NBO (b) and the corresponding evolution of $\epsilon_i(t)$ (d).

the dependence of $T_{p,i}$ on n_i, the dynamics of the oscillators strongly depend on the number of spikes per burst. In summary, $v_i(t)$ expresses the spiking dynamics, while $y_i(t; n_i)$ describes the difference between BOs and NBOs through the fact that n_i is different among them, and enables the existence of the resting time $T_{s,i}$.

Going in-depth on the dynamical description, during the active stage, the dynamic $v_i(t)$ is given by two terms: the first ruling the charging stage, identified by $\epsilon_i(t) = 1$, and the second one regulating the discharging term identified by $\epsilon_i(t) = 0$. The binary variable $\epsilon_i(t)$ determines thus if the active oscillator is charging or discharging. The temporal evolution of $\epsilon_i(t)$ is given by:

$$\epsilon_i(t^+) = \epsilon_i(t) - \epsilon_i(t)H(v_i(t) - V_u) + (1 - \epsilon_i(t))H(V_l - v_i(t)), \tag{2}$$

which indicates that $\epsilon_i(t)$ switches to 0 when $v_i(t)$ reaches the upper threshold V_u, and switches to 1 when $v_i(t)$ reaches the lower threshold V_l, and keeps this value for $T_{s,i}$ time units, i.e., for all $t^+ \in [t, t + T_{s,i}]$.

In Fig. 2, we report the temporal behavior of $v_i(t)$ for both an isolated BO (Fig. 2(a)) and NBO (Fig. 2(b)) oscillator, together with the binary function $\epsilon_i(t)$ and the associated flashing event (Fig. 2(c) for BO and Fig. 2(d) for NBO) denoted by a rectangular mark displayed each time a flash is emitted.

Let C_m, resp. C_f, be the sets identifying the BOs, resp. the NBOs. When $N = |C_m| + |C_f|$ oscillators interact with each other, the dynamical equations are modified as follows. Once an oscillator $i \in C_m$, in the active stage, receives signals from other units, then the derivative of $v_i(t)$ increases. The opposite holds true for NBOs, namely once $i \in C_f$ in the active stage, the signals from the group decrease the derivative of $v_i(t)$. This different behavior can be cast into the following form for $i = 1, \ldots, N$:

$$\begin{aligned} \dot{v}_i(t) &= \left[\frac{\ln 2}{T_{c,i}}\left(V_M - v_i(t)\right)\epsilon_i(t) - \frac{\ln 2}{T_{d,i}} v_i(t)\left(1 - \epsilon_i(t)\right) + \gamma_i(t)\right] z_i(t; n_i) \\ z_i(t; n_i) &= H(t - \Delta\Phi_i) + \sum_{k=1}^{M_i'(t)} \left[H(t - t_{k,i} - T_{s,i}) - H(t - t_{k,i})\right], \end{aligned} \tag{3}$$

where $\gamma_i(t)$ is the coupling term, $z_i(t; n_i)$ determines the oscillator stage when coupled ($z_i(t; n_i) = 0$ during the resting stage, $z_i(t; n_i) = 1$ during the active stage) and $t_{k,i}$ is the time instant at which, for the k-th time, the following conditions are satisfied:

$$\textbf{(i)} \quad \text{mod}\left(\eta_i(t), n_i\right) = 0\ , \ \textbf{(ii)}\ v_i(t) = V_l \text{ and } \textbf{(iii)}\ \epsilon_i(t) = 0\ . \tag{4}$$

with $\eta_i(t)$ being the number of peaks reached by $v_i(t)$ in the time interval $[0, t]$, and $n_i = n_m$ if $i \in C_m$ and 1 otherwise. The term $M_i'(t)$ indicates the number of times these three conditions are satisfied in the time interval $[0, t]$. Notice that in this case, the oscillator stage is not determined by the function $y_i(t; n_i)$ since, due to the coupling, the complete cycle of the oscillator no longer has a constant duration. The coupling term is expressed as:

$$\gamma_i(t) = \theta_i \sigma_i \sum_{j=1}^{N} a_{ij}\left(1 - \epsilon_j(t)\right), \tag{5}$$

where σ_i is the coupling strength, a_{ij}, $i, j = 1, \ldots, N$, are the entries of the adjacency matrix of the interaction graph, and θ_i is a term that assumes the value of 1 if $i \in C_m$, -1 otherwise. This models the fact that the coupling term, when non-zero and when the oscillator is in the active stage, affects the rate of change of $v_i(t)$. Specifically, when i is in the charging stage, $T_{c,i}$ decreases if i is a BO, and increases otherwise. Conversely, when i is in the discharging stage, $T_{d,i}$ increases if i is a BO, and decreases otherwise.

Notice that the coupling is defined in terms of the variable $\epsilon(t)$, in particular when $\epsilon_j(t) = 0$, i.e., the j-th oscillator is flashing, the latter influences the i-th one. Let us observe that in the rest of this work, we assume the interaction graph to be undirected and fully connected, namely each robot can potentially communicate with all the others.

In the mathematical model, two scenarios have been considered: one where the oscillators are stationary and interact through an unweighted adjacency matrix, and another where the oscillators are mobile and interact through a weighted time-varying interaction graph, with the weights depending on the distance between agents [15]. In both cases, the coupling strength is uniform across all agents, i.e., $\sigma_i = \sigma$ for all i. In this work, we focus on a scenario where the units are stationary and interact through an unweighted adjacency matrix, but the coupling strength depends on the type of oscillator: $\sigma_i = \sigma_m$ if $i \in C_m$ and $\sigma_i = \sigma_f$ if $i \in C_f$.

In the following we will focus on the configurations of oscillators given by:

1. *Mutually Coupled NBOs*: In this case, N NBOs interact among themselves.
2. *Mutually Coupled BOs*: This configuration considers interacting N BOs.
3. *Leader (BO) - Follower (NBO)*: In this configuration, both BOs and NBOs are present. The BOs are identical but do not interact with other units ($\sigma_m = 0$), whereas the NBOs interact among themselves and are influenced by the BOs ($\sigma_f \neq 0$). Notice that the condition $\sigma_m = 0$ implies that the dynamics of the BOs is described by Eq. (1).
4. *Mutually Coupled BOs and NBOs*: Here, both BOs and NBOs interact with each other and their behavior is governed by Eq. (3).

In [30] it is shown that: BOs can synchronize among them, NBOs cannot for any values of σ_i, and when BOs and NBOs are mingled, BOs can still synchronize and their synchronization induces the response of NBOs.

4 Robotic Implementation

This section aims to introduce the implementation of the mathematical model (3) in the robots and the technical details we had to consider and resolve.

During the experiments, the robots remain stationary and, based on the values of the parameters (θ_i, σ_i, $T_{c,i}$, $T_{d,i}$, $T_{s,i}$, V_M, n_i, $\Delta\Phi_i$) loaded from a file stored on their memory, they are assigned either to BO, i.e., "male", or to NBO, i.e., "female". At each time step $t_h = ht_s$ ($h = 1, 2, \ldots$), each robot updates its state variable by performing an integration step of Eq. (3), by using the 4th-order Runge-Kutta method with a fixed step size of Δt. The robot then updates the value of $\epsilon_i(t_h)$ by using Eq. (2), and based on this value, it controls a green LED (activated if $\epsilon_i(t_h) = 0$, deactivated otherwise) and broadcast its current value of $\epsilon_i(t_h)$. Both $v_i(t_h)$ and $\epsilon_i(t_h)$ are recorded in a file stored inside the memory of each robot, that will be transmitted via Wi-Fi to a computer at the end of each experiment, to post-process the produced data.

To prevent time drift during integration, before starting each experiment, the computer sends a file to the robots containing an absolute reference time, t_{ref}. This ensures that, at the beginning of the experiment, the robots, equipped with a universal clock, start their integration process once $t \geq t_{ref}$. As a result, the robots will perform the integration steps at times $t = t_{ref} + ht_s$ for $h = 1, \ldots, T/t_s$, where T denotes the total duration of the experiment.

In the mathematical model, the interaction graph is fully connected. On the other hand, in the experiment the neighbor detection relies on local communication provided by the range & bearing board, with a transmission power set to enable communication between robots up to a distance of 40 [cm], to minimize battery consumption. Given that the communication system is not immune to message loss and message corruption, the neighborhood of the robot i is defined as follows:

$$\mathcal{N}_i(t_h) = \{j : \text{a valid message from } j \text{ is received by } i \text{ in } t' \in]t_{h-1}, t_h]\}$$

where "valid" refers to messages that are not corrupted due to interference or that do not come from previous time steps or by the robot itself. To ensure the reliability of communication, the robots broadcast messages containing information that allows the receiving robots to check their validity. Specifically, the message, consisting of two bytes, is structured as follows: the most significant byte contains the robot ID, i, which helps prevent the receiving robot from considering the same neighbor information multiple times during an integration step; the least significant byte is divided into 4-bit segments: the upper segment containing $r_i(t_h) = \text{mod}(h, 5)$, and the lower one containing $\epsilon_i(t_h)$.

Each time a robot i receives a message from another unit j, it performs a set of checks, including:

- verifying that the message is not corrupted (for example, if $\epsilon_j(t_h)$ is different from 0 or 1 the message is considered corrupted;
- ensuring that the message comes from the current time step, i.e., $r_i(t_h) = r_j(t_h)$;
- checking that the sender j does not already belong to $\mathcal{N}_i(t_h)$.

If a message does not meet all these conditions, it will be neglected by the receiver. As a result, there is no guarantee that the adjacency matrix describing the interaction graph among the robots is time-constant and undirected. Consequently, the value of a_{ij} in Eq. (3) becomes time-varying, with $a_{ij}(t_h) = 1$ if and only if $j \in \mathcal{N}_i(t_h)$.

The elementary time step, t_s, must be chosen large enough to allow the robots to detect their neighbors but at the same time it cannot be too large, as the use of the range & bearing board can significantly drain the battery. To determine this value, we followed the same methodology adopted in [36]. Specifically, we conducted experiments by using $N = 2$ robots communicating at a distance $d = 15$ [cm] and we tested different values of t_s. During these experiments, robots continuously sent messages, and, at intervals of t_s, they recorded a value of 1 in a file if a message was received within the interval, or 0 otherwise. These files were then post-processed to compute the probability of message reception. The outcomes of these experiments are shown in Fig. 3(a), which depicts the likelihood of receiving messages as a function of t_s. Based on these findings, we select $t_s = 500$ [ms] which provides a message reception probability $p = 0.85$. Let us notice that this probability refers to the case in which communication occurs between two robots. Indeed, when the number of robots, N, increases, the value of p decreases, see Fig. 3(b) where we set $t_s = 500$ [ms]. We also investigated the impact of the distance among the robots on the communication probability. For this purpose, we conducted experiments by using $N = 2$ robots and an elementary time step of $t_s = 500$ [ms]; as shown in Fig. 3(c), the probability remains constant for distances from $d = 10$ [cm] to $d = 35$ [cm], after which it starts to decrease, reaching zero at $d = 45$ [cm]. Let us observe that this behavior is expected, given that the transmission power of the robots is configured to allow communication up to a maximum distance of approximately 40 [cm].

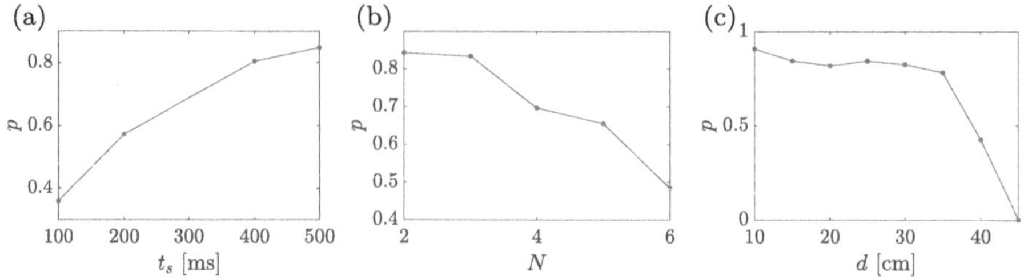

Fig. 3. Probability of receiving a non-corrupted message, p, as function of the elementary time step t_s (a), the number of robots N (b), and the distance among robots d (c). In panels (a) and (c), $N = 2$, whereas in panels (b) and (c), $t_s = 500$ [ms].

5 Experimental Results

In this section, we present the results obtained from the experiments conducted by using the robotic implementation described in the previous section. Specifically, we carried out four types of experiments: the first shows that the NBOs are not able to synchronize in the absence of BOs; the second demonstrates that BOs can achieve synchronization; the third characterizes the response of the NBOs to the BOs synchronization in a leader-follower configuration; and the fourth one investigates the NBOs response to BOs synchronization in a mutually coupled BO-NBO configuration.

We use the synchronization error, computed separately for BOs and NBOs, to evaluate the synchronization among the oscillators:

$$\delta_k(t) = \frac{1}{|C_k|} \sum\nolimits_{\substack{i,j \in C_k \\ j \neq i}} |v_i(t) - v_j(t)| \quad \text{for } k \in \{m, f\},$$

$\delta_k(t)$ is small if the BOs, resp. NBOs, exhibit at time t close values of v_i.

The following parameters, based on previous studies involving real fireflies [31], were used in all experiments: $T_{c,m} = 0.5$ [s], $T_{d,m} = 0.2$ [s], $T_{s,m} = 5.8$ [s], $n_m = 6$, $T_{c,f} = 6$ [s], $T_{d,f} = 0.1$ [s], $T_{s,f} = 3.9$ [s], and $\Delta t = 0.05$ [s]. Additionally, $V_M = 9$, resulting in $V_l = 3$ and $V_u = 6$. For the ODE integration, we used $\Delta t \neq t_s$, resulting in two distinct time scales: numerical and physical.

In the first experiment, lasting $T = 1000$ [s], $N = 2$ NBOs interact at a distance $d = 20$ [cm]. Each oscillator is initialized to a different initial condition and, at each time step t_h, it integrates the ODE (3) with $\sigma_f = 2$. Figure 4 shows the dynamics of the oscillators and the corresponding flashes, revealing that, despite starting from very similar initial conditions, the two oscillators do not synchronize. This demonstrates that, as in the mathematical model, NBOs cannot synchronize on their own, independently from the coupling strength value.

For the second experiment, a group of $N = 6$ BOs is arranged in a hexagonal formation with a radius $d_r = 20$ [cm], and interact with each other with a coupling strength $\sigma_m = 0.05$. The experiment lasts $T = 800$ [s] and, as shown in Fig. 5(a), the system reaches synchronization despite a reception probability

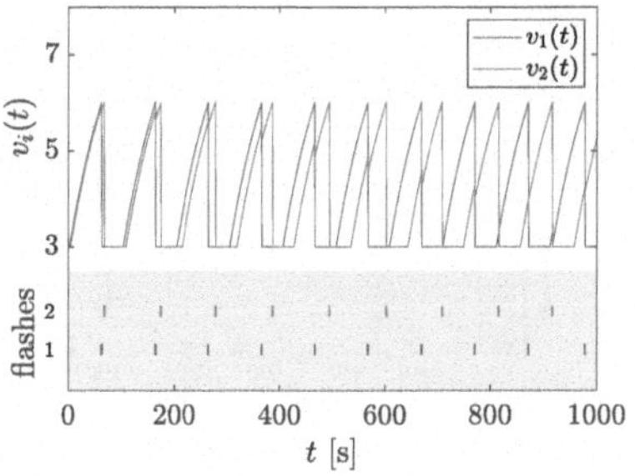

Fig. 4. Experimental results: 2 NBOs. Dynamics and corresponding flashes of two NBOs interacting at a distance of $d = 20$ [cm]. Despite starting from very close initial conditions, the NBOs do not achieve synchronization.

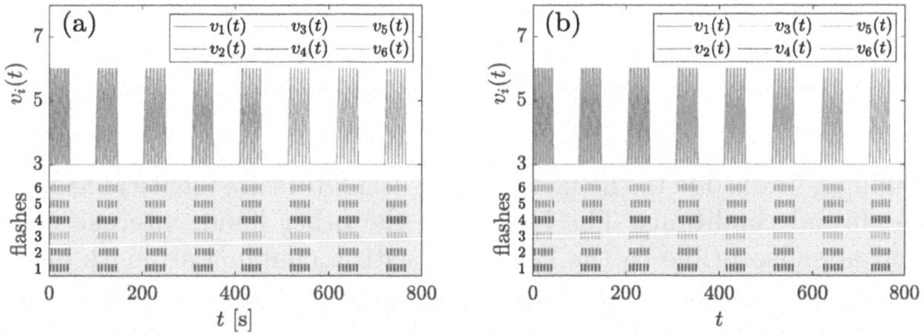

Fig. 5. Experimental and numerical results: 6 interacting BOs. (a) Dynamics of $N = 6$ interacting robots, each exhibiting a bursting oscillator, arranged in a hexagonal formation with a radius $d_r = 20$ [cm]. (b) Results obtained from a simulation in which a probability of message reception $p = 0.5$ is considered. In both cases, a coupling strength $\sigma = 0.05$ has been used.

estimated to be ~ 0.5, according to the results shown in Fig. 3(a). We compare this result with a numerical simulation where we introduced the parameter p to emulate the message loss characterizing the local communication provided by the range & bearing board. Specifically, at each time step, the coupling term becomes $\gamma_i'(t) = \theta_i \sigma_i \sum_{j=1}^{N} \omega_{ij}(t)\,(1 - \epsilon_j(t))$, where $\omega_{ij}(t)$ takes the value 1 with probability p and 0 otherwise. We can observe (Fig. 5(b)) that the numerical results agree with the experimental ones.

The third set of experiments involves $N = 8$ oscillators, consisting of 6 BO and 2 NBO units. In particular, the BOs are arranged in a hexagonal formation, with the NBOs placed at 3 $[cm]$ from the center of the hexagon, facing each other. In each experiment, the distance between the center and the vertices of the hexagon, denoted by d_r, is kept constant but varies across experiments, being incrementally increased from one experiment to the next. In each experiment, the BOs start from the same initial condition and do not interact with any other units ($\sigma_m = 0$). As a result, they maintain synchronization throughout the entire experiment. Notice that, in this case, synchronization does not arise

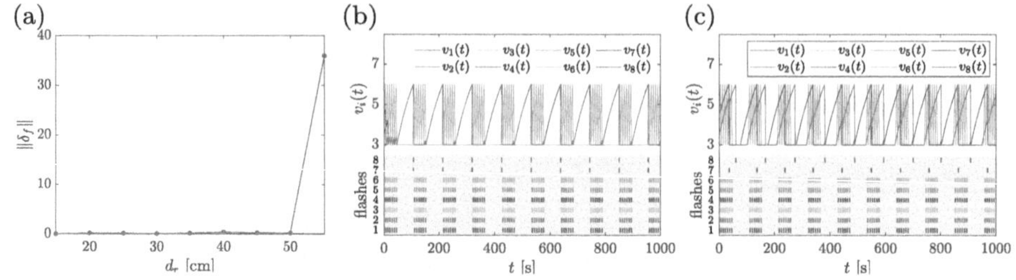

Fig. 6. Experimental results: 6 BOs (Leaders) and 2 NBOs (Followers). The experiments use the configuration in Fig. 1(b), with BOs arranged in a hexagon formation with a radius d_r and NBOs placed 3 cm from the center of the hexagon, facing each other. (a) Synchronization error of the NBOs as a function of the hexagon radius d_r. (b) Dynamics and corresponding flashes of the oscillators for $d_r = 20$ [cm], showing that the NBOs respond to synchronization after a short time. (c) Dynamics and corresponding flashes of the oscillators for $d_r = 55$ [cm], showing that NBOs do not synchronize.

from coupling, but is due to the fact that the oscillators are identical as they have the same initial conditions. The NBOs, on the other hand, interact with each other and are susceptible to the BO state, with a coupling strength of $\sigma_f = 2$. Figure 6(a) shows the total synchronization error among the NBOs computed as the norm of $\delta_f(t)$ over the time interval $\left[\frac{4}{5}T, T\right]$, i.e., we remove an initial transient. We can appreciate a low synchronization error for many values of d_r, i.e., up to $d_r = 50$ [cm]. For $d_r = 55$ [cm], however, the synchronization error is high, indeed the NBOs, being very far from the BOs, do not detect the latter. The dynamics of the oscillators and corresponding flashes for $d_r = 20$ [cm] and for $d_r = 55$ [cm] are shown in Fig. 6(b) and Fig. 6(c), respectively. We notice that, for $d_r = 20$ [cm], the NBOs respond to the synchronization and synchronize between themselves in a short time.

The fourth experiment also includes $N = 8$ oscillators: 6 BOs arranged in a hexagonal formation with radius $d_r = 25$ [cm] and 2 NBOs placed at 2 [cm] from the center facing each other. We hereby consider a scenario in which BOs and NBOs start from different initial conditions and interact among themselves. Specifically, the BOs interact with a coupling strength of $\sigma_m = 0.5$, while the NBOs have a coupling strength of $\sigma_f = 2$. The response to synchronization in NBOs occurs once the BOs synchronize among themselves, i.e., about at $t \sim 1000$ [s] or 10 bursts of the BOs. Figure 7 shows the time evolution of the oscillators state variables and the corresponding flashes.

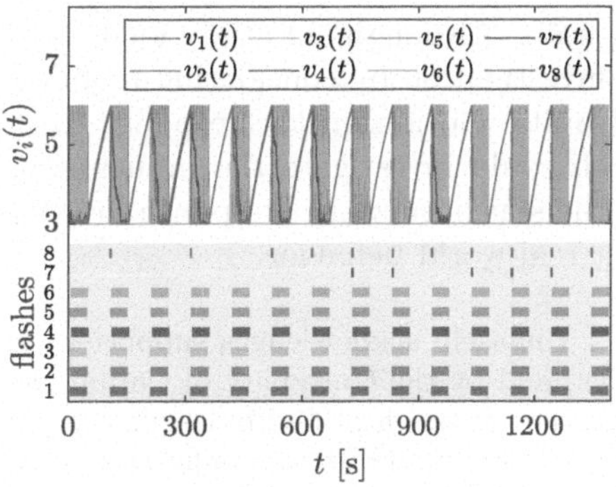

Fig. 7. Experimental results: 6 BOs and 2 NBOs. The experiment employs the same configuration used for Fig. 6, with $d_r = 25$ [cm] and the NBOs placed at 2 [cm] from the center. The BOs interact with a coupling strength $\sigma_m = 0.5$ and the NBOs with $\sigma_f = 2$. Each oscillator starts from different initial conditions. The figure shows the time evolution of the oscillators state variables and their corresponding flashes.

6 Conclusions

In this work, we used a team of e-puck2 robots to implement and validate the mathematical model discussed in [30], describing the response to synchronization in fireflies. The model considers a heterogeneous system composed of two types of oscillators: NBOs, representing females unable to synchronize alone, and BOs, representing males capable of achieving synchronization by their own. Specifically, the model focuses on the phenomenon of response to synchronization, where NBOs synchronize as a result of their interactions with synchronous BOs, demonstrating how heterogeneity can induce coordinated behavior in multi-agent systems.

For robotic implementation, we considered two populations of robots: one representing BOs and the other representing NBOs. Such implementation is fully distributed as the robots integrate their own dynamics and locally interact through the range & bearing board. The experiments explored various configurations. In more detail, using a configuration consisting only of NBOs in absence of BOs, we demonstrated that, despite starting from nearly identical initial conditions, they are unable to synchronize on their own. We also tested interacting BOs, confirming their ability to achieve synchronization. Additionally, we investigated both leader-follower (BOs acting on NBOs) and interacting BOs and NBOs configurations to study the response of the NBOs to the BO synchronization, proving that the NBOs can synchronize in the presence of the BOs. Through our experiments, we reproduced the main features of the mathematical model, proving its robustness against real-world factors, including message loss during communication or corrupted messages.

The validation of this model offers insight into practical applications involving units that are not capable of synchronizing alone. Indeed, the introduction of additional units capable of synchronizing could induce synchronization among the non-synchronizing units. Although our study focused on static robots, future work will extend this investigation by considering mobile robots, allowing us to explore the impact of units' motion on the response to synchronization. Additionally, future research could investigate the influence of varying the number of NBOs and BOs on the synchronization response, by studying how closely the observed results reflect real-world behavior.

Acknowledgments. C. Tomaselli acknowledges support by the Namur Institute for Complex Systems (naXys) and by the University of Catania for her visiting period at the University of Namur. The research of C. Tomaselli and L.V. Gambuzza has been framed and supported by University of Catania under the PIA.CE.RI. project, starting grant for RTDb. G. M. Ramírez-Ávila acknowledges the European Union's Horizon 2020 research and innovation program under the Marie Sklodowska-Curie grant agreement No 101034383.

Disclosure of Interests. The authors have no competing interests to declare that are relevant to the content of this article.

References

1. Aihara, I., et al.: Complex and transitive synchronization in a frustrated system of calling frogs. Phys. Rev. E **83**(3), 031913 (2011)
2. Backwell, P.: Synchronous waving in fiddler crabs: a review. Curr. Zool. **65**(1), 83–88 (2019)
3. Balanov, A., Janson, N., Postnov, D., Sosnovtseva, O.: Synchronization: From Simple to Complex. Springer, Berlin (2007)
4. Boccaletti, S., Pisarchik, A.N., del Genio, C.I., Amann, A.: Synchronization: From Coupled Systems to Complex Networks. Cambridge University Press, Cambridge (2018)
5. Buck, J., Buck, E.: Flash synchronization as tool and as enabler in firefly courtship competition. Am. Nat. **116**(4), 591–593 (1980)
6. Buck, J.B.: Synchronous flashing of fireflies experimentally induced. Science **81**(2101), 339–340 (1935)
7. Carlson, A.D., Copeland, J., Raderman, R., Bulloch, A.: Response patterns of female Photinus macdermotti firefly to artificial flashes. Anim. Behav. **25**, 407–413 (1977)
8. Copeland, J., Moiseff, A.: Flash precision at the start of synchrony in Photuris frontalis. Integr. Comp. Biol. **44**(3), 259–263 (2004)
9. Cratsley, C.K.: Flash signals, nuptial gifts and female preference in Photinus fireflies. Integr. Comp. Biol. **44**(3), 238–241 (2004)
10. De Cock, R., Faust, L., Lewis, S.: Courtship and mating in Phausis reticulata (coleoptera: lampyridae): male flight behaviors, female glow displays, and male attraction to light traps. Fla. Entomol. **97**(4), 1290–1307 (2014)
11. Dunning, D.C., Byers, J.A., Zanger, C.D.: Courtship in two species of periodical cicadas, Magicicada septendecim and Magicicada cassini. Anim. Behav. **27**, 1073–1090 (1979)
12. Ercsey-Ravasz, M., SÁrközi, Z., Néda, Z., Tunyagi, A., Burda, I.: Collective behavior of electronic fireflies. Eur. Phys. J. B **65**(2), 271–277 (2008)

13. Faust, L.F.: Natural history and flash repertoire of the synchronous firefly Photinus carolinus (coleoptera: Lampyridae) in the great smoky mountains national park. Florida Entomol. 208–217 (2010)
14. Faust, L.F.: Fireflies, Glow-worms, and Lightning Bugs: Identification and Natural History of the Fireflies of the Eastern and Central United States and Canada. University of Georgia Press, Athens (2017)
15. Fujiwara, N., Kurths, J., Díaz-Guilera, A.: Synchronization in networks of mobile oscillators. Phys. Rev. E **83**(2), 025101 (2011). https://doi.org/10.1103/PhysRevE.83.025101
16. Hartbauer, M., Römer, H.: A novel distributed swarm control strategy based on coupled signal oscillators. Bioinspir. Biomimetics **2**(3), 42 (2007)
17. Kousaka, T., Kawakami, H., Ueta, T.: Synchronization of electric fireflies by using square wave generators. IEICE Trans. Fundam. Electron. Commun. Comput. Sci. **81**(4), 656–663 (1998)
18. Lloyd, J.E.: Fireflies of melanesia: bioluminescence, mating behavior, and synchronous flashing (coleoptera: Lampyridae). Environ. Entomol. **2**(6), 991–1008 (1973)
19. Maquitico, Y., Coronado, J., Luna, A., Vergara, A., Cordero, C.: Deceptive seduction by femme fatale fireflies and its avoidance by males of a synchronous firefly species (coleoptera: Lampyridae) (2024)
20. McCrea, M., Ermentrout, B., Rubin, J.E.: A model for the collective synchronization of flashing in Photinus carolinus. J. R. Soc. Interface **19**(195), 20220439 (2022)
21. Michaelidis, C.I., Demary, K.C., Lewis, S.M.: Male courtship signals and female signal assessment in Photinus greeni fireflies. Behav. Ecol. **17**(3), 329–335 (2006)
22. Mirollo, R.E., Strogatz, S.H.: Synchronization of pulse-coupled biological oscillators. SIAM J. Appl. Math. **50**(6), 1645–1662 (1990)
23. Moiseff, A., Copeland, J.: Firefly synchrony: a behavioral strategy to minimize visual clutter. Science **329**(5988), 181 (2010)
24. Nelson, S., Carlson, A.D., Copeland, J.: Mating-induced behavioural switch in female fireflies. Nature **255**(5510), 628–629 (1975)
25. Nityananda, V., Balakrishnan, R.: Leaders and followers in katydid choruses in the field: call intensity, spacing and consistency. Anim. Behav. **76**(3), 723–735 (2008)
26. Peleg, O.: A new chapter in the physics of firefly swarms. Nat. Rev. Phys. **6**(2), 72–74 (2024)
27. Perez Diaz, F.: Firefly-inspired synchronization in swarms of mobile agents. Ph.D. thesis, University of Sheffield (2016)
28. Pikovsky, A., Rosenblum, M., Kurths, J., Synchronization, A.: A universal concept in nonlinear sciences. Self **2**, 3 (2001)
29. Rabha, M.M., Sharma, U., Goswami, A., Gohain Barua, A.: Bioluminescence emissions of female fireflies of the species luciola praeusta. J. Photochem. Photobiol. B Biol. **170**, 134–139 (2017)
30. Ramírez-Ávila, G.M., Kurths, J.: Unraveling the primary mechanisms leading to synchronization response in dissimilar oscillators. Eur. Phys. J. Spec. Top. 2487–2506 (2016). https://doi.org/10.1140/epjst/e2016-60033-5
31. Ramírez-Ávila, G.M., Deneubourg, J.L., Guisset, J.L., Wessel, N., Kurths, J.: Firefly courtship as the basis of the synchronization-response principle. EPL (Europhys. Lett.) **94**(6), 60007 (2011)
32. Ramírez-Ávila, G.M., Kurths, J., Deneubourg, J.L.: Fireflies: a paradigm in synchronization, pp. 35–64. Springer, Cham (2018)

33. Ramírez-Ávila, G.M., Kurths, J., Depickère, S., Deneubourg, J.L.: Modeling fireflies synchronization, pp. 131–156. Springer, Cham (2019)
34. Sarfati, R., Hayes, J.C., Peleg, O.: Self-organization in natural swarms of Photinus carolinus synchronous fireflies. Sci. Adv. **7**(28), eabg9259 (2021)
35. Tomaselli, C., Guastella, D.C., Muscato, G., Minati, L., Frasca, M., Gambuzza, L.V.: Synchronization of moving chaotic robots. IEEE Robot. Autom. Lett. **9**(7), 6496–6503 (2024)
36. Tomaselli, C., Guastella, D.C., Muscato, G., Frasca, M., Gambuzza, L.V.: A multi-robot system for the study of face-to-face interaction dynamics. IEEE Robot. Autom. Lett. (2023)
37. Trianni, V., De Simone, D., Reina, A., Baronchelli, A.: Emergence of consensus in a multi-robot network: from abstract models to empirical validation. IEEE Robot. Autom. Lett. **1**(1), 348–353 (2016)
38. Vencl, F.V., Blasko, B.J., Carlson, A.D.: Flash behavior of female Photuris versicolor fireflies (coleoptera: Lampyridae) in simulated courtship and predatory dialogues. J. Insect Behav. **7**(6), 843–858 (1994)

Attractive-Repulsive Challenge in Swarmalators with Time-Dependent Speed

Steve J. Kongni[1,4](✉), Thierry Njougouo[2,3,4], Gaël R. Simo[4,5], Patrick Louodop[1,4,6,7], Robert Tchitnga[1,4], and Hilda A. Cerdeira[7,8]

[1] Research Unit Condensed Matter, Electronics and Signal Processing, University of Dschang, Dschang, Cameroon
[2] IMT School for Advanced Studies Lucca, Lucca, Italy
[3] Faculty of Computer Science and Namur Institute for Complex Systems (naXys), University of Namur, Namur, Belgium
[4] MoCLiS Research Group, Dschang, Cameroon
KongniSteve@Yahoo.fr
[5] Laboratory of Electrotechnics, Automatics and Energy, Higher Technical Teachers, Training College (ENSET) of Ebolowa, University of Ebolowa, Ebolowa, Cameroon
[6] Potsdam Institute for Climate Impact Research (PIK) Member of the Leibniz Association, Potsdam, Germany
[7] Instituto de Física Teórica, São Paulo State University (UNESP), São Paulo, Brazil
[8] Epistemic ME, São Paulo, Brazil

Abstract. We examine a network of entities whose internal and external dynamics are intricately coupled, modeled through the concept of "swarmalators" as introduced by O'Keeffe et al. [1]. We investigate how the entities' natural velocities impact the network's collective dynamics and path to synchronization. Specifically, we analyze two scenarios: one in which each entity has an individual natural velocity, and another where a group velocity is defined by the average of all velocities. Our findings reveal two distinct forms of phase synchronization—static and rotational—each preceded by a series of states where attractive or repulsive interaction dominates the movement of the entities. This interaction phase, which depends sensitively on initial conditions, allows for selective modulation within the network. By adjusting initial parameters, we can isolate specific entities to experience attractive-repulsive interactions distinct from the group, prior to the onset of full synchronization. This nuanced dependency on initial conditions offers valuable insights into the role of natural velocities in tuning synchronization behavior within coupled dynamic networks.

Keywords: Swarmalators · attractive repulsive interactions · Rotational phase synchronization

---

Supplementary Information The online version contains supplementary material available at https://doi.org/10.1007/978-3-031-93631-9_18.

T. Carletti et al. (Eds.): WIVACE 2024, CCIS 2532, pp. 223–235, 2025.
https://doi.org/10.1007/978-3-031-93631-9_18

1 Introduction

The study, understanding, and synthesis of swarm and oscillator dynamics have recently attracted considerable attention from researchers. This research primarily aims to enhance our understanding of, and to replicate, the behaviors exhibited by various living systems, including birds, fish, frogs, and social insects. Over the last decade, a new class of systems known as mobile systems has emerged, characterized by the coupling of internal and spatial dynamics. These systems, referred to as swarmalators by O'Keeffe et al. [1], merge the dynamics of swarming with those of oscillators. Numerous studies in this domain have highlighted phenomena such as external sinusoidal excitation [2], phase transitions [3,4], and various other interactions [5–13].

Building on the two-dimensional model of swarmalators, Sar et al. developed a one-dimensional version, demonstrating that random pinning can induce chaotic behavior within the system [13]. Furthermore, researchers have explored this system in various configurations within one-dimensional and two-dimensional spaces. However, much of the existing research has concentrated on the swarmalator model primarily through pairwise interactions. Recently, the focus has shifted toward incorporating higher-order interactions within swarmalator systems to better approximate the dynamics of specific real-world scenarios [14].

To the best of our knowledge, all existing research on swarmalator dynamics has primarily been aimed at identifying new dynamics and transitions toward synchronization, while often assuming that particles do not possess self-propulsion velocities. In this way by not considering memory of their previous directions of movement, the movement of the oscillators in real space has been restricted to rotation in a circle. The attractive and repulsive interactions described in the model by O'Keeffe et al. [1] play a critical role in the spatial mobility of entities. This attractive and repulsive interactions were also argued in many studies where some of them highlighted the existence of explosive transition toward synchronization due to the presence or not of repulsive coupling [3,17]. However, incorporating a self-propulsion velocity for each entity in the swarmalators dynamics, would more accurately reflect the behavior observed in many living systems, capturing their individual responses to external factors and adaptations to group movement.

In this work, we incorporate a non-zero self-propulsion speed to investigate its impact on the transition toward a synchronized state. Following this introduction, we present the studied model in Sect. 2. In Sect. 3, we analyze the numerical results, focusing first on the effects of individual self-propulsion speed, followed by the effects of group self-propulsion speed. We conclude with a summary of our findings.

2 Model

Let us consider a model, which has spurred extensive research into swarmalator dynamics, encompassing diverse configurations, coupling mechanisms, and emergent behaviors [1,2,5,7,8]. This system is described by Eqs. 1 and 2, which capture the core interactions underlying these dynamics.

$$\dot{r}_i = v_i + \frac{1}{N}\sum_{j\neq i}^{N} F_{att}(r_{ij})\, W(\theta_{ij}) - F_{rep}(r_{ij}), \tag{1}$$

$$\dot{\theta}_i = w_i + \frac{K}{N}\sum_{j\neq i}^{N} H_{att}(\theta_{ij})\, G(r_{ij}), \tag{2}$$

with $i, j = 1, \ldots, N$, where N denotes the total number of swarmalators; θ_i represents the phase of the internal dynamics of each entity, and $r_i = (x_i, y_i) \in \mathbb{R}^2$ denotes the spatial coordinate of the ith swarmalator. The parameters v_i and w_i correspond to the velocity and natural frequency of each entity, respectively. The spatial attractive and repulsive interactions between entities in the network are represented by the functions F_{att} and F_{rep}, respectively, while H_{att} represents the phase interaction between them [1,2]. The influence of the internal dynamics on the oscillators' movement is represented by the functions W and G, which also capture the reciprocal effect of movement on internal dynamics. The model described by Eqs. 1 and 2 can be reformulated as follows:

$$\dot{r}_i = \alpha V(t) + \frac{1}{N}\sum_{j\neq i}^{N} \underbrace{\frac{r_{ij}}{\|r_{ij}\|}\left(A + J\cos(\theta_{ij})\right)}_{\textit{Attractive term}} - \underbrace{B\frac{r_{ij}}{\|r_{ij}\|^2}}_{\text{Repulsive term}}, \tag{3}$$

$$\dot{\theta}_i = w_i + \frac{K}{N}\sum_{j\neq i}^{N} \frac{\sin(\theta_{ij})}{\|r_{ij}\|}. \tag{4}$$

Here, $r_{ij} = r_j - r_i$ and $\theta_{ij} = \theta_j - \theta_i$. The constant α is chosen within $[0, 1]$; $V(t)$ is defined by Eqs. 5 and 6 depending on the considered hypothesis defined below; A and B are set to 1; K represents the phase coupling strength, and the interaction between spatial and phase dynamics is modulated by $A + J\cos(\theta_j - \theta_i)$. J measures the influence of phase similarity on spatial attraction. Previous studies on swarmalators often simplify their models by assuming both the natural velocity v_i and natural frequency w_i are zero as in Eq. 1. In contrast, we will set $w_i = 0$ while allowing $v_i \neq 0$ and equal to $v_i = \alpha V(t)$ in our analysis (Eqs. 5 and 6).

Assuming zero intrinsic speed fails to accurately describe the dynamics of many living systems. For instance, in a flock of birds or a school of fish, the intrinsic speed of individuals may vary based on their internal states or external factors, such as the search for food or the presence of a predator. In these scenarios, a member may adjust its speed to pursue food or evade threats, leading to changes in direction, potential shifts in leadership, and variations in the intrinsic speeds of the group members.

Building on this example, we propose incorporating the effect of each entity's individual speed as it returns to the group after an external disturbance. For simplicity, we model each entity's speed as dependent on its speed from the previous moment (see Eq. 6), representing a memory state of its intrinsic speed.

Furthermore, interactions among entities within their collective dynamics have enabled us to define an additional form of speed, the "group speed" v_g (see Eq. 5). This group speed characterizes each entity's tendency to return to the collective movement pattern established prior to an external disturbance, facilitating coordinated movement and minimizing collision risk.

The two hypotheses formulated above allow us to examine the influence of a group speed v_g and an individual speed distribution v_p, as defined by Eqs. 5 and 6.

- Delayed group velocity

$$V(t) = v_g = \langle \dot{r}_i(t - \Delta t) \rangle_{i=1 \dots N} \tag{5}$$

- Delayed individual velocity

$$V(t) = v_p = \dot{r}_i(t - \Delta t) \tag{6}$$

$\langle \dots \rangle$ represents the time average, and Δt denotes the constant time delay, set equal to the integration step ($\Delta t = dt = 0.1$) for simplicity. Initially, all entities begin with zero velocity.

We will then analyze how these different choices of natural speeds affect the overall dynamics of the swarmalator network, as well as their potential influence on the network's progression toward phase synchronization.

3 Numerical Results

To analyze our model's dynamics, we simulated a network of 50 nodes using the fourth-order Runge-Kutta method with a time step of $dt = 0.1$. Initial spatial positions is randomly assigned from a uniform distribution over $[-1, 1]$, and internal phases is initialized within $[-\pi, \pi]$. Each solution is computed over 10^6 iterations. In this section, we examine the effects of both individual and group speed distributions on the network dynamics and further investigate how the choice of initial conditions influences the balance between attractive and repulsive interactions.

3.1 Effect of Group Velocity to the Route to Phase Synchronized State.

Let us consider in this initial case that the natural velocity at time t is uniform and corresponds to the average of velocities of all elements at time $t - \Delta t$, as described by Eq. 5. The dynamics of the system are governed by Eqs. 3 and 4.

Based on this assumption, Fig. 1 presents the transition from the static wing state to a static phase-synchronized state. The figure illustrates this progression through snapshots, emphasizing the repulsive and attractive interactions that drive one agent to behave as a solitary wave or "soliton". In these snapshots, we observe various transitional states preceding static phase synchronization. These transitions reveal that synchronization is achieved through a balance of competing forces: repulsive interactions that foster the formation of an active solitary node (AN) and attractive interactions that pull this node back toward the group, resulting in stable phase synchronization.

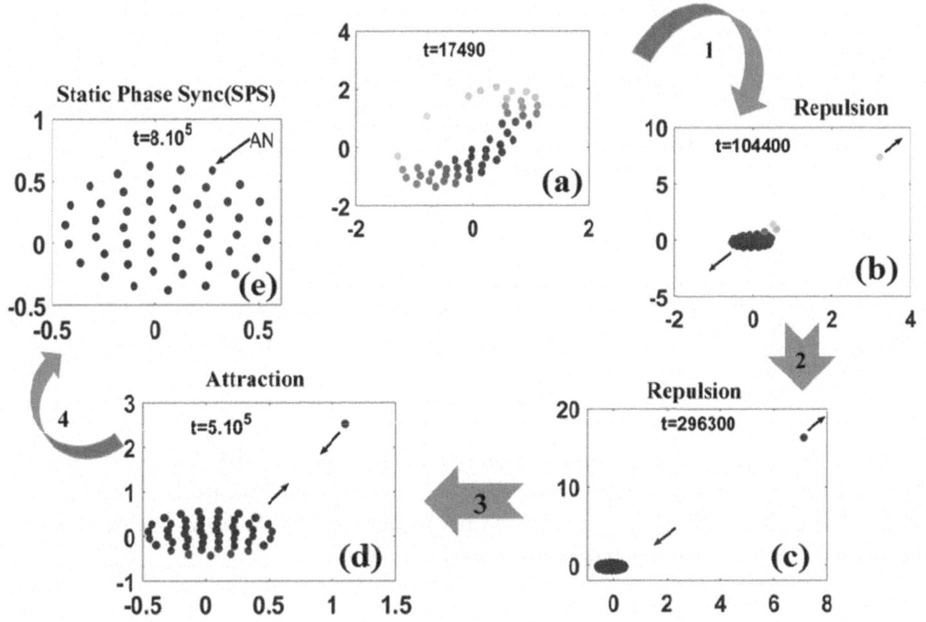

Fig. 1. Snapshots of states showing the evolution toward synchronization throught the attractive-repulsive challenge with the existence of active node (AN) when the delayed group velocity is applied. (a) Static wing state; (b, c) repulsion between active node (AN) and the static sync group; (d) attraction between AN with the static sync group; (e) Static phase sync (SPS). Plotted for $J = 1, K = 0.0004$ and $\alpha = 0.3$. The black arrow indicate the sens of displacement of both active node (AN) and the group (see movie 1 in the supplementary material [15]).

Figure 1(a) illustrates the creation of a solitary node at time $t = 17490$, driven by the repulsive interaction. When this repulsive force dominates over the attractive interaction, the solitary node moves away from the rest of the group, as seen in the transition from stages 1 to 2 (i.e., Figs. 1(b) and 1(c)). Stage 3, highlighted in Fig. 1(d), demonstrates the return of the solitary node to the group. This stage is characterized by the predominance of the attractive interaction over the repulsive one between the active node and the other elements in the network (see Fig. 3).

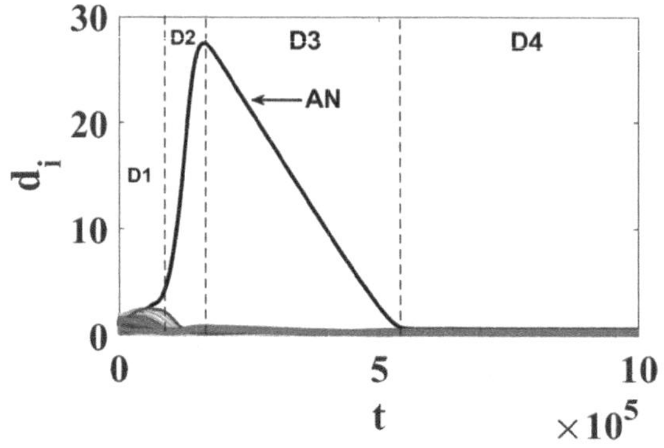

Fig. 2. Time evolution of the mean radius of all the nodes when the delayed group velocity is applied. The black curve represents the active node (AN). Plotted for $J = 1, K = 0.0004$ and $\alpha = 0.3$.

To better illustrate the challenge of attractive and repulsive interactions, we represent in Fig. 2 the relative distance of mutual interaction, d_i (see Eq. 7), between the elements of the network. This figure highlights four distinct regions (D1 to D4) corresponding to stages 1 to 4 of Fig. 1.

$$d_i(t) = \sqrt{x_i^2(t) + y_i^2(t)} \tag{7}$$

At the initial moment in region D1, the elements are distributed randomly, resulting in a relative distance d_i whose average value is so small (but not equal to zero). During this time, the attractive and repulsive interactions largely compensate for each other ($\mathrm{F_{rep}} \simeq \mathrm{F_{att}}$), with a slight dominance of the attractive interaction occurring after $t = 28970$ ($\mathrm{F_{att}} \geq \mathrm{F_{rep}}$), as indicated by the blue curve in Fig. 3.

In region D2 (stage 2 in Fig. 1(b,c)), we observe the onset of dominance by the repulsive interaction of the active element, represented by the red curve in Fig. 3, where $\mathrm{F_{rep}} > \mathrm{F_{att}}$. This shift is characterized by an increase in the relative distance d_i of the active element from the rest of the group (see Fig. 2). The change in concavity of the relative distance observed in region D3 suggests an imminent shift in dominance between the two interaction forces, similar to what we noted in D2. Specifically, the repulsive force continues to prevail in D3, maintaining the relationship $\mathrm{F_{rep}} > \mathrm{F_{att}}$, as illustrated in the zoomed-in view of Fig. 3. However, approaching region D4, we witness a tendency for the active node (AN) to be attracted toward the rest of the group, as depicted in Fig. 1(d). This indicates a transition to a dominance of the attractive force over the repulsive force, where $\mathrm{F_{att}} > \mathrm{F_{rep}}$.

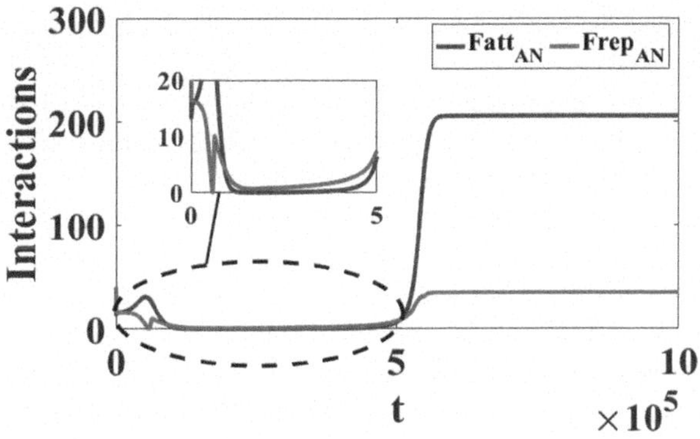

Fig. 3. Time evolution of the mean of the interactions suggested by the active node (AN) when the delayed group velocity is applied. The red color indicates the repulsive interaction while the blue one represents the attractive interaction between the active node and the other elements. Plotted for $J = 1, K = 0.0004$ and $\alpha = 0.3$. (Color figure online)

In addition to the evolutions illustrated in Figs. 1 and 2, it is important to highlight that during the transition toward static phase synchronization, the active node demonstrates distinct internal phase dynamics across regions D1 to D3 (see Fig. 4(a)). Notably, the entire group maintains a consistent phase difference of 2π between regions D2 and D3, which is indicative of phase synchronization. However, the active node only achieves a phase difference of either zero or 2π starting from region D4. Conversely, the evaluation of the internal phase energy H_i [16] between the entities reveals a direct relationship with the relative distance of interaction d_i between the active node and the group. Specifically, as the distance d_i increases, the internal phase energy H_i also increases, and conversely, it decreases as d_i decreases. Notably, the internal phase energy stabilizes

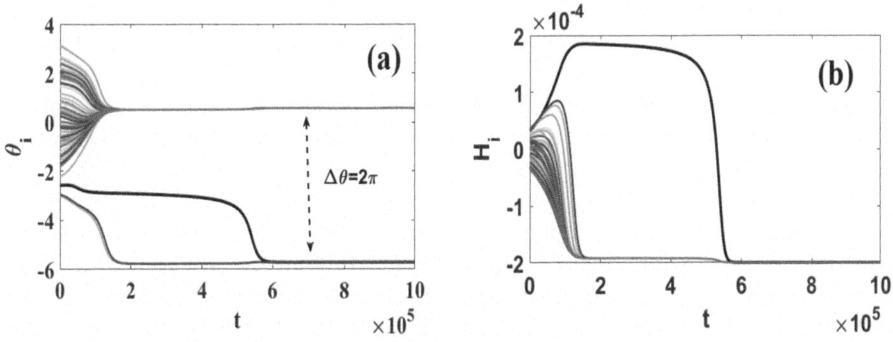

Fig. 4. Time evolution showing transition to static phase synchronization. (a) The internal phase dynamic θ_i of each element. (b) The internal phase energy H_i of each element. Plotted for $J = 1, K = 0.0004$ and $\alpha = 0.3$.

once synchronization is achieved among the internal phases of the entities, as illustrated in Fig. 4(b).

The increase in internal phase energy H_i with the relative distance d_i between the active node and the group can be attributed to several factors. As the distance increases, the internal phase energy often rises due to the dominance of repulsive forces over attractive ones, resulting in a loss of cohesion and stability within the system. This greater separation leads to increased phase differences, which disrupts synchronization and raises the energetic costs associated with rejoining the group. Additionally, the effective strength of interactions typically diminishes with distance, creating energy barriers that the active node must overcome to synchronize again. Overall, these dynamics reflect the underlying physical principles governing the interactions within the system, where both individual and collective behaviors significantly influence the energy landscape.

3.2 Distributed Velocity Effect on the Transition to Phase Synchronization

Let us now examine the impact of a delayed individual speed distribution on the transition to synchronization. Given that the function $V(t)$ in Eq. 3 is defined by Eq. 6, we find that similar to the effects of delayed group speeds, this delayed distribution significantly influences not only the transition to synchronization but also the type of synchronization achieved. The snapshots in Fig. 5 show the different transitional states leading to the synchronization state, where we observe a rotational behavior among the synchronized entities. Thus, by generating the initial conditions as seen in Fig. 5(a), we witness the formation of a compact group and a solitary or active node (AN), characteristic of a dominant repulsive interaction (Fig. 5(b)). In Figs. 5(c) and (d), we observe a repulsive effect between the active node and the group of entities (shown by the opposite black arrow). This repulsion is accompanied by the dominance of the repulsive interaction force over the attractive one ($F_{rep} > F_{att}$), as seen in Fig. 6(a) within region R2.

This dominance is also accompanied by an increase in the interaction distance of the active node relative to the other elements in region R2 of Fig. 6(b), followed by a decrease in this distance as it approaches region R3. Unlike in the case of delayed group speed, the return of the active node here depends not only on a dominant attractive interaction ($F_{att} > F_{rep}$) in region R3 but also on the presence of a constant spatial phase difference, $\Delta\phi_{ij} = |\phi_j - \phi_i|$, between the active node and the oscillating group. This search for a constant phase difference can be observed in the continuous movement of the active node in the opposite direction to the rotating group (indicate by the blue arrow in Fig. 5(e) and (f)). However, in Fig. 5(g), we observe an attraction and penetration of the active node into the rest of the group, with a tangential orientation to the constant rotational movement, leading to the achievement of rotating phase synchronization (RPS) as shown in Fig. 5(h).

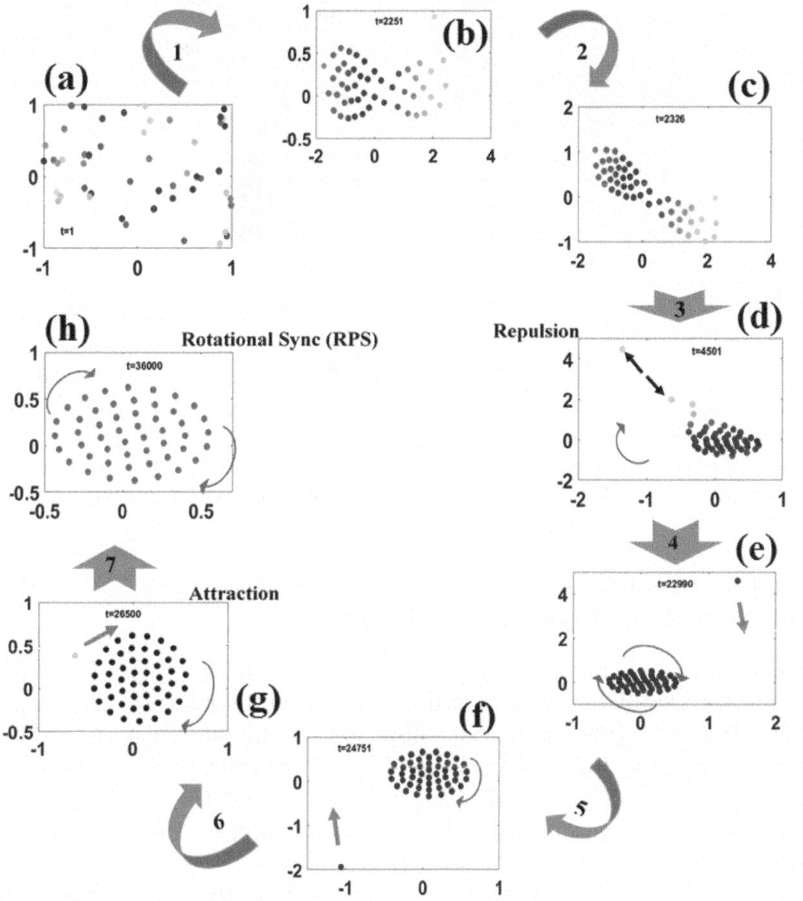

Fig. 5. Snapshots of states showing the evolution toward synchronization throught the attractive-repulsive challenge with the existence of active node (AN) when the delayed individual velocity is applied. (a) Initial distribution of entities; (b) Repulsion with creation of active node (AN); (c, d) Repulsion between active node (AN) and the dynamic sync group; (e, f, g) Attraction between AN with the dynamic sync group; (h) Rotational phase sync (RPS). Plotted for $J = 1, K = 0.0008$ and $\alpha = 1$, the bleu arrow indicate the sens on displacement of the active node (AN) while the red arrow indicate the sens of rotation of the group (see movie 2 in the supplementary material [15]).

We can conclude that achieving a synchronized state, whether under delayed individual speeds or delayed group speeds, requires navigating a simultaneous tension between attractive and repulsive interactions.

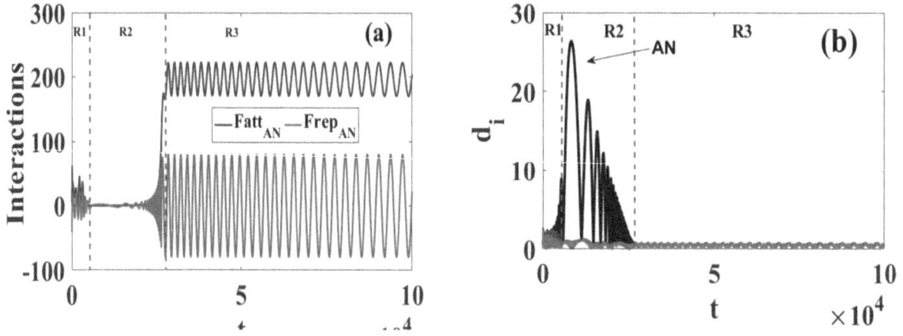

Fig. 6. Time evolution showing the transition to a static phase synchronization when the delayed individual velocity is applied. (a) Interactions, the red color indicates the repulsive interaction while the blue one represents the attractive one between the active node and the other elements; (b) The mean radius of the nodes (the black curve is for the active node (AN)). Plotted for $J = 1, K = 0.0008$ and $\alpha = 1$. (Color figure online)

3.3 Initial Condition Dependence on the Existence of Attractive-Repulsive Challenge

The synchronization transitions observed above for different speed distributions underscore the significant impact of non-zero speed distribution on swarmalator dynamics. However, these transitions appear to be highly sensitive to the choice of initial conditions. Specifically, for a given set of initial conditions, we found that the occurrence of synchronization transitions-and the resulting balance of attractive and repulsive interactions-depends strongly on the initial phase configuration of the entities.

Alternatively, it has been observed that modifying the initial phase condition of the active node alone can successfully adjust its behavior. In other words, this approach enables effective propagation of the phenomenon to the desired element. However, changes to the initial spatial conditions have no impact on the occurrence or progression of this transition toward synchronization.

To extend our observation, we applied perturbations ϵ_1 and ϵ_2 to the initial spatial conditions and internal phases, as given in Eqs. 8 and 9.

$$r_i = r_i(0) + \epsilon_1 \tag{8}$$

$$\theta_i = \theta_i(0) + \epsilon_2 \tag{9}$$

By simultaneously varying ϵ_1 and ϵ_2 in the context of delayed group speed, we find that specific conditions can reveal the presence or absence of an attractive-repulsive dynamic between the active node and the network as a whole. Figure 7 illustrates the domains of existence (E) or non-existence (NE) of synchronization transitions under this dynamic, represented in cyan and blue, respectively. Our observations indicate that perturbations in the spatial variable have minimal

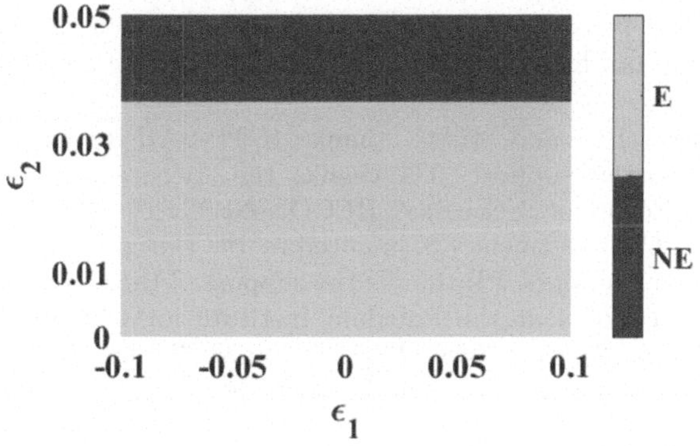

Fig. 7. Domain of existence (E) and non-existence (NE) of a transition to synchronization with attractive-repulsive challenge. Plotted for $J = 1; K = 0.0004$ and $\alpha = 0.3$.

influence on the presence of interaction conflicts. Specifically, applying perturbations within the range $\epsilon_1 \in [-0.1, 0.1]$ does not alter the synchronization transition. However, perturbations in the internal phase variable ϵ_2 have a notable impact: as ϵ_2 increases from 0 to 0.05, it promotes a synchronization transition featuring an attractive-repulsive dynamic, provided that $\epsilon_2 \leq 0.04$. Beyond this threshold, when $\epsilon_2 > 0.04$, this transition ceases to exist.

4 Conclusion

In this study, we investigated the transition to synchronization in a swarmalator system influenced by non-zero individual speeds, an area that has received less attention in previous research. We focused on how these non-zero speeds affect synchronization dynamics, addressing two distinct formulations. First, we examined the scenario where the speed of an element at time t is dependent on the average speed at time $t - \Delta t$, as defined by Eq. 5. Our findings indicate that the transition to synchronization is characterized by a consecutive emergence of attractive and repulsive interactions. This dynamic results in an increase in the internal phase energy of the active node (AN) and culminates in a static phase synchronization state. In our second approach, we explored the dependency of an entity's speed at the present time t on its speed at the previous time $t - \Delta t$, as described by Eq. 6. This formulation introduces a memory component to the behavior of the node. Here, we identified a novel state of dynamic synchronization termed Rotational Phase Synchronization (RPS), which is preceded by a transition that also highlights the attractive-repulsive challenge. Furthermore, we emphasized the significance of initial conditions on the type of transition to synchronization, establishing a domain of existence or non-existence for this behavior. Investigating how initial conditions can influence the number of entities exhibiting this synchronization could provide deeper insights into the dynamics of living systems, such as those observed in flocks of birds or schools of fish.

This line of inquiry presents a promising avenue for future research, as it may enhance our understanding of collective behaviors in biological systems.

Acknowledgments. PL and HAC thank ICTP-SAIFR and FAPESP grant 2021/14335-0 for partial support. TN thanks the "Reconstruction, Resilience and Recovery of Socio-Economic Networks" RECON-NET EP_FAIR_005 - PE0000013 "FAIR" - PNRR M4C2 Investment 1.3, financed by the European Union - NextGenerationEU for the partial support. PL thanks the support of the Deutscher Akademischer Austausch Dienst (DAAD) at the Potsdam Institute for Climate Impact Research (PIK) under the Grant Number (91897150). SJK, TN, PL and MoCLiS research group thank ICTP for the equipment, donated under the letter of donation Trieste 12th August 2021.

Disclosure of Interests. The authors have no competing interests to declare that are relevant to the content of this article.

References

1. O'Keeffe, K.P., Hong, H., Strogatz, S.H.: Oscillators that sync and swarm. Nat. Commun. **8**(1), 1–13 (2017)
2. Lizarraga, J., Aguiar, M.: Synchronization and spatial patterns in forced swarmalators. Chaos **30**(5), 053112 (2020)
3. Kongni, S.J., et al.: Phase transitions on a multiplex of swarmalators. Phys. Rev. E **108**(3), 034303 (2023)
4. Kongni, S.J., Njougouo, T., Louodop, P., Tchitnga, R., Ferreira, F.F., Cerdeira, H.A.: Expected and unexpected routes to synchronization in a system of swarmalators. Phys. Rev. E **110**(6), L062301 (2024)
5. O'Keeffe, K.P., Evers, J.H.M., Kolokolnikov, T.: Ring states in swarmalator systems. Phys. Rev. E **98**, 022203 (2018)
6. O'Keeffe, K., Bettstetter, C.: A review of swarmalators and their potential in bio-inspired computing. In: Micro-and Nanotechnology Sensors, Systems, and Applications XI, vol. 10982, p. 109822E. International Society for Optics and Photonics (2019)
7. Hong, H., Yeo, K., Lee, H.K.: Coupling disorder in a population of swarmalators. Phys. Rev. E **104**(4), 044214 (2021)
8. Sar, G.K., Chowdhury, S.N., Perc, M., Ghosh, D.: Swarmalators under competitive time-varying phase interactions. New J. Phys. **24**(4), 043004 (2022)
9. Barciś, A., Bettstetter, C.: Sandsbots: robots that sync and swarm. IEEE Access **8**, 218752–218764 (2020)
10. Ghosh, S., Sar, G.K., Majhi, S., Ghosh, D.: Antiphase synchronization in a population of swarmalators. Phys. Rev. E **108**(3), 034217 (2023)
11. Hong, H., O'Keeffe, K.P., Lee, J.S., Park, H.: Swarmalators with thermal noise. Phys. Rev. Res. **5**(2), 023105 (2023)
12. Ceron, S., O'Keeffe, K., Petersen, K.: Diverse behaviors in non-uniform chiral and non-chiral swarmalators. Nat. Commun. **14**(1), 940 (2023)
13. Sar, G.K., Ghosh, D., O'Keeffe, K.: Pinning in a system of swarmalators. Phys. Rev. E **107**(2), 024215 (2023)
14. Anwar, M.S., Sar, G.K., Perc, M., Ghosh, D.: Collective dynamics of swarmalators with higher-order interactions. Commun. Phys. **7**(1), 59 (2024)

15. This supplementary file show the movies describing the attractive-repulsive challenge in both cases observed in the snapshots of fig.1 and fig.5. http://www.insertlinkhere.com
16. Hong, H.: Active phase wave in the system of swarmalators with attractive phase coupling. Chaos Interdisc. J. Nonlinear Sci. **28**(10), 103112 (2018)
17. Sharma, A.: Explosive synchronization through attractive-repulsive coupling. Chaos Solitons Fractals **145**, 110815 (2021)

A "Plug and Play" Table for Composable Molecular Neural Networks Built with DNA

Ethan Collinson(✉) and Harold Fellermann

Interdisciplinary Computing and Complex Biosystems Research Group, School of Computing, Newcastle University, Newcastle upon Tyne, UK
e.collinson@newcastle.ac.uk

Abstract. DNA nanotechnology has evolved over the past three decades into a promising field, providing programmable molecular building blocks for various applications in biotechnology and synthetic biology. We developed a modular "plug and play" table to explore and optimise various neural network components and evaluated their performance through simulations. We employ chemical reaction network models to simulate these components, revealing insights into their compatibility and performance metrics. Our findings highlight the impact of architectural choices on network functionality and suggest pathways for future advancements. Ultimately, this research emphasises the potential of scalable, *in-vivo* biological neural networks capable of online learning, opening new avenues for applications in biotechnology, molecular diagnostics, and synthetic biology.

1 Introduction

Since its inception over 30 years ago [1,5], DNA nanotechnology has blossomed into a vibrant research domain. It offers programmable building blocks for molecular construction, computing, sensing, and actuation, with a wide range of promising applications in biotechnology, smart materials, environmental monitoring, molecular diagnostics, and therapeutics [6,10,15,21].

Complementing the toolkit of Synthetic Biology with nucleic acid nanotechnology offers a broad spectrum of potential advantages. For instance, as detailed below, DNA nanocircuits can implement arbitrary chemical reaction systems that are not limited by the on/off dynamics of typical gene regulation mechanisms [16]. These circuits can consist of dozens to hundreds of independently interacting strands, far surpassing the number of transcription factors available to synthetic biologists. Moreover, because the function of these circuits arises from the interactions of nucleic acids, they are expected to impose significantly less burden on host cells compared to mechanisms that necessitate the translation of regulatory proteins [18].

Therefore, it is crucial to explore methods that enable computation within biological systems. Biocompatible models of computation, inspired by biological systems, are designed to integrate seamlessly with living organisms and

T. Carletti et al. (Eds.): WIVACE 2024, CCIS 2532, pp. 236–249, 2025.
https://doi.org/10.1007/978-3-031-93631-9_19

biological processes. These paradigms replicate essential features of biological systems, such as self-organisation, adaptability, and fault tolerance [2]. Investigating these approaches can uncover areas where molecular computing outperforms traditional computing models and identify opportunities for effectively utilising cellular computation [8].

Artificial neural networks (ANNs) have emerged as a cornerstone of modern machine learning, finding numerous real-world applications [20]. Inspired by the biological networks of human brains, ANNs are constructed based on various mathematical models of neurons [9]. Today, deep learning - employing ANNs with multiple hidden layers, known as deep ANNs - stands as one of the most successful approaches in machine learning. This advancement has significantly enhanced applications across various fields, including speech and visual object recognition, drug discovery, and genomics [11]. While most implementations of ANNs have been conducted *in silico*, there have been notable attempts to explore unconventional, non-silicon-based computing platforms for these networks [17].

In this work, we explore the current implementations and future potentials of molecular neural networks as biologically relevant models of computation. These are artificial neural networks constructed using engineered biological components, designed for use in a wide array of synthetic biology applications. Unlike previous studies, our work uniquely focuses on systematically deconstructing molecular neural networks into their fundamental components to identify key design principles. We decided to break down the current implementations of molecular neural networks and performed an investigation to find an idealised molecular neural network design. Within this work, we created a modular "plug and play" table based on seminal molecular neural network motifs to determine whether we could improve the performance of current molecular neural networks. By taking a novel modular approach, we provide insights into optimising molecular neural networks for greater efficiency and scalability. We explore their future possibilities as well as the challenges that need to be addressed to successfully develop a scalable *in vivo* biological neural network capable of online learning.

2 Chemical Perceptron Designs

A perceptron is the basic unit of a neural network and an algorithm used for supervised learning of binary classifiers [13]. The perceptron can take in numerical inputs and weights. The inputs and weights are multiplied together to create the weighted sum. The weighted sum is then compared to the threshold value and activates depending on if the threshold value is exceeded or not. This allows perceptrons to act as binary classifiers which can separate specific inputs into two different groups. Perceptrons can be implemented in multiple different ways; chemical perceptrons use distinct reagents and mixtures to achieve a similar output to computational perceptrons. Most chemical perceptrons' usual output is by using fluorescence or pH because it is easy to observe and measure. The implementation of perceptrons in biochemical networks is extremely varied, with multiple different design approaches used.

There are a wide range of designs for chemical multipliers, with varying degrees of complexity and effectiveness. In its simplest terms, an input can interact with a gate which acts as a multiplier. This reaction approximates a scalar multiplication between input and output equilibria, and has simple reaction kinetics (See Fig. 1(a)). Conversely a simple threshold rectifier consists of a simple threshold gate and reporter gate (See Fig. 2(f)). This reaction uses the hill equation to model sigmoidal behaviour. This allows for the concentration of the rectifier gate to determine, the concentration at which the network would report a result.

The seminal "seesaw" gate [12] uses a toehold exchange reaction where the binding of an input strand releases an output strand and simultaneously opens a second toehold. This second toehold is accessible to a fuel strand which re-releases the input strand back into solution, so that the input acts as a catalyst to the overall reaction (See Fig. 1(b)). The output strand in turn can engage in downstream seesaw reactions. Basic seesaw gates can be combined larger gates to implement particular functions: in particular, the inclusion or omission of threshold gates and fuel strands determines the gate function (See Fig. 2(g)). Amplifying gates require both fuel and threshold to catalytically release output when the input exceeds the threshold gate concentration. Integrating gates require no fuel and threshold, and output the sum of all inputs. Both the multiplier and rectifier subcomponent of the seesaw network was analysed.

Cherry et al. [7] have expanded the above work by introducing pairwise annihilation gates within the rectifier step. These reactions amplify the molecular majority species to the expense of the minorty species (See Fig. 2(j)). The authors utilised these gates to implement a winner-takes-all (WTA) network in the form of a three-layer, fully connected artificial neural network. The output layer displays the classification outcomes for handwritten digits from '1' to '9', highlighting the possibility to embed learning within molecular neural networks. Therefore, it would be interesting to compare the effectiveness of this rectifier compared to their initial rectifier design, as well as the other rectifier designs.

Another molecular network is the DNA switching circuit (DSC) architecture [19]. The input interacts with the gate, resulting in the generation of an output strand (See Fig. 1(e)). Subsequently, this output strand propagates and reaches the next gate. Gates undergo state changes based on the presence of the output strand. Consequently, the ON and OFF states are toggled, effectively representing Yes and No. This flow of the output strand mirrors the passage of a current through the gate of the DSCs. This multiplier method may show interesting results when combined with other architectures.

Another specific architecture that is worth investigating is the weight-loop and weight-race perceptron designs [3]. The weight-loop perceptron operates by transforming weights into output species while preserving the weights' concentrations through backup copies. Inputs are inputted into the system, and catalyse reactions that consume weights and fuel and produces outputs alongside backup weights. After output production, weights are then restored from backups, ensuring no mass is lost. From a functional standpoint, the perceptron operates in a sequential manner: it processes an input, generates an output, restores weights, and ultimately executes a cleanup phase (See Fig. 1(c)). In the weight-race perceptron model, a departure from the weight-loop approach is seen through the reversal of roles for inputs and weights. Here, weights catalyse the input-to-output reactions directly instead of inputs catalysing weight transformations (See Fig. 1(d)). This modification eliminates direct weight comparisons, opting instead for a competitive catalyst-based mechanism where weights vie against each other in input-to-output reactions. This system relies on rapid annihilation of output species, simultaneous weight racing, and comparable rate function shapes (See Fig. 2(i)).

Samaniego et al. [14] put forth an interesting architecture based upon biomolecular neural networks. Theoretically the system demonstrated that biochemical networks, incorporating mechanisms like sequestration or phosphorylation reactions, can effectively mimic ideal ANNs for both linear and nonlinear classification tasks (See Fig. 2(h)). The biomolecular reaction network is capable of performing the perceptron function at equilibrium, which may show interesting results when combined with other architectures. These circuits exhibit potential for reversibility, thereby opening avenues for expanding functionalities within established biological circuits.

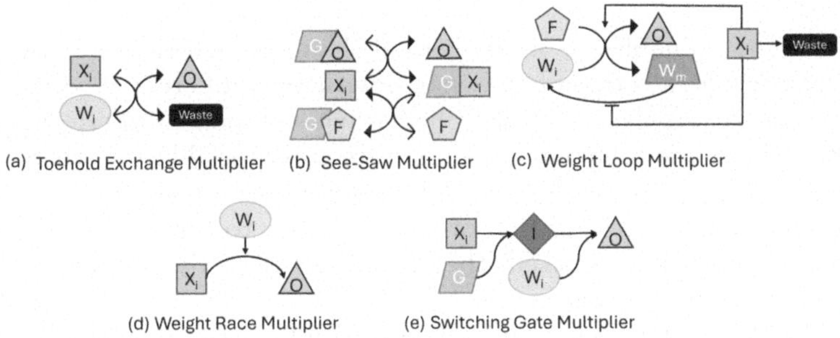

Fig. 1. Summary figure of modular molecular neural network multipliers. The coloured shapes represent unique species within the system. Each system has many different species that can act in different ways. Species with the same colour and shape provide similar functions across each modular design.

3 Chemical Reaction Network Models

For all multiplier and rectifier components within the last section we have developed chemical reaction network models of the desired reaction pathways. Mass action kinetics were applied to derive ordinary differential equations (ODE's) of these systems. ODE's were integrated using SciPy version 1.12.0.

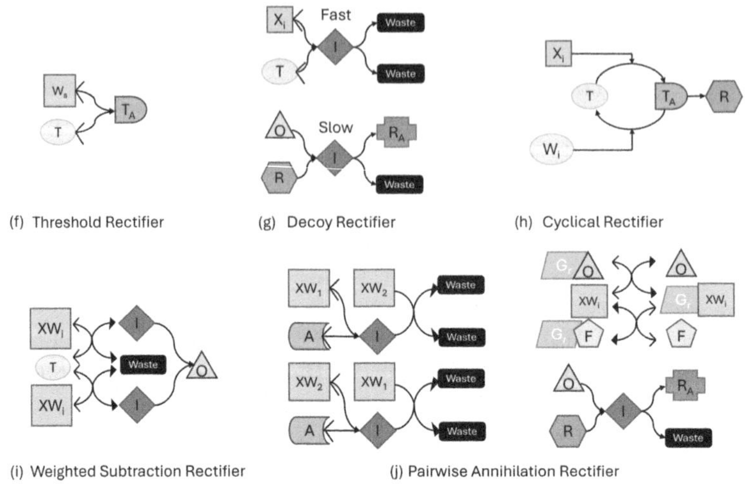

Fig. 2. Summary figure of modular molecular neural network rectifiers. The coloured shapes represent unique species within the system. Each system has many different species that can act in different ways. Species with the same colour and shape provide similar functions across each modular design. (Color figure online)

This simulation enabled us to analyse all the molecular neural network components, measuring the error of the system with each different modular component. Through this analysis, we identified which components performed better than others. Some modular components were incompatible due to fundamental differences in their initial designs. Additionally, certain components required modifications, such as the addition of more species, to function together. As a result, while these system worked, they could not be truly considered modular due to these necessary changes.

To fairly compare the components, we decided to create a standardised result network. Specifically, we developed a two-input network capable of encoding both positive and negative weights. This network would only produce an output when the weighted input exceeded the rectifier's threshold value. Under specific initial conditions, the network is expected to generate certain results. By observing how the modular components cause deviations from these expected values, we can determine the effectiveness of each component in affecting the network's performance. Additionally, this network enabled us to compare input species and determine which output species would be produced. This provided another method to evaluate how the network's performance deviated from the expected results. These metrics were crucial in gauging the compatibility of component subfunctions based on their ability to produce the anticipated output.

The "plug and play" table references the proposed neural network (NN) designs. The "seesaw" gate, introduced by Qian et al. [12], uses a toehold exchange reaction for catalytic input and downstream reactions. This was expanded by Cherry et al. [7] with pairwise annihilation gates to implement a winner-takes-all (WTA) network in a complex, three-layer artificial neural networks. The DNA switching circuit (DSC) architecture by Wang et al. [19] features input interactions generating output strands that propagate to subsequent gates. The weight-loop and weight-race perceptron designs by Banda et al. [3] involve weight transformations to output species while maintaining weight concentrations, with variations in input and weight roles. Finally, Samaniego et al. [14] proposed biomolecular neural networks using sequestration or phosphorylation reactions to mimic ideal ANNs, performing perceptron functions at equilibrium with potential for reversibility and expanded functionalities. We decided to use these networks because they represent some of the most seminal work within the field, additionally these architectures are likely to be robust enough to be broken down and used as modular components within potential composable molecular neural networks.

4 The "Plug and Play" Table

We created the "plug and play" table to systematically analyse and optimise molecular neural networks by breaking down current implementations into their fundamental components (See Fig. 3). This modular approach allows for flexibility in experimenting with different network configurations, facilitating a better understanding of how specific changes impact performance. By leveraging famous molecular neural network motifs, we aimed to improve upon established designs and identify the most effective configurations. In addition to these established designs, we attempted to create the simplest molecular neural network. This simple network prioritised the minimum number of chemical reactions and species, based from our own designs.

We decided to break down these networks into specific subfunctions, primarily the multiplier subfunction (Fig. 1) and the rectifier subfunction (Fig. 2). The multiplier subfunction is crucial for applying weights to input signals, allowing the network to integrate multiple inputs with varying degrees of influence to enable different processing pathways. On the other hand, the rectifier subfunction serves as an activation function, introducing non-linearity into the network, which is essential for modeling complex relationships that linear functions cannot capture. This subfunction also helps in thresholding signals, ensuring that only significant inputs propagate through the network, similar to the role of rectified linear units (ReLU) in ANNs. By incorporating these subfunctions, the molecular neural network gains the ability to effectively weight and integrate inputs while introducing necessary non-linear dynamics and signal thresholding, contributing to the development of a robust and efficient system capable of complex information processing.

Throughout our investigation into different multiplier configurations, we explored their performance in tandem with various rectifier modules aimed at enhancing stability, accuracy, and overall computational efficiency.

Perceptron Components	Toe Hold Exchange Multiplier	See-saw Multiplier	Weight Loop Multiplier	Weight Race Multiplier	Switching Gate Multiplier
Threshold Rectifier	✓	-	-	-	✓
Decoy Rectifier	✓	✓	✓	✓	✓
Cyclical Rectifier	✓	×	-	-	×
Weighted Subtraction Rectifier	✓	✓	✓	✓	✓
Pair Wise Annihilation Rectifier	✓	✓	✓	✓	-

Fig. 3. Summary table of modular molecular neural network subfunctions. Ticks represent the modular components function together within a network. Dashes represent that the modular components function together within a network, but require modifications. Crosses represent that the modular components would not work within a network.

Initial analysis of the toehold exchange multiplier started by trying to find the simplest and smallest number of chemical species and reactions required to build a multiplier component. In its simplest terms, an input can interact with a gate which acts as a multiplier. This reaction approximates a scalar multiplication between input and output equilibria, and has simple reaction kinetics. This specific multiplier design was investigated mathematically. By checking the multiplier design by calculating the maths by hand can help improve the validity of future model simulations. The toehold exchange multiplier demonstrated robust performance when paired with the threshold rectifier, and showed a lower bound to the number of species and reactions required. The decoy rectifier showed superior threshold capabilities, likely due to reporter and threshold interactions. Integration of the cyclical rectifier with the toe hold exchange multiplier succeeded. This multiplier design allowed the cyclical rectifier to find the boundary conditions within the network to show a stable state. Specifically we can see how the system rectifies with different initial input conditions and specific weights. The 3D plot in 4(c) shows the sigmoidal nature of the rectifier when viewed in the format as well. The top-down view also shows the boundary in which the input with overcome the threshold and show a response. This is what we want to see when determining if these modular components work with a molecular network. Additionally, both the weighted subtraction rectifier and pairwise annihilation rectifier worked and showed the desired performance with the toehold exchange multiplier.

The See-saw multiplier functioned effectively with the decoy rectifier, weighted subtraction rectifier, and pairwise annihilation rectifier. These rectifiers were most similar to designs used throughout the established literature networks. The pairwise annihilator rectifier is of particular interest within this

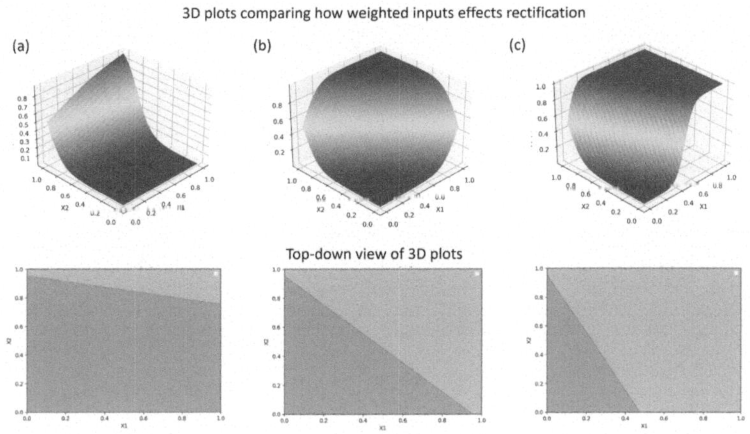

Fig. 4. 3D plots comparing how weighted inputs effects the rectification of a molecular network. **Plot (a):** The weight is set to 0.1. The rectifier doesn't threshold until the combined inputs are large. **Plot (b):** The weight is set to 0.5. The rectifier requires smaller concentration of input present to threshold. **Plot (c):** The weight is set to 0.9. The rectifier thresholds very quickly with small concentrations of input. There is also a period when input values are similar to threshold values and full thresholding behaviour isn't seen. This is due to chemical reactions not being instantaneous.

modular design of network because it inherently allows for handling negative weights and multiple inputs within these networks. Figure 5 shows the two input network that can compare the concentration of input species. This network produces a different output, dependent on the initial conditions of the system. From this we see that the greater the difference between two input concentrations the faster the "major" input produces the desired output. This behaviour holds everywhere, except in the zero-measure space defined by the concentration of inputs being equal and the desired output of the pairwise annihilation circuit is not specified. However, the cyclical rectifier was incompatible with this multiplier, and integrating it with the threshold rectifier required modifications. The cyclical rectifier relies on boundary conditions to operate, whereas the See-saw multiplier sets the system state based directly on initial conditions. As a result, the system lacked a space to reach a stable state. For the threshold rectifier, a minor modification involving the addition of more species enabled functionality; however, this adjustment compromised its complete modularity.

The Weight Loop Multiplier functioned successfully, with the decoy rectifier performing as intended. Both the weighted subtraction rectifier and the pairwise annihilation rectifier achieved the desired performance metrics. The decoy, weighted subtraction, and pairwise annihilation rectifiers worked because they fit within the perceptron's cycle of transforming, restoring, and maintaining weights. Each successfully processed inputs, produced outputs, and preserved weight concentrations as required. This is depicted in Fig. 6, where the effects of the multiplier and rectifier are analysed in a two-input network, capturing both

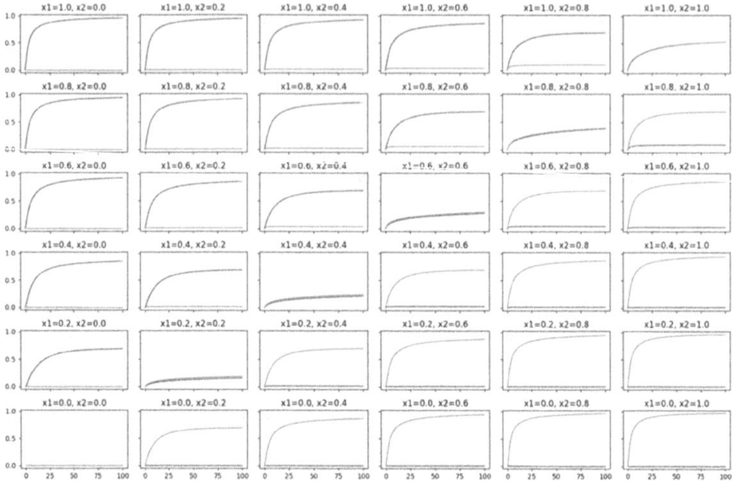

Fig. 5. Repeating simulations for different input concentrations between 0 and 1. Results show that the design is successfully emulating the all-or-none law across all concentrations where the input concentrations aren't equal.

positive and negative weights. As shown, the network produces an output only when the weighted input exceeds the rectifier's threshold, illustrating the influence of the multiplier and rectifier on output generation under standardised conditions. However, the cyclical rectifier and threshold rectifier modules required modifications to function with this multiplier. The cyclical and threshold rectifier did not fully align with the operational specifications of the weight-loop perceptron. Their designs required adjustments to effectively integrate into the perceptron's sequential processing cycle and ensure optimal performance.

The Weight Race Multiplier successfully functioned with the decoy rectifier, while the weighted subtraction rectifier and pairwise annihilation rectifier both demonstrated the desired performance outcomes. Figure 7 demonstrates the distinct effects of a multiplier and a rectifier in a two-input network. It shows how the network encodes both positive and negative weights and identifies the conditions under which the weighted input surpasses the rectifier's threshold, resulting in an output. The figure emphasises the similarities among modular components across various systems, as comparable outcomes have been observed when contrasting the weight loop multiplier with the weighted subtraction rectifier. This supports our assertion that these components are modular and can be utilised to construct composable molecular neural networks. However, the cyclical and threshold rectifier modules required adjustments to fully align with the multiplier's operational specifications. The cyclical and threshold rectifier modules required adjustments to optimise their performance. These modifications were necessary to ensure that all components functioned cohesively within the system.

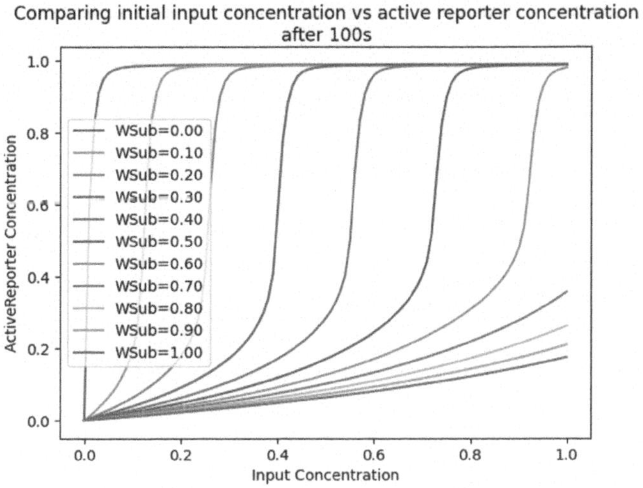

Fig. 6. The distinct effects of a multiplier and rectifier within a two-input network, capable of encoding both positive and negative weights, are examined. The network generates an output only when the weighted input surpasses the rectifier's threshold. Using standardised initial conditions, the influences of the multiplier and rectifier on the output are analysed under varying initial conditions.

The Switching Gate multiplier successfully demonstrated the effectiveness of the decoy rectifier, as well as the weighted subtraction rectifier and threshold rectifier, all of which performed as desired. The comparison between the Switching Gate Multiplier and Decoy Rectifier with an ideal standardised perceptron in Fig. 8 reveals how different modular components influence output within a standardised two-input network. This network evaluates how variations in modular components affect performance, with ideal perceptron responses (in orange) produced when the weighted inputs surpass the rectifier's threshold, leading to a consistent binary output. However, simulated perceptron responses (in blue) display deviations from the ideal, which reflect variations in component behavior. These deviations highlight how differences in input species and subfunction compatibility can significantly impact the network's overall performance. However, the pairwise annihilator modules required modifications to meet the operational specifications of the multiplier, thus compromising its modularity. Unfortunately, the cyclical rectifier did not function as intended as the cyclical rectifier relies on boundary conditions to operate, whereas this multiplier also sets the system state based directly on initial conditions.

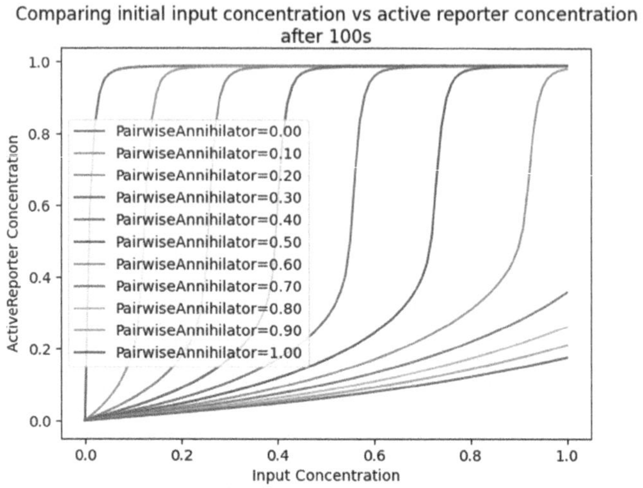

Fig. 7. The differential effects of a multiplier and rectifier within a two-input network capable of encoding both positive and negative weights. The network produces an output only when the weighted input exceeds the rectifier's threshold. Under standardised initial conditions, the multiplier's and rectifier's impacts on the output are observed under different initial conditions.

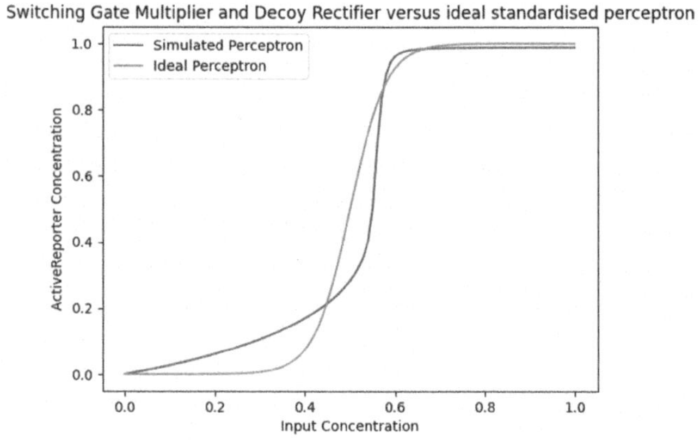

Fig. 8. Comparison of Switching Gate Multiplier and Decoy Rectifier to an ideal standardised perceptron. The standardised two-input network evaluates the effect of different modular components on the output. Ideal perceptron responses (in orange) occur when weighted inputs exceed the rectifier's threshold, producing the expected binary output. Simulated perceptron responses (in blue) demonstrate deviations from ideal outputs based on component variation, illustrating the impact of input species and subfunction compatibility on performance. (Color figure online)

5 Discussion

In summary, our study highlights how architectural design choices significantly impact network performance. These findings underscore the importance of thoughtful design and the potential for further development to enhance these designs and achieve improvements. It is essential to deconstruct the elements that contribute to the scalability of molecular neural networks. The "plug and play" table represents an advancement in the study of molecular neural networks by providing a structured framework for analysing and trying to optimise various network configurations. By systematically evaluating well-established molecular neural network motifs and introducing novel designs, our work facilitates deeper insights into how alterations in specific components can affect overall network performance.

The investigation into different multiplier configurations demonstrated their varied performances in conjunction with various rectifier modules. The analysis of the toehold exchange multiplier provided a foundational understanding of how minimal chemical reactions can achieve effective scalar multiplication. This foundational design offers a low-complexity solution that can be foundational for future model simulations. Our findings indicate that pairing the toehold exchange multiplier with the threshold rectifier yields strong performance, particularly in demonstrating the network's capability to stabilise and accurately respond to varying input conditions. Additionally, the exploration of the see-saw multiplier revealed its compatibility with established rectifier designs, especially the pairwise annihilator rectifier. This design's ability to handle negative weights and multiple inputs is particularly noteworthy, as it expands the network's functional capabilities. However, challenges arose when integrating the cyclical rectifier with the see-saw multiplier, underscoring the importance of initial conditions in establishing stable states within the network. Overall, the seesaw architecture offered a balanced and efficient approach to addressing challenges in various applications with the decoy rectifier playing a role as a highly effective component, despite its relative simplicity. The decoy rectifier was one of the most reliable components as its simple nature allowed for a wide range of versatility. As complexity grew within the molecular neural network, pairwise annihilation rectifier became effective in reducing the number of species required in complex networks. This combination ensured robust performance and resilience, which made the seesaw architecture a preferred choice for simple and complex molecular neural networks.

Our exploration into molecular neural networks reveals both the promising current implementations and the vast future potentials of this innovative technology [4]. Through dissecting existing designs and attempting to find an idealised modular framework, we aimed to enhance the efficiency and functionality of these networks. Moving forward, the development of scalable *in vivo* biological neural networks capable of online learning presents a significant challenge yet offers unprecedented opportunities for synthetic biology applications. Addressing these challenges will be crucial in unlocking the full transformative potential of molecular neural networks in the realm of biotechnology and beyond.

Acknowledgments. All authors have received funding from the United Kingdom Engineering and Physical Sciences Research Council (EPSRC). All authors have contributed equally to the conceptualisation, research and writing of the paper. None of the authors declares any competing interests.

References

1. Adleman, L.M.: Molecular computation of solutions to combinatorial problem. **266**(5187), 1021–1024. https://doi.org/10.1126/science.7973651
2. Amos, M., Dittrich, P., McCaskill, J., Rasmussen, S.: Biological and chemical information technologies. **7**, 56–60. https://doi.org/10.1016/j.procs.2011.12.019. https://www.sciencedirect.com/science/article/pii/S1877050911006843
3. Banda, P., Teuscher, C., Lakin, M.R.: Online learning in a chemical perceptron. **19**(2), 195–219. https://doi.org/10.1162/ARTL_a_00105
4. Braccini, M., Collinson, E., Roli, A., Fellermann, H., Stano, P.: Recurrent neural networks in synthetic cells: a route to autonomous molecular agents? **11**
5. Chen, J., Seeman, N.C.: Synthesis from DNA of a molecule with the connectivity of a cube. **350**(6319), 631–633. https://doi.org/10.1038/350631a0
6. Chen, Y.J., Groves, B., Muscat, R.A., Seelig, G.: DNA nanotechnology from the test tube to the cell. **10**(9), 748–760. https://doi.org/10.1038/nnano.2015.195
7. Cherry, K.M., Qian, L.: Scaling up molecular pattern recognition with DNA-based winner-take-all neural networks. **559**(7714), 370–376. https://doi.org/10.1038/s41586-018-0289-6
8. Grozinger, L., et al.: Pathways to cellular supremacy in biocomputing. **10**(1), 5250. https://doi.org/10.1038/s41467-019-13232-z. https://www.nature.com/articles/s41467-019-13232-z
9. Hasson, U., Nastase, S.A., Goldstein, A.: Direct-fit to nature: an evolutionary perspective on biological (and artificial) neural networks. Neuron **105**(3), 416–434 (2020)
10. Jung, C., Ellington, A.D.: Diagnostic applications of nucleic acid circuits. **47**(6), 1825–1835. https://doi.org/10.1021/ar500059c
11. LeCun, Y., Bengio, Y., Hinton, G.: Deep learning. Nature **521**(7553), 436–444 (2015). https://doi.org/10.1038/nature14539
12. Qian, L., Winfree, E.: Scaling up digital circuit computation with DNA strand displacement cascades. **332**(6034), 1196–1201. https://doi.org/10.1126/science.1200520
13. Rosenblatt, F.: Perceptron simulation experiments. **48**(3), 301–309. https://doi.org/10.1109/JRPROC.1960.287598
14. Samaniego, C.C., Moorman, A., Giordano, G., Franco, E.: Signaling-based neural networks for cellular computation. In: 2021 American Control Conference (ACC), pp. 1883–1890. https://doi.org/10.23919/ACC50511.2021.9482800. ISSN 2378-5861
15. Simmel, F.C., Yurke, B., Singh, H.R.: Principles and applications of nucleic acid strand displacement reactions. **119**(10), 6326–6369. https://doi.org/10.1021/acs.chemrev.8b00580
16. Soloveichik, D., Seelig, G., Winfree, E.: DNA as a universal substrate for chemical kinetics. **107**(12), 5393–5398. https://doi.org/10.1073/pnas.0909380107. http://www.pnas.org/content/107/12/5393

17. Stepney, S.: Programming unconventional computers: dynamics, development, self-reference. Entropy [electronic only] **10** (2012). https://doi.org/10.3390/e14101939
18. Stone, A., Youssef, A., Rijal, S., Zhang, R., Tian, X.J.: Context-dependent redesign of robust synthetic gene circuits. Trends Biotechnol. (2024). https://doi.org/10.1016/j.tibtech.2024.01.003. https://www.cell.com/trends/biotechnology/abstract/S0167-7799(24)00003-9
19. Wang, F., et al.: Implementing digital computing with DNA-based switching circuits. **11**(1), 121. https://doi.org/10.1038/s41467-019-13980-y
20. Xu, Y., et al.: Artificial intelligence: a powerful paradigm for scientific research. Innovation **2**(4), 100179 (2021). https://doi.org/10.1016/j.xinn.2021.100179
21. Zhang, J., Huang, Y., Sun, M., Wan, S., Yang, C., Song, Y.: Recent advances in aptamer-based liquid biopsy. **5**(5), 1954–1979. https://doi.org/10.1021/acsabm.1c01202

Educational Robotics: From Computational to Complexity Thinking

Onofrio Gigliotta[1(✉)], Michela Ponticorvo[1], Erica Chinzer[1,2], Clara Nobile[1], Diana Giorgini[3], and Alessandra Vitanza[3]

[1] University of Naples, Federico II, Naples, Italy
onofrio.gigliotta@unina.it
[2] University of Macerata, Macerata, Italy
[3] ISTC-CNR, Rome, Italy

Abstract. Since the seminal work of Seymour Papert, Educational Robotics (ER) has spread across various educational curricula. ER embodies Papert's constructionism which stresses the importance of learning by making things. According to Papert and Piaget children construct knowledge through an interactive process with the surrounding environment. ER has been widely utilized to foster Computational Thinking (CT), a strategy of problem solving that borrows concepts from computer science. Ironically, Papert warned us about the risk of embracing a specific (programming) language, as it can favor one way of thinking over another. Computer science, however, is a very young discipline, and humans have been coping with complex problems for ages without the possibility to resort to it. In this paper, we describe two bioinspired approaches which have used robotics to teach how to evolve and understand the behavior of groups of robots. The idea we support is that by incorporating Complexity Thinking (CxT) into ER, students not only learn how to decompose tasks (as in CT) but also how to navigate and adapt to complex, decentralized systems, similar to the behavior observed in nature. In this way, the aim is to enrich the learning experience by fostering a deeper understanding of the underlying dynamics.

Keywords: Evolutionary Robotics · Swarm Robotics · Educational Robotics

1 In the Beginning was the Turtle

1.1 The Turtle

Educational Robotics (ER) began with the pace of a robotic turtle (see Fig. 1). Designed and developed by Seymour Papert and colleagues at MIT in the '70s, the Turtle was thought of as a 'transitional object' [1]. In Papert's mind, transitional objects or systems are tangible things able to capture ideas: "the creation of a transitional object, the Turtle, that can exist in the child's environment and make contact with the ideas" [1]. In this sense, the tangible system—the robot—embodies ideas, spanning from math to physics and the like. Moreover, the role of the Turtle was to complement the educational programming language LOGO (derived from LISP [2]), developed by Papert and colleagues to foster a philosophy of education based on Piaget's Constructivism, which later evolved into Constructionism.

T. Carletti et al. (Eds.): WIVACE 2024, CCIS 2532, pp. 250–260, 2025.
https://doi.org/10.1007/978-3-031-93631-9_20

In fact, Papert had the chance to collaborate with the renowned psychologist Jean Piaget—one of the leading proponents of Constructivism. This theory posits that knowledge acquisition occurs through an active process where learners construct new concepts based on their prior knowledge and direct experiences [3]. Papert advanced this perspective by introducing Constructionism, which asserts that the creation of tangible artifacts by learners not only serves as a motivational factor but also enables them to observe the concrete outcomes of their efforts and reflect on the strategies employed in problem-solving. In this context, the use of technology, particularly educational robotics, emerges as an effective means of supporting the learning process [3, 4].

The theory of Constructionism and the learning by design approach, which emphasize the significance of experiential and active learning, are closely linked to the principles of Constructivism [5, 6]. Papert [1] also stressed the importance of fostering computational thinking from early childhood, not only to meet the growing demands of technological literacy in modern society but also to structure cognitive processes, stimulate creativity, and enhance abstract reasoning skills.

1.2 The Turtle and Its Successors

Turtle robots, however, represented a significant investment. The Terrapin Inc. model featured in BYTE Magazine (Vol. 3, No. 7, p. 71, 1978) was sold for $500, approximately $2,420 today. Fortunately, in the late '90s, Lego developed and distributed a more affordable educational robotic kit, based on the work of Papert, and non-incidentally named Lego Mindstorms after the successful book of the educational scientist [1]. Soon after, educational robotics kits flourished in the market with solutions varying in price and complexity, finding fertile soil in STEM (Science, Technology, Engineering and Math) education. Lego kits have been effectively used to demonstrate mathematical concepts [7], engineering [8], computer programming, geospatial concepts, and more [9]. Robotics represents an interdisciplinary field able to appeal to and excite students [10]. These aspects explain why this field has attracted so much attention in education.

1.3 Hard and Soft Skills

Educational robotics offers a powerful, hands-on approach to teaching complex hard and soft skills such as programming and computational thinking. By engaging students in the analysis, design, and programming processes inherent in traditional robotics, this discipline fosters active learning and deeper understanding. These hard skills are crucial for future success and should be nurtured from the earliest stages of education [11]. However, to ensure a comprehensive and interdisciplinary learning experience, it is essential to integrate these technical skills with the development of *soft skills*. The concept of *soft skills* is multidimensional, encompassing a broad set of transversal abilities that involve emotional, behavioral, and cognitive aspects. Moreover, the use of robotics in collaborative settings fosters the development of essential skills such as communication, teamwork, and leadership [12–16]. These abilities play a critical role in managing interpersonal dynamics and problem-solving, distinguishing them from the technical competencies that define *hard skills* [13, 14]. While the transmission of technical knowledge through educational robotics remains a primary objective, this paper

Fig. 1. Seymour Papert with a turtle robot

takes a broader perspective, exploring the role of educational robotics as a tool to teach how to manage complex systems and the related uncertainty by harnessing their self-organizing properties. In an era marked by distributed intelligence and complexity, this soft skill, i.e., the ability to leverage the peculiar dynamics of such systems, becomes crucial [17].

1.4 Fostering Complexity Thinking

By putting the emphasis on coding, many ER applications focus on fostering Computational Thinking (CT), a problem-solving strategy based on the concepts fundamental to computer science [18]. Hence, CT can be viewed as a soft skill notwithstanding the overlap with computer programming. In a world full of digital technologies, on one hand, we believe that CT is a powerful tool for understanding and learning; on the other hand, we strongly believe that thinking devices should embrace complexity, as the world is also crowded with complex systems. Denning and Tedre [19] emphasize that CT requires specific domain knowledge for its effective application. Conversely, CxT enables users of ER based on complex systems to simultaneously acquire management skills and domain knowledge, mirroring Piaget's concept of learners as "little scientists".

Holland [20] defines a complex system as an interconnected set of parts able to exhibit a property called emergence: the behavior of the system is more than the sum of the behavior of the single parts.

Complexity Thinking (CxT) is based on concepts related to physical and social sciences, including dynamic systems, self-organization, adaptive systems, emergence, and others. Examples include the behavior of social groups, swarms of insects, school of fish, weather patterns, evolution, etc.

In this paper, we present two applications developed to foster CxT: in the first one, we present an Evolutionary Robotics platform to link the digital and physical world:

Breedbot and its epigons [21, 22]; in the second, the power of a robot swarm as an educational tool [23].

2 Evolutionary Robotics

Nature is a forge of problems and solutions, and very often evolution is at the origin of both. Evolutionary Robotics (EvR) is inspired by the Darwinian principle of selective reproduction with mutation of the fittest [24]. The fittest individuals produce more offspring than others. This principle, captured by Genetic Algorithms, can be applied iteratively to the design of specific behaviors. Humans, long before they understood the mechanisms of natural evolution, found a way to harness its power by artificially selecting the fittest animals for their own purposes: to produce more milk and wool, or to become more docile.

Evolutionary Robotics (EvR) facilitates the investigation of the interaction between artificial organisms and their environment, as well as the emergent behaviors that arise from this dynamic relationship. It is based on the premise that organisms, through continuous interaction with their surroundings, autonomously develop the capacity to structure their actions in response to adaptive feedback from the environment. The initial step involves equipping artificial organisms with a sensorimotor system, comprising sensory modules, actuators, and a control architecture—typically an artificial neural network—that integrates and regulates their interactions [25].

EvR is grounded in the principles of Genetic Algorithms [26]. Each evolutionary experiment begins with an initial population of robotic agents, whose defining characteristics—such as control architecture and morphological traits—are encoded within a randomly generated artificial genotype. Each genotype undergoes a decoding process, mapping it onto its corresponding phenotype. The resulting robotic agents, whether physical or simulated, are then deployed within an artificial environment, where their emergent behavior is assessed according to a predefined fitness function, established by the experimenter.

The fittest individuals are allowed to generate the next population of robots. Reproduction may occur through either asexual or sexual mechanisms. In both cases, the Genetic Algorithm duplicates the genotypic information of the parent robot (asexual reproduction) or recombines the genetic material from two parent robots (sexual reproduction). Throughout this process, genetic operators introduce stochastic modifications (mutations) into the offspring's genotypes, fostering genetic diversity. The progeny of the first generation constitutes the second generation of robots, initiating the next cycle of evolutionary progression [25].

This iterative evolutionary process is repeated over multiple generations, progressively refining the behavioral capabilities of the robotic agents until they exhibit the target behaviors specified by the experimenter. The methodology of Evolutionary Robotics can be schematically represented as an iterative optimization process that harnesses artificial evolution to achieve adaptive robotic behavior [25].

2.1 Breedbot and Its Epigons

From an educational perspective, robots can be conceptualized as tools for developing complex skills related to STEM disciplines [27]. In this context, robots are meticulously designed and programmed, constituting the culmination of an engineering process. Alternatively, robots can be envisioned analogously to human beings, as organisms—systems that must dynamically adapt to their environment and respond to both external and internal stimuli, thereby representing complex systems that must be governed [28]. To facilitate the development of transversal competencies, including the mastery of complexity, Breedbot has been employed.

Breedbot [11] and its epigons, Brainfarm and Teach E-AI 2C Robotic Farm [29] are integrated software/hardware platforms able to connect simulated environments to real counterparts. These platforms are designed for the "breeding" of robots and aimed at fostering the development of skills in managing complex systems by allowing users to set up a custom world, with obstacles and walls and to train/breed robots without writing a single line of code. Indeed, no computer programming skills are required. Each robot is controlled by an artificial brain (a neural network) whose parameters are encoded into a genetic string. Users can iteratively select individuals with the behavior they prefer the most. Once satisfied, users can download the evolved artificial brain into the real robot. This evolutionary process not only allows users to design better control systems but also helps them uncover the power of the transformational ideas behind the scientific theory of evolution.

The Brainfarm laboratory (Fig. 2, left) represents an innovative approach to robotics, moving away from traditional programming approaches in favor of artificial intelligence methodologies that lead to emergent solutions. This platform, developed on the basis of Breedbot [11], an evolutionary robotics application for a robot built with Lego Mindstorms and equipped with custom IR sensors, facilitates the training and evolution of virtual robots, which can subsequently be tested in physical environments. The software is equipped with a 3D interface reminiscent of a video game and permits either automated or user-guided evolution of robotic agents [30]. The system comprises a population of nine wheeled robots, each equipped with infrared sensors for obstacle detection and dual motors for wheel speed regulation.

Users can choose between a simple feed-forward neural network and more sophisticated architectures to regulate the robots' behavior. The parameters of the network (e.g., synaptic weights) are encoded into a genetic string, which serves as the foundation for the evolutionary process. This process begins with a randomly generated population of genetic strings, subjected to iterative cycles of selection, reproduction, and mutation. Users may assume the role of breeders, selecting the most effective or preferable agents and thereby guiding the evolutionary trajectory. This process culminates in a new generation of robots, closely resembling the selected predecessors but incorporating specific modifications. While the robots' morphology remains static throughout, the parameters of the artificial neural networks are dynamically adjusted, refining the artificial brain in terms of connectivity. Students typically validate their optimized robots by transferring the simulated control systems to physical robots in real-world environments [31].

This platform has been employed to convey evolutionary biology concepts [32]. However, it has also proven highly effective in developing transversal competencies,

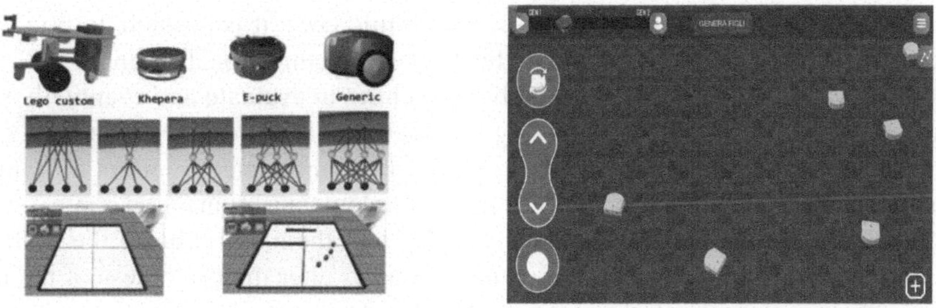

Fig. 2. *Left*: Brainfarm empowers users to determine the physical configuration (i.e., the robot's morphology), design the control architecture (i.e., the artificial neural network), configure the environment, select the mode of learning (phylogenetic or ontogenetic), and ultimately deploy the optimized neural controller onto a corresponding physical robot [30]. *Right*: Teach E-AI 2C Robotic Farm

enabling learners to effectively manage complex systems. The Brainfarm laboratory has enhanced students' abilities to articulate ideas effectively, foster constructive interpersonal relationships, and develop collaborative solutions [31], Teach E-AI 2C Robotic Farm (Fig. 2, left) it is the latest version with a revamped graphic design and a new supported robot, the Thymio.

3 Educational Robot Swarms

A single ant cannot build a bridge, but a swarm can. Collective intelligence, in fact, emerges from the complex interactions among individual intelligences [33]. In educational contexts, this concept can be translated into innovative learning experiences, where multiple robots can support the learning of complex concepts such as cooperation and collective decision-making [34–36]. By interacting with robots, learners can observe how collective intelligence works in practice. For example, in [23], Thymio robots are used to illustrate two fascinating educational case studies that highlight the principles of cooperation: 1) the house-hunting behavior of robot bees, which illustrates how bees as a swarm make joint decisions for the benefit of the group, and 2) the stick-pulling task, in which robots have to cooperate to move an object together, demonstrating the importance of synchronized action and teamwork. These examples are interesting not only because they enhance students' understanding of robotic systems but also because they offer valuable lessons on the dynamics of collaboration and problem-solving in real-world scenarios.

3.1 House-Hunting Robots

When searching for a new nest, bees engage in a fascinating process of collective evaluation: when multiple sites are discovered they must collectively assess different solutions. To move to the new nest, bees must ultimately agree on a single solution. Collective consensus decision making is, indeed, a ubiquitous problem in nature [37], spanning from ants to humans, and it plays a crucial role in survival and, more broadly, in efficiency.

In this context, using Thymio robots [38] that mimic bees, it is possible to provide an innovative way to explore and illustrate this concept in educational settings.

In the developed demo, 8 Thymio robot bees have to evaluate and eventually select the best nest from 4 available options (Fig. 3). Each nest is represented by a light whose brightness, reflecting the *quality* of the nest, can be manipulated by users to influence the robots' decision-making process. The robots implement a bioinspired control system based on the real behavior of bee swarms, allowing them to replicate the collective decision-making process observed in nature. As users alter the brightness of the nests, they can observe how these changes affect the collective decision of the group. This dynamic interaction allows the illustration of principles of collective intelligence, such as spatial heterogeneities and the formation of consensus clusters, and enables students to grasp key concepts and gain insights into how groups can reach informed decisions despite the limited rationality of individual agents.

Moreover, the Thymio robots provide a hands-on, tangible learning experience that makes abstract concepts of decentralized control and swarm intelligence more accessible and engaging. By examining how the robots arrive at a consensus, learners not only understand the mechanics of collective decision-making but also explore broader concepts of cooperation, problem-solving, and the power of distributed intelligence in real-world scenarios. Educational parallelism is a natural result.

Fig. 3. Demonstration of house-hunting robots

3.2 Stick-Pulling in Robots

The cooperative pulling paradigm requires two or more animals to cooperate in order to solve a problem that cannot be accomplished by a single individual. This concept has been widely studied in both natural and artificial systems to understand the underlying mechanisms of coordination and mutual assistance [39]. In this example, the setup simulates a benchmark robotics scenario where many sticks are randomly distributed in the environment, and robots may find them through a random walk. An effective strategy consists in pulling a stick whenever a robot finds one and then waiting for another robot to assist in completing the task, that is, to collaborate. However, it is clear that the success of this strategy is highly dependent on timing and coordination. So, the *waiting time* is a critical parameter that directly influences the possibility and level of cooperation among robots. To explore these dynamics, this parameter is a variable that users can manipulate to affect behavior. A short waiting interval can hinder collaboration, as the first robot may abandon position before another arrives, leading to failed attempts. Conversely, a longer interval ensures more opportunities for cooperation but may reduce overall efficiency by increasing idle time. Finding the right balance is essential, but it depends on spatial and task-specific variables.

In this demonstration, some Thymio robots act as Thymio-Beacons and represent the stick (i.e., a collaboration site), while others are Thymio-Workers and actively seek the sites (Fig. 4). When a sufficient number of Thymio-Workers ($n = 2$) gather at a given location (i.e., near a Thymio-Beacon), the *freedom criterion* is met, and the Thymio-Beacon turns green, indicating a successful collaboration. Meanwhile, the Thymio-Workers continue searching for Thymio-Beacons and wait for a designated time period before resuming their exploration to find new sites. This waiting period is crucial, and it can be set by the users to influence the outcomes. By variating it, users can observe how different strategies emerge and how they impact cooperation, efficiency and success rate.

From an educational point of view, this setup allows learners to experiment with fundamental principles of task allocation, coordination constraints, and decision-making in distributed systems. The hands-on, interactive modality based on interactions helps make abstract concepts more tangible by enabling students to analyze real-world behaviors and reinforcing both their understanding of technical concepts and their problem-solving skills.

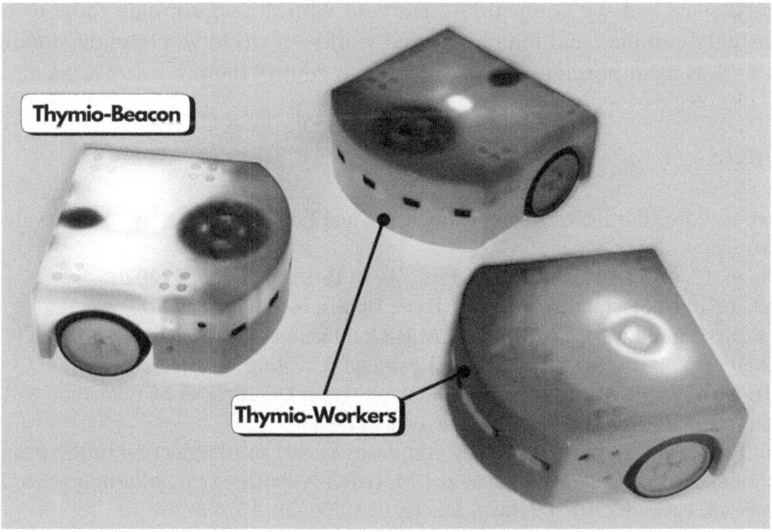

Fig. 4. Demonstration of stick-pulling

4 Conclusion

Computational Thinking is a powerful tool for solving problems using principles borrowed from computer science. However, in a world increasingly crowded with complexity—where interactions between artificial intelligent agents, decentralized decision-making, and emergent complex behaviors are becoming more prevalent—teaching how to manage and harness this complexity represents a complimentary and powerful resource alongside CT. In this near future, we will likely manage an even more complex ecosystem by relying on CxT rather than CT. While CT focuses on the decomposition of problems into simpler steps, CxT provides the tools needed to understand and manage the unpredictable dynamics that emerge from complex systems. This concept is particularly evident in the context of recent artificial intelligence techniques. Consider, for example, how people these days try to manipulate prompts for Large Language Models, an intensive trial-and-error search.

As demonstrated through the robotics educational experiences discussed here, learning how to navigate complexity is crucial. For instance, students engaging with collective intelligence principles begin to comprehend how decentralized systems self-organize, how emergent behaviors arise from simple rules, and how collaboration can be optimized despite discrepancies. These lessons extend beyond robotics and have implications for both hard and soft life skills. In an era of multi-agent systems and AI-driven ecosystems, mastering the art of complexity management will be a paramount skill.

The educational robotics experiences described in this paper show how autonomous robots can be very easily approached from learners, in Papertian terms the floor is very low. Users interact with those robots as if they were animals and by interacting they can discover regularities and finally build knowledge as novel scientists.

In conclusion, we consider CxT a complementary skill to CT. Only by embracing both perspectives can we equip future learners with the appropriate tools to navigate an increasingly complex and interconnected world—a world where understanding how systems work is as important as knowing how to control them.

References

1. Papert, S.: Mindstorms: Children, Computers, and Powerful Ideas. Basic Books, Inc., USA (1980)
2. Solomon, C., et al.: History of logo. Proc. ACM Program. Lang. **4** (2020)
3. Piaget, J.: To Understand is to Invent. Basic Books, New York (1974)
4. Papert, S., Harel, I.: Constructionism. Ablex Publishing Corporation, New York (1991)
5. Bonwell, C.C., Eison, J.A.: Active Learning: Creating Excitement in the Classroom. ASHE-ERIC Higher Education Report, Washington DC: School of Education and Human Development, George Washington University (1991)
6. Cope, B., Kalantzis, M.: The things you do to know: an introduction to the pedagogy of multiliteracies. In: Cope, B., Kalantzis, M. (eds.) A Pedagogy of Multiliteracies. Palgrave Macmillan, London (2015). https://doi.org/10.1057/9781137539724_1
7. Ortiz, A.M.: Examining students' proportional reasoning strategy levels as evidence of the impact of an integrated LEGO robotics and mathematics learning experience. J. Technol. Educ. (2015). https://doi.org/10.21061/jte.v26i2.a.3

8. Kaloti-Hallak, F., Armoni, M., Ben-Ari, M.: The effect of robotics activities on learning the engineering design process. Inform. Educ. **18**, 105–129 (2019). https://doi.org/10.15388/infedu.2019.05
9. Benitti, F.B.V.: Exploring the educational potential of robotics in schools: a systematic review. Comput. Educ. **58**, 978–988 (2012). https://doi.org/10.1016/j.compedu.2011.10.006
10. Robotics in Education: Proceedings of the RiE 2023 Conference. Springer, Cham (2023). https://doi.org/10.1007/978-3-031-38454-7
11. Learning Robots discover and teach AI with the AlphAI robot. https://learningrobots.ai/. Accessed 15 Apr 2024
12. Merdan, M., Lepuschitz, W., Koppensteiner, G., Balogh, R.: Robotics in Education. Research and Practices for Robotics in STEM Education. Springer, Heidelberg (2016)
13. Touloumakos, A.K.: Expanded yet restricted: a mini review of the soft skills literature. Front. Psychol. **11**, 2207 (2020) https://doi.org/10.3389/fpsyg.2020.02207. PMID: 33013574; PMCID: PMC7500090
14. Caudron, S.: The hard case for soft skills. Workforce **78**(7), 60–64 (1999)
15. Strange, W., Bacolod, M., Blum, B.: Urban interactions: soft skills versus specialization. J. Econ. Geography **9**, 227–262 (2009). https://doi.org/10.1093/jeg/lbn057
16. Rubinacci, F., Ponticorvo, M., Passariello, R., Miglino, O.: Robotics for soft skills training. Res. Educ. Media **9**, 20–25 (2017). https://doi.org/10.1515/rem-2017-0010
17. Thurlow, N., et al.: The Flow System Guide. Independently Published (2019)
18. Voogt, J., Fisser, P., Good, J., Mishra, P., Yadav, A.: Computational thinking in compulsory education: towards an agenda for research and practice. Educ. Inf. Technol. **20**, 715–728 (2015)
19. Denning, P.J., Tedre, M.: Computational thinking: a disciplinary perspective. Inform. Educ. **20**, fasc. 3, 361–390 (2022)
20. Holland, J.H.: Complexity: A Very Short Introduction. Oxford University Press, Oxford (2014)
21. Miglino, O., Gigliotta, O., Ponticorvo, M., Nolfi, S.: Breedbot: an evolutionary robotics application in digital content. Electron. Libr. **26**, 363–373 (2008)
22. Miglino, O., Gigliotta, O., Ponticorvo, M., Nolfi, S.: Breedbot: an edutainment robotics system to link digital and real world. Lecture Notes in Computer Science (including subseries Lecture Notes in Artificial Intelligence and Lecture Notes in Bioinformatics), vol. 4693, pp. 74–81 (2007). https://doi.org/10.1007/978-3-540-74827-4_10
23. Vitanza, A., Rossetti, P., Mondada, F., Trianni, V.: Robot swarms as an educational tool: the Thymio's way. Int. J. Adv. Rob. Syst. **16**, 1729881418825186 (2019). https://doi.org/10.1177/1729881418825186
24. Nolfi, S., Bongard, J., Husbands, P., Floreano, D.: Evolutionary robotics. In: Springer Handbooks, pp. 2035–2068. Springer, Cham (2016). https://doi.org/10.1007/978-3-319-32552-1_76
25. Ponticorvo, M., Walker, R., Miglino, O.: Evolutionary Robotics as a Tool to Investigate Spatial Cognition in Artificial and Natural Systems (2006) https://doi.org/10.4018/9781599041117.ch007
26. Mitchell, M.: An Introduction to Genetic Algorithms. MIT Press, Camrbdige (1996)
27. Eguchi, A.: RoboCupJunior for promoting STEM education, 21st century skills, and technological advancement through robotics competition. Robot. Auton. Syst. **75**, Part B 692–699 (2016) https://doi.org/10.1016/j.robot.2015.05.013. ISSN 0921-8890
28. Marchetti, A., Massaro, D. (eds.): Robot sociali e educazione. Interazioni, applicazioni e nuove frontiere, Raffaello Cortina Editore, Milano 240 (2023). https://hdl.handle.net/10807/249828

29. Nobile, C., Marocco, D., Ponticorvo, M., Gigliotta, O.: Teaching embodied artificial intelligence to children (Teach E-AI 2C): an educational proposal for young learners. In: 2024 IEEE International Conference on Metrology for eXtended Reality, Artificial Intelligence and Neural Engineering (MetroXRAINE), pp. 879–883 (2024)
30. Gigliotta, O., Giancarlo, P., Massimiliano, S.: BrainFarm: Cervelli Robotici a Portata Di Mouse. BrainFarm: Cervelli Robotici a Portata Di Mouse (2011)
31. Rubinacci, F., Ponticorvo, M., Gigliotta, O., Miglino, O.: Breeding robots to learn how to rule complex systems. In: Merdan, M., Lepuschitz, W., Koppensteiner, G., Balogh, R. (eds.) Robotics in Education. Advances in Intelligent Systems and Computing, vol. 457. Springer, Cham (2017). https://doi.org/10.1007/978-3-319-42975-5_13
32. Miglino, O., Rubinacci, F., Pagliarini, L., Lund, H.: Using artificial life to teach evolutionary biology. Cogn. Process. **5**, 123–129 (2004). https://doi.org/10.1007/s10339-004-0009-z
33. Bonabeau, E., Dorigo, M., Theraulaz, G.: Swarm Intelligence: From Natural to Artificial Systems. Oxford University Press, Oxford (1999)
34. Mitnik, R., Recabarren, M., Nussbaum, M., et al.: Collaborative robotic instruction: a graph teaching experience. Comput. Educ. **53**(2), 330–342 (2009)
35. Vitanza, A., Rossetti, P., Trianni, V.: Robotica educativa e decisioni di gruppo. Pedagogika.it: "La Robotica Educativa" 24–31 (2019)
36. Vitanza, A., Trianni, V.: La robotica di sciame come nuovo paradigma educativo per l'insegnamento delle competenze trasversali. In: Grimaldi R., a cura di, La società dei robot, Mondadori Università, Milano, pp. 275–285 (2022)
37. Reina, A., Valentini, G., Fernández-Oto, C., Dorigo, M., Trianni, V.: A design pattern for decentralised decision making. PLoS ONE **10**, 1–18 (2015)
38. Mondada, F., Bonani, M., Riedo, F., et al.: Bringing robotics to formal education: the Thymio open-source hardware robot. IEEE Robot. Autom. Mag. **24**(1), 77–85 (2017)
39. Ijspeert, A.J., Martinoli, A., Billard, A., et al.: Collaboration through the exploitation of local interactions in autonomous collective robotics: the stick pulling experiment. Auton. Robot. **11**(2), 149–171 (2001)

An Ecological Model of Cancer Development and Control

Elpida Tzafestas[(✉)]

Laboratory of Cognitive Science, Department of History and Philosophy of Science, National and Kapodistrian University of Athens, University Campus, Ano Ilisia, 15771 Athens, Greece
etzafestas@phs.uoa.gr

Abstract. Cancer may be viewed as a collapse of equilibrium in a cellular ecology, where one subpopulation of atypical cells grows abnormally to the detriment of the others and drives the system to premature death. We present an abstract agent-based model of a cellular ecology comprising two distinct cell types that are interconnected in a way that allows a sustainable population equilibrium. We show that some kinds of genetically controlled mutations may cause cancer and how accumulation of deleterious mutations in atypical cells may lead to aggressive cancer. We also show the effects of treatment with a toxic nutrient which affects atypical cells more than typical ones. In that case, we show that cancer may be constrained or healed depending on the toxicity of the nutrient, the age of cancer attack, the time elapsed before cancer identification and the duration of the treatment. These results are in accordance with the properties of real cancerous systems and thus our model may be extended to support further research on cancer development in cellular ecologies and means of fight against it. An additional Hayflick-like limit of reproduction is finally shown to lead to more robust cellular behavior and to resistance to accumulation of deleterious mutations.

Keywords: Cancer development · Cellular ecology · Toxic cancer treatment · Hayflick limit

1 Introduction

Cancer is the phenomenon where a subpopulation of cells within a multicellular organism grows abnormally fast compared to the other cells and may potentially lead the organism to premature death. The phenomenon is attributed to a number of genetic anomalous or deleterious mutations that a cell may carry or develop. Hanahan and Weinberg [1, 2] have summarized a set of 'hallmarks of cancer' that are mutations found in cancerous cells. These hallmarks have been simulated with an agent-based model by Abbott and colleagues [3]. A number of mathematical studies, based for the most part on differential equations or on hybrid approaches, also abound [4, 5].

Most of these models define and study a single cell population and its growth rates. Instead, we regard the organism as an ecology, where various 'species' of cells interact and coexist in equilibrium. This equilibrium may be perturbed by so-called atypical cells, whose behavior is abnormal. The cells' behavior depends not only on genetic factors,

T. Carletti et al. (Eds.): WIVACE 2024, CCIS 2532, pp. 261–275, 2025.
https://doi.org/10.1007/978-3-031-93631-9_21

but also on the environment they are in and on the way their metabolism works. Many phenomena related to cancer bear striking similarities with ecological ones, like tumor metastasis with species migration, eradication of cancer with extinction of species and others. Most notably, it appears that a tumor microenvironment has a protective role and functions as an ecological niche for the cancer cells by allowing specific types of mutualisms reminiscent of those in ecology (cf. Reynolds et al. [6] and Somarelli [7]). In that spirit, we study some genetically controlled factors that affect the interactions between cells and how mutations in some factors may lead to cancer. This aligns with the observation that not all mutations are deleterious and lead to cancer and not all types of cells have such cancerous potential. A number of controls (treatment regimes) that act on the metabolic function of atypical cells may allow us to heal such a system, by bringing it back to equilibrium. With our abstract cellular ecology model we purport to reproduce the phenomenon of cancer in a way that will allow for further insights of ecological inspiration concerning on the one hand the nature of cancer and how it develops and on the other hand new means with which it might be possible to cure it. Our approach is also loosely inspired by the breakthrough work of Kaneko and colleagues [8, 9] that study interactions and differentiation in cellular populations and by philosophical investigations that propose synthetic views of the phenomenon [10]. Section 2 presents the basic model of the ecology of cells and Sect. 3 describes the simulation setup. Section 4 introduces the atypical cells and shows which gene mutations are deleterious and the effect of the accumulation of mutations. Section 5 studies toxic treatment, while Sect. 6 introduces a cellular reproduction limit. Finally, Sect. 7 concludes with a discussion of the results and their future perspectives.

2 Model of an Ecology of Biological Cells

We have defined a population of cells of two types that are interdependent in a way that allows them to stay at equilibrium. This is therefore an ecological equilibrium that corresponds to a multicellular organism, where the cell types play the role of different species. The cells move randomly in the environment, they gain energy by consuming chemical substances found in it and they produce other substances when their energy exceeds a production level. Type 1 cells can consume external chemicals of type Nutrition 1 and Nutrition 2 while type 2 cells can consume external chemicals of type Nutrition 2 only. Moreover, type 1 cells can consume chemicals of type Food 2 that are produced by type 2 cells while the latter cells can consume chemicals of type Food 1 that are produced by type 1 cells. All cells gain 1 unit of energy for each piece of chemical consumed. Figure 1 shows the relations between the two cell types.

All chemicals are characterized by a real value that defines their type: for example, chemicals of type Nutrition 1 have a value in the interval [−3, 3]. Cells have appropriate receptors that recognize substances if they fall within certain bounds: for example, receptors for Nutrition 1 have *low and high thresholds* in the intervals [−3, −1.5] and [1.5, 3], respectively. These *thresholds* are mutated on cell division: with a certain *mutation probability* any of these values is randomly reinitialized within a *mutation range*. The ranges of the intervals, together with the values for the rest of the parameters presented next, have been tuned so as to allow a large but not too large space of normal, typical behavior, that is, to ensure that a system reaches equilibrium by default (Sect. 3), but that the equilibrium may be perturbed by atypical cells (Sect. 4).

Cells reproduce when their energy exceeds a reproduction level (*energy to divide*). Newly born cells inherit all the mother cell's parameters except *max age* and chemical *recognition thresholds*, that are reinitialized in the appropriate ranges. A mechanism of apoptosis is additionally defined: type 1 cells decompose into one residue piece of Nutrition 1, one residue piece of Nutrition 2 and one residue piece of Food 1, while type 2 cells decompose into one residue piece of Nutrition 1, one residue piece of Nutrition 2 and one residue piece of Food 2. Decomposition happens 50 steps after death has occurred. Finally, cells may starve when no suitable chemicals are found in their environment: this happens after a certain period of time has elapsed without consumption (*starveAge*).

The above are summarized in Fig. 1 and Table 1. Figure 1 shows the relations between the two cell types while Table 1 gives all the parameters used together with all the parameters for the atypical cells that are presented later. As with nutrition intervals, these parameters are selected and their typical and atypical values are tuned so that the typical case can sustain the cellular ecology, while the atypical case can allow for faster cellular growth and can thus present a potential for abnormal development and cancer.

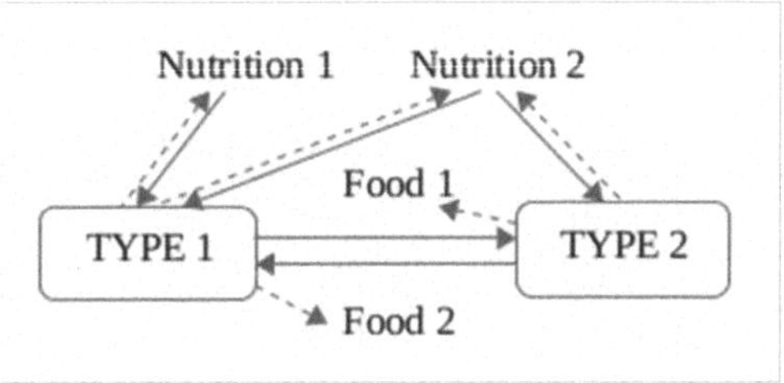

Fig. 1. Pictorial description of the relation and interactions between the two cell types. Solid arrows represent outputs from cell type to cell type, while dashed arrows represent residual output during cell decomposition (after death).

The algorithm representing the metabolic function of the cells is as follows:

IF *age* > <u>*max age*</u> OR food age > <u>*starve age*</u>, THEN die.

IF *energy* > <u>*energy to divide*</u>, THEN reproduce.

Consume as many chemicals from the environment as allowed by <u>*speed*</u> (speed functions as a probability and is decreased by one on each consumption)

For each consumed chemical increase *energy* and a *production counter*. Reset *food age*

IF *production counter* > <u>*energy to produce*</u>, THEN produce chemicals with <u>*prob. to produce*</u>

3 Simulation Setup and Control Simulation Experiment

The simulation is individual-based and uses a 2D grid of 30 × 30 positions and 60 cells at start that are randomly dispersed in the environment. All positions may hold at most one cell. The environment is initialized randomly with 500 pieces of Nutrition 1 and

Table 1. Basic parameters of the two cell types, the corresponding atypical values and the chemical parameters. (*) Atypical cells may behave either as type 1 with atypical values or as generalists that can consume and produce all allowed chemicals (see later text).

Cells	Type 1 cells (T1)	Type 2 cells (T2)	Atypical values (AT)
Max age	250–500	250–500	500–1000
Starve age	400	350	500
Energy to divide	5	4	3
Energy to produce	2	2	3
Prob. to produce	0.9	0.9	0.1
Speed	0.9–1.1	0.9–1.1	0.9–3.1
Mutation prob	0.1	0.1	0.5
Mutation range	0.1	0.1	0.2
Nutrition 1	IN, DEC	DEC	IN (*)
Nutrition 2	IN, DEC	IN, DEC	IN (*)
Food 1	OUT, DEC	IN	IN, OUT (*)
Food 2	IN	OUT, DEC	IN, OUT (*)
Chemicals	**Range**	**Low threshold**	**High threshold**
Nutrition 1	[−3, 3]	[−3, −1.5]	[1.5, 3]
Nutrition 2	[11, 17]	[11, 12.5]	[15.5, 17]
Food 1	[18, 24]	[18, 19.5]	[22.5, 24]
Food 2	[4, 10]	[4, 5.5]	[8.5, 10]

1000 pieces of Nutrition 2. In the following simulation experiments, there is no nutrition renewal, periodic or random. Rather the cellular ecology manages to sustain itself by interaction. Unless otherwise stated, the following simulations are performed for 10,000 cycles per run and all results are averages of 100 runs. As with the parameters of the cell model, the simulation time is abstract and does not correspond to real time. If anything, it can be seen as metabolic time unit.

The described configuration allows the cell population to reach a dynamic equilibrium, as shown in Fig. 2. Note that our cells are potentially immortal in the sense of being able to divide forever without a Hayflick-type limit [11], but this constraint is relaxed and studied in Sect. 6. Thus a normal population of cells will remain in this dynamic equilibrium forever. The number of cells after 100,000 steps (average of 100 runs) is from 350 to 400 for each type, so that the whole population fills about 80–85% of the space, which presents itself as a Malthusian-type equilibrium where the "carrying capacity" of the environment has been reached. Because the cells decompose during apoptosis, Nutrition 1 and 2 also stabilize at around 50–60 and 20–30 pieces respectively throughout equilibrium. The carrying capacity enforces the ecological condition, in the sense that abnormal "tumor" development cannot show as atypical growth in excessive numbers but as growth to the detriment of the typical cells.

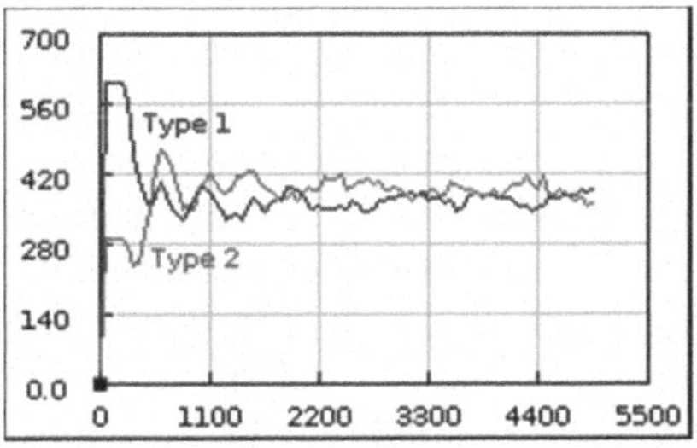

Fig. 2. Development of an ecological equilibrium between the two cell types. (x axis) time, (y axis) numbers of cells of types 1 and 2.

4 Atypical Genes

In any ecological equilibrium, a slightly modified species may perturb the equilibrium and drive one or more of the common species into extinction. Likewise, a modified or 'atypical' cell species in the above environment may lead the system to cancer, i.e. to abnormal development of one cell type to the detriment of the others and finally to organismic death. We have tested 'atypical' values for the eight parameters of Table 1 (*max age* to *mutation range*) as well as an 'atypical cell type' that is generalist and can consume anyone of the four chemicals (nutrition 1 and 2, food 1 and 2), can produce both types of food (provided it has already consumed some of it) and does not decompose to meaningful residues.

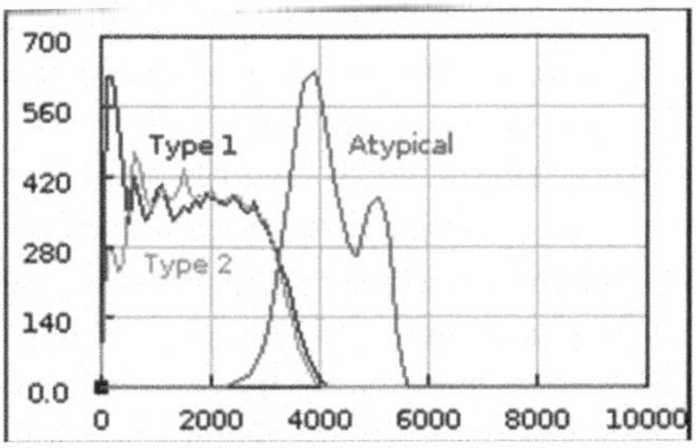

Fig. 3. Development of cancer by introduction of a single highly aggressive atypical cell (carrying all possible mutations) at t = 2000, after the initial population has stabilized. (x axis) time, (y axis) numbers of cells of types 1, 2 and atypical.

Figure 3 shows how cancer can develop in such a system: after the system has stabilized at t = 2000, a single highly aggressive cell is introduced, that is atypical in all the parameters. The figure shows how the population of the atypical cell grows while the two normal cell type populations collapse and finally the whole system dies. Since the cells are immortal, i.e. without a division limit, some other types of equilibria can arise at short-term (10,000 steps): equilibria where there is only one remaining cell type, typical or atypical (this again is abnormal and corresponds to death of the organism) and equilibria where two or three cell types survive: either atypical cells with one of the other types (generally type 2, because of the structure of the cellular ecology) or atypical cells together with type 1 and type 2 cells, as in Fig. 4. We call these equilibria respectively: **immortalized** equilibrium (atypical cells only), **atypical** equilibrium (atypical cells with either type 1 or type 2 cells) and **irregular** equilibrium (atypical, type 1 and type 2 cells). The latter may be acceptable as a form of chronic cancer, if it persists throughout life. However, most populations eventually stabilize after a long period in one of two conditions: either a **regular** healthy equilibrium between the cells of type 1 and type 2, or death.

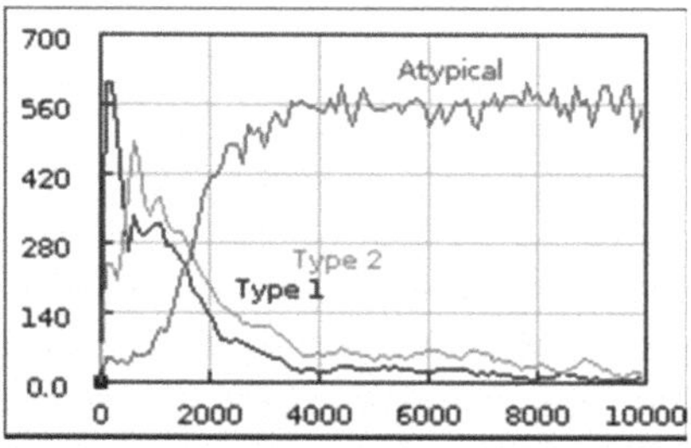

Fig. 4. Development of an irregular equilibrium (chronic cancer), where cells of both types are in equilibrium with atypical ones. In this simulation experiment the atypical cell was introduced at start, i.e. at t = 0. (x axis) time, (y axis) numbers of cells of types 1, 2 and atypical.

We have run simulations with heritable 'gene mutations' for the atypical values: if the corresponding gene is mutated, the value of the parameter is the atypical rather than the typical one (all typical and atypical parameter values are given in Table 1). In Tables 2, 3 and 4 we report simulations with each one of the genes. All simulations are run for 10,000 steps to unravel the tendencies of the system, because at this stage many populations have not yet stabilized, so we observe the transitive phenomenon. The atypical cell "attacks" (i.e. the emergence of cells with one atypical gene) take place generally at t = 2000, when the cellular population has stabilized, as Fig. 2 shows Atypical equilibria are anyway very rare and are counted here together with irregular ones. Table 2 shows that the only gene mutations that can lead to cancer are the 'max age' gene mutation and the 'energy to divide' gene mutation, and in these cases we obtain 30 to 70 percent of potentially chronic cancer and no deaths after 10,000 steps. Table 3 shows that if the 'atypical cell type' gene mutation is present we have significant amounts of deaths and chronic cancers after 10,000 steps for the same gene mutations as before, for example we obtain only 17% and 46% healthy survival (regular equilibrium) for the 'max age' and 'energy to divide' gene mutations, respectively. These values are significantly lower than the corresponding values in Table 2 when the 'atypical cell type' gene mutation is absent.

Table 2. Results with atypicity genes for mutated type 1 cells. Introduction of a single atypical 'max age' cell at t = 2000. Percentages of deaths, immortalized equilibria, irregular and regular equilibria after 10,000 steps. Averages of 100 runs.

	Deaths	Immortalized	Irregular	Regular
Max age	0	0	70%	30%
Starve age	0	0	2%	98%
Energy to divide	0	0	33%	67%
Energy to produce	0	0	0	100%
Prob. to produce	0	0	0	100%
Speed	0	0	2%	98%
Mutation prob.	0	0	2%	98%
Mutation range	0	0	1%	99%

Table 3. Results with atypicity genes for mutated 'atypical type' cells, that metabolize all chemicals of type 'nutrition 1', 'nutrition 2', 'food 1' and 'food 2' and produce all chemicals of type 'food 1' and 'food 2' (these are shown in Table 1). Introduction of a single atypical 'max age' cell at t = 2000. Percentages of deaths, immortalized equilibria, irregular and regular equilibria after 10,000 steps. Averages of 100 runs. Compare with Table 2.

	Deaths	Immortalized	Irregular	Regular
Max age	26%	32%	25%	17%
Starve age	8%	10%	12%	70%
Energy to divide	28%	15%	11%	46%
Energy to produce	4%	12%	8%	76%
Prob. to produce	1%	14%	12%	73%
Speed	6%	7%	11%	76%
Mutation prob.	4%	11%	4%	81%
Mutation range	9%	13%	6%	72%

Table 4 additionally shows that if the atypical cell type gene mutation is introduced very early in the system, the system develops young cancer which is much more lethal than at a later age: no survival for the same two mutations and only 29% and 24% potentially chronic cancers, respectively.

Finally, we have studied the effect of a low gene mutation probability at cell division time. Table 5 shows the results for several values of this probability. As the probability increases, more gene mutations are accumulated and the death rates increase while the healthy or chronic outcomes decrease. For example, if the mutation probability is 0.0000001 (1E−7), we obtain healthy survival and chronic cancer in 67% and 26% of the cases, whereas if the probability is one order of magnitude larger (0.000001 = 1E−6)

Table 4. Same as Table 3, but with introduction of a single atypical 'max age' cell at t = 0, i.e. at birth. Percentages of deaths, immortalized equilibria, irregular and regular equilibria after 10,000 steps. Averages of 100 runs. Compare with Tables 2 and 3.

	Deaths	Immortalized	Irregular	Regular
Max age	57%	14%	29%	0
Starve age	28%	13%	43%	16%
Energy to divide	67%	9%	24%	0
Energy to produce	22%	13%	22%	43%
Prob. to produce	4%	36%	15%	45%
Speed	42%	5%	13%	40%
Mutation prob	33%	17%	35%	15%
Mutation range	39%	14%	28%	19%

Table 5. Results for various probabilities of gene mutation. Percentages of deaths, immortalized equilibria, atypical, irregular and regular equilibria after 10,000 steps. Results for atypical and irregular equilibria are given separately to show the progression of cancers. Averages of 100 runs.

	Deaths	Immortalized	Atypical	Irregular	Regular
5xE-8	4%	3%	0	10%	83%
1xE-7	3%	3%	1%	26%	67%
1xE-6	13%	13%	3%	62%	9%
3xE-6	34%	31%	4%	31%	0
5xE-6	42%	37%	11%	10%	0
7xE-6	59%	28%	9%	4%	0
9xE-6	66%	29%	4%	1%	0
1xE-5	60%	32%	6%	0	0
5xE-5	86%	14%	0	0	0
0.0001	95%	5%	0	0	0

we obtain 9% and 62%, respectively. Note that at death time, an atypical cell has on average 3 to 4 gene mutations.

All the above results align conceptually with the usual tendencies we see in real cancers: some mutations are more aggressive than others (for example, only some mutations are aggressive in prostate cancer [12]), attacks early in life are more threatening than attacks at later stages [13] and the more deleterious mutations accumulate the higher the death toll [14]. Note that although the atypical genes in our model affect parameters related to the actual metabolism, such as 'energy to divide' (Warburg and others have insisted on the metabolic origins of cancer [15–17]), this setup is not incompatible with signaling theories of cancer [18], because defective genes may control metabolism indirectly just by signaling metabolic actions and hyper-activating metabolism and growth rather than by changing directly the metabolic pathways (point made by [16]).

5 Treatment with Toxicity

Next, we have defined and studied one type of chemical control of cancer by the introduction of a toxic nutrient, provided that it affects atypical cells more than typical ones. More specifically we define toxic Nutrition 1 as infecting a cell in a way that changes its metabolic cycle and makes it more difficult for it to produce and to reproduce, thus it restrains growth potential:

Without infection (cf. Table 1)
(Type 1 cells) *energy to divide* = 5, *energy to produce* = 2
(Type 2 cells) *energy to divide* = 4, *energy to produce* = 2
On first infection
energy to divide = 10, *energy to produce* = 6
On second infection
energy to divide = 15, *energy to produce* = 8

A cell may be infected by a toxic piece of Nutrition 1 three times before it gets killed by toxicity and an infection lasts at most 10 steps before the cell recovers its previous metabolic values. As long as a cell is in infected state, it runs the risk of dying with a certain probability (*virulence*). The toxic element affects atypical cells more that typical ones (type 1 cells) when the *virulence* for type 1 cells is less than for atypical cells. Another type of useful toxicity that will be considered in future work will be to have more drastic changes for atypical than typical cells, independently of *virulence.*

Table 6 gives comparative results for various regimes of treatment with toxicity (*virulence* 0.1 for type 1 cells, 0.5 for atypical cells). Toxic treatment is implemented as repeated introduction of 2 toxic pieces of Nutrition 1 randomly in the environment and for as long as the treatment takes place. In this and the following Tables, all simulations are run for up to 100,000 steps or until they have reached death, a regular equilibrium or immortality of one of the cell types. This happens in the majority of cases well before 30,000 steps. Note that the first row in all Tables represents the long term equilibrium in the untreated case. For example, in the case of Table 6, the first row may be compared to the equivalent short-term value (after 10,000 steps) given in the first row of Table 3 ('max age' gene mutation): some of the early irregular equilibria resolve to self-healed regular ones, but most of them finally lead to death. Finally, diagnosis time and treatment durations are given values that make sense: later diagnosis need a much longer treatment duration to reach lower survival rates.

For the specific 'max age' gene attack at t = 2000, Table 3 shows that if the cancer is left untreated it leads to 17% immediate spontaneous recovery and 25% potentially 'chronic' cancer that may allow it to be discovered. If discovered after 1000 steps have elapsed, it can lead to 76% recovery if treated for a period of 3000 steps and to lower degrees of recovery if treated for significantly longer or significantly shorter periods (5000 steps or 1000 steps, in this simulation). This is a general property of toxic treatment: there is generally an optimal range of treatment duration where survival is maximum and less than that has lower effect while more than that kills the system more often. Sometimes there is a threshold after which the effect of the toxic treatment does not change for better or for worse. Any treatment is however better than no treatment at all.

Table 6. Comparative results for various treatment regimes. Single cancer 'max age' gene attack at t = 2000. The indicated times are the time elapsed after attack before the cancer is diagnosed and treatment is started and the duration of the treatment. Treatment virulence 0.1 for typical cells, 0.5 for atypical cells. Percentages of deaths, immortalized equilibria, atypical, irregular and regular equilibria. Averages of 100 runs for up to 100,000 steps.

Diagnosis time	Treatment duration	Deaths	Immortalized	Atypical	Irregular	Regular
None	None	74%	4%	0	0	22%
1000	1000	41%	3%	0	0	56%
1000	3000	23%	1%	0	0	76%
1000	5000	38%	0	0	0	62%
2000	1000	55%	2%	0	0	43%
2000	5000	40%	6%	0	0	54%
2000	8000	43%	5%	0	0	52%

Table 7 gives the results for the same system but with higher toxicity of the treatment (virulence 0.5 for type 1 cells, 0.9 for atypical cells). As expected, for higher toxicities or for later diagnoses, we need longer treatments to obtain more or less the same survival rates. Finally, Table 8 gives the results for the case of a young cancer (appearing earlier in life). This has a lower default recovery rate if untreated but needs a shorter treatment period to obtain a comparable or better prognosis, i.e. subsequent survival rate, as the higher age cancer (3000 steps instead of 5000 steps to achieve 54% rather than 44% survival rate). In general, to obtain optimal results, the duration of treatment should take into account the age of first attack, the time elapsed before diagnosis and the toxicity of the treatment.

Table 7. Same as Table 6, but treatment virulence is 0.5 for typical cells, 0.9 for atypical cells. Percentages of deaths, immortalized equilibria, atypical, irregular and regular equilibria. Averages of 100 runs for up to 100,000 steps.

Diagnosis time	Treatment duration	Deaths	Immortalized	Atypical	Irregular	Regular
None	None	74%	4%	0	0	22%
1000	1000	53%	1%	0	0	46%

(*continued*)

Table 7. (*continued*)

Diagnosis time	Treatment duration	Deaths	Immortalized	Atypical	Irregular	Regular
1000	3000	27%	2%	0	0	71%
1000	5000	24%	0	0	0	76%
3000	2000	58%	9%	0	0	33%
3000	5000	44%	12%	0	0	44%
3000	8000	48%	14%	0	0	38%

Table 8. Same as Tables 6 and 7, but cancer 'max age' gene attack at t = 500 (younger cancer). Treatment virulence is 0.5 for typical cells, 0.9 for atypical cells. Percentages of deaths, immortalized equilibria, atypical, irregular and regular equilibria. Averages of 100 runs for up to 100,000 steps.

Diagnosis time	Treatment duration	Deaths	Immortalized	Atypical	Irregular	Regular
None	None	75%	7%	0	0	18%
2000	1000	52%	2%	0	0	46%
2000	3000	39%	7%	0	0	54%
2000	5000	55%	3%	0	0	42%

6 Reproduction Limit

Finally, we have investigated a Hayflick-like reproduction limit, where a cell cannot reproduce or divide more than a certain number of times. Hayflick [11] has found that a normal cell cannot divide indefinitely but only up to about 60 times, after which period it enters the so-called senescence state. The molecular mechanism that is generally accepted as the cause for this behavior is telomere shortening and consequent DNA damage.

We have investigated arbitrary division limits that lead to different lifespans. When the division limit for a cell is exhausted, the cell can no longer divide and dies when its maximum age is reached. The results given next are obtained with simulations where the division limit was set to 350. This limit has been chosen to allow a long lifespan for the organism (the cellular ecology); this lifespan is between 51,000 and 53,000 steps. As Fig. 5 shows, the organism is stable for a long period and collapses somewhat abruptly after overshooting in the end. All simulations with the division limit are therefore run for 48,000 steps which is a little before natural collapse of the system, so that, if a system survives to this age from cancer attack, it can be thought to have survived in general.

We have repeated the simulations of the previous sections with the division limit and the results are qualitatively similar but with some notable differences. Firstly, as shown in Table 9, the cellular ecology is more robust for some gene attacks. Secondly, as shown

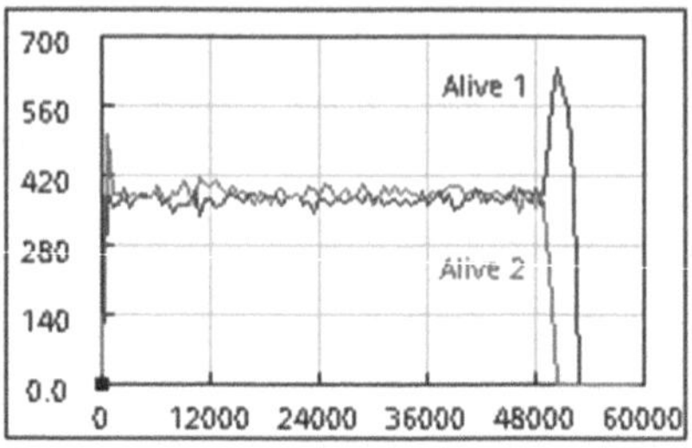

Fig. 5. Collapse of the ecological equilibrium and organismic death with a division limit. (x axis) time, (y axis) numbers of cells of types 1, 2 and atypical.

in Table 10, cumulative gene damage becomes lethal at very high, rather unrealistic, mutation rates, compared to the limitless case. Both these manifestations of robustness to deleterious mutations are due to the fact that, the cells that do not divide after a certain limit, do not have the chance to pass the deleterious mutations to their offspring. Thus, less damage is found in the subsequent cell generations when there is a limit to cellular division than when there is no such limit. Note that with the division limit and the resulting finite lifespan of organisms, a 'young' or 'adult' cancer can be defined more precisely and unlike the previous case that had no limit. The simulation experiments have indeed shown that very early cancers have dramatic effects, in parallel with Table 4. We have chosen to study the attacks at 'middle age' around 30,000. However, no age of the mortal organism can be directly compared to the immortal one and attacks at 2000 as before are really too early for the mortal system and give survival rates comparable but slightly better that the limitless case. Of course, what makes the organism so vulnerable at early age remains to uncover.

Table 9. Comparison of survival rates (regular equilibria at end of simulation) with and without division limit for some of the gene attacks of Tables 2, 3 and 4. All simulations with mutated 'atypical type' gene, as in Tables 3 and 4. Averages of 100 runs.

Attack at 30,000	Survival (no division limit)	Survival (division limit = 350)
Max age	14%	11% (+14% chronic)
Energy to divide	64%	69% (+1% chronic)
Starve age	69%	74% (+1% chronic)
Prob. to produce	44%	69% (+3% chronic)

We have also investigated toxic treatment of cancers as in Sect. 5 above. Again, as Table 11 shows, treatment drastically improves survival rates but, beyond a certain treatment duration, it has no additional effect (this is the case of attack at 2000 in the table) or it has even an inverse effect (this is the case of attack at 30,000 in the table). The effect is also more pronounced for early attacks (at 2000 in the table) compared to late attacks (at 30,000 in the table).

Table 10. Comparison of survival rates (regular equilibria at end of simulation) with and without division limit for various probabilities of gene mutation as in Table 5. Averages of 100 runs. Middle column taken from last column of Table 5.

	Survival (no division limit)	Survival (division limit = 350)
5xE-8	83%	99%
1xE-7	67%	100%
1xE-6	9%	99%
5xE-6	0	98%
1xE-5	0	88%
5xE-5	0	60%
1xE-4	0	33%
0.001	0	0
0.01	0	0

Table 11. Comparative results for toxic treatment regimes of a 'max age' gene attack with division limit. Treatment virulence 0.1 for typical cells, 0.5 for atypical cells. Percentages of deaths, immortalized equilibria, atypical, irregular and regular equilibria. Averages of 100 runs.

Attack at 2000 Diagnosis time	Treatment duration	Deaths	Immortalized	Atypical	Irregular	Regular
None	None	87%	0	0	0	13%
500	500	48%	5%	0	0	47%
500	1000	29%	4%	0	0	67%
500	5000	33%	4%	0	0	63%
Attack at 30,000 Diagnosis time	Treatment duration	Deaths	Immortalized	Atypical	Irregular	Regular
None	None	42%	32%	4%	0	22%
500	500	24%	33%	0	3%	40%
500	1000	32%	39%	0	1%	28%
500	3000	33%	38%	1%	0	28%
3000	8000	48%	14%	0	0	38%

Taken together, all these results show that a Hayflick-like limit is advantageous to a system where deleterious mutations may emerge and accumulate. Since some of these may lead to cancer, a Hayflick-like limit is a way to bypass and avoid cancer development and would, for this reason, be evolutionarily promoted.

7 Discussion

We have presented an abstract agent-based model of a cellular ecology that is in equilibrium but can be invaded by an 'atypical' cell carrying one or more mutations that signal a change in its metabolic and reproductive activity in a way that perturbs the ecological equilibrium and drives the whole cell population to premature death. Our results are qualitative and their objective is to show the breadth of emergent phenomena that may be obtained with this and other similar cellular ecology models. Our model and its results spawn a number of observations.

- Not all mutations may be causes of cancer, for example the 'energy to produce' gene mutation has no effect, unless the generalist 'cell type' gene mutation exists. And some mutations may control the effect of others, such as the 'cell type' gene mutation, mentioned above.
- That said, all these mutations are deleterious in the specific ecology but probably not in a different one. Then we may think of introducing factors that are not toxic but that 'cooperate' with atypical cells to keep them in balance, for example by providing specific inputs to the atypical cells that are competitive with input from typical cells or by inserting specific cellular organisms that can establish with them a parallel interaction network compared to the one shown in Fig. 1. These are ideas for future work in the line of the perspective and the many options discussed in [19, 20].
- In Table 5 we have seen how the probability of gene mutation affects cancer development, namely that higher gene mutation probabilities and accumulation of mutations have a higher potential for cancer development. But the prognosis changes dramatically if there exists even a very slight probability of gene repair. For example, the highest mutation probability of 0.0001 accompanied by a one-order-of- magnitude smaller repair probability (0.00001) in the limitless case leads to 32% immediate recovery and 56% chronic cancer at the short term compared to 95% deaths without gene repair. This is an indication that treatment with a factor that is not toxic but that triggers gene repair may yield very significant results, especially by constraining atypical cells without affecting so much the typical cells.
- Further modeling based on the Hayflick limit may show why the stem cells are of such importance to normal development and what the effect of cancer stem cells [21, 22] may be.
- One further concern is that the cells modeled as above are largely homogeneous, while we know that real tumours tend to attain and conserve higher heterogeneity than the cells of the normal tissue. We plan to introduce differentiation mechanisms such as the ones introduced by the Kaneko group [9] and mutations that control differentiation to study the effect of diversity on the resistance to cancer development. Regular somatic differentiation introduces and sustains diversity which is known to show more resistance to attacks, but for the same reason a diverse cancer cell population may be more aggressive and may also escape more easily toxic treatment.

All these ideas show that viewing development of cancer as collapse of an ecological equilibrium in a population of cells allows not only a valid alternative understanding of the phenomenon but also inspiration for a number of interventions and treatments of non-toxic chemical nature.

Disclosure of Interests. The author has no competing interests to declare that are relevant to the content of this article.

References

1. Hanahan, D., Weinberg, R.A.: The hallmarks of cancer. Cell **100**, 57–70 (2000)
2. Hanahan, D., Weinberg, R.A.: Hallmarks of cancer: the next generation. Cell **144**(5), 646–674 (2011)
3. Abbott, R.G., Forrest, S., Pienta, K.J.: Simulating the hallmarks of cancer. Artif. Life **12**, 617–634 (2006)
4. Domschke, P., Trucu, D., Gerisch, A., Chaplain, M.A.J.: Mathematical modelling of cancer invasion: implications of cell adhesion variability for tumour infiltrative growth patterns. J. Theor. Biol. **361**, 41–60 (2014)
5. Greene, J.M., Levy, D., Funsg, K.L., Souza, P.S., Gottesman, M.M., Lavi, O.: Modeling intrinsic heterogeneity and growth of cancer cells. J. Theor. Biol. **367**, 262–277 (2015)
6. Reynolds, B.A., Oli, M.W., Oli, M.K.: Eco-oncology: applying ecological principles to understand and manage cancer. Ecol. Evol. **10**, 8538–8553 (2000)
7. Somarelli, J.A.: The hallmarks of cancer as ecologically driven phenotypes. Front. Ecol. Evol. (Sec. Population, Community, and Ecosystem Dynamics) **9**, 99–1128 (2021)
8. Furusawa, C., Kaneko, K.: Emergence of rules in cell society: differentiation, hierarchy, and stability. Bull. Math. Biol. **60**, 659–687 (1998)
9. Furusawa, C., Kaneko, K.: Chaotic expression dynamics implies pluripotency: when theory and experiment meet. Biol. Direct **4**, 17 (2009)
10. Bertolaso, M.: Philosophy of Cancer: A Dynamic and Relational View. Springer, Dordrecht (2016)
11. Hayflick, L.: The illusion of cell immortality. Br. J. Cancer **83**, 841–846 (2000)
12. Darst, B.F., Saunders, E., et al.: Germline sequencing analysis to inform clinical gene panel testing for aggressive prostate cancer. JAMA Oncol. **9**(11), 1514–1524 (2023)
13. https://projects.research-and-innovation.ec.europa.eu/en/horizon-magazine/five-things-know-about-childhood-cancer
14. Takeshima, H., Ushijima, T.: Accumulation of genetic and epigenetic alterations in normal cells and cancer risk. Precis. Oncol. **3**(7) (2019)
15. Warburg, O.: On the origin of cancer cells. Science **123**, 309–314 (1956)
16. Vander Heiden, M.G., Cantley, L.C., Thompson, C.B.: Understanding the Warburg effect: the metabolic requirements of cell proliferation. Science **324**, 1029–1033 (2009)
17. Schwartz, L., Supuran, T., Claudiu, O., Alfarouk, F.: The Warburg effect and the hallmarks of cancer. Anti-Cancer Agents in Medicinal Chemistry (Formerly Current Medicinal Chemistry - Anti-Cancer Agents) **17**(2), 64–170 (2017)
18. Sever, R., Brugge, J.S.: Signal transduction in cancer. Cold Spring Harb. Perspect. Med. **5**, a006098 (2015)
19. Kroemer, G., McQuade, J.L., Merad, M., André, F., Zitvogel, L.: Bodywide ecological interventions on cancer. Nat. Med. **29**(29), 59–74 (2023)
20. Pienta, K.J., McGregor, N., Axelrod, R., Axelrod, D.E.: Ecological therapy for cancer: defining tumors using an ecosystem paradigm suggests new opportunities for novel cancer treatments. Transl. Oncol. **1**(1), 158–164 (2008)
21. Lobo, N.A., Shimono, Y., Qian, D., Clarke, M.F.: The biology of cancer stem cells. Annu. Rev. Cell Dev. Biol. **23**, 675–699 (2007)
22. Zacijek, G., Wilensky, U.: NetLogo Tumor model (1998). http://ccl.northwestern.edu/netlogo/models/ Tumor Center for Connected Learning and Computer-Based Modeling, Northwestern University, Evanston, IL

Stability of Input Representations in Biological Neural Network Reservoirs

Trym A. E. Lindell[1(✉)], Ola H. Ramstad[1], Ioanna Sandvig[2], Axel Sandvig[2], and Stefano Nichele[1,3]

[1] Oslo Metropolitan University, 0130 Oslo, Norway
trymlind@oslomet.no
[2] Norwegian University of Science and Technology, 7491 Trondheim, Norway
[3] Østfold University College, 1757 Halden, Norway

Abstract. Biological neural networks operate with significantly lower energy costs than their artificial counterparts. Harnessing this efficiency could drastically reduce the massive energy consumption of artificial intelligence systems. One method of doing this involves using biological neural networks grown on micro-electrode arrays (MEA) *in vitro* as reservoirs within reservoir computing systems.

However, involving biology in computing does not come without challenges. *In vitro* neural networks develop and change over time, which may affect their representation of, and dynamic range to, input. This can be problematic when an input encoder is calibrated to a given dynamic range, and a readout decoder is trained based on a set of representations that are no longer produced by the network.

In this study we explore the stability of input representations and dynamic range in biological neural networks. We use a series of input consisting of inter-stimulation intervals to determine the networks' dynamic range to stimulation timing at different days *in vitro*. To assess the stability of the representations of input we also train classifiers based on the response on one Day *In Vitro* (DIV) and test on responses from two other days *in vitro*. Our results show that networks are generally unstable both in dynamic range and in the stability of their representations.

Keywords: Micro Electrode Array · Biological Neurons · Reservoir Computing

1 Introduction

1.1 Motivation

Biological systems have evolved under strict energy constraints which have given rise to organisms that can perform highly complex computations at extremely low energy costs when compared to today's artificial computing systems. This is problematic as excessive energy use can have negative environmental impact, particularly if the energy is sourced from fossil fuels. This has motivated the construction of hybrid bio-artificial computing systems. Current implementations

T. Carletti et al. (Eds.): WIVACE 2024, CCIS 2532, pp. 276–291, 2025.
https://doi.org/10.1007/978-3-031-93631-9_22

are capable of controlling simple robotics [1], playing pong [17] and predicting chaotic time series [4]. While these applications show the potential of biological computing, some unique challenges arise due to biology being "messier" than silicon. While we can fix the weights and architecture of artificial neural networks, biological neural networks are far more unstable: They go through developmental stages which change their dynamics [7,27,30], they can change their responses due to different stimulation protocols [4,5,17,33] and, even in mature animals, are in a constant flux due to representational drift [8,26]. This dynamic behaviour may result in specific challenges when we try to utilize such biological systems for practical computations that require an interface with otherwise static in-silico systems. If, for example, the representation of a computation changes after a decoder has been constructed to translate the biological computations into the desired machine or human usable outputs, the decoder may no longer produce the correct output. This may require continuous re-training of the decoder. Furthermore, changes in the overall dynamics of the network may affect its ability to encode input. If input is encoded as the timing between stimulation pulses, the networks ability to produce different responses to different variations of stimulation timing may change, i.e., the upper and lower bounds of its dynamic range may change over time. This would require updating the encoder such that input can be properly encoded.

As research is ongoing into the feasibility of using biological neural networks as alternative computing substrates to artificial neural networks, it becomes increasingly important to asses the stability of their computations as this will impact their energy efficiency through the need for decoder/encoder retraining/calibration and their practical usability for computational tasks.

In this study, we probe 6 *in vitro* (in glass, meaning outside of the organism) neural networks at *days in vitro* 19, 29, and 35. We used stimulation timing as the encoding method. This consisted of injecting pairs of stimulation pulses with different relative delays into the networks. The same delays were used across all three days, and changes in dynamic range was assessed by identify changes in the lower and upper bound of delays that could be correctly classified. Representation changes were explored by training classifiers on spike responses from one DIV and testing on the two other.

The paper is organized as follows: We first introduce relevant background related to reservoirs computing, *in vitro* neural networks grown on MEA and the different stability challenges that can affect their computations. This is followed by the methods, results and discussion and conclusion.

1.2 Background

Reservoir Computing. Reservoir computing is an umbrella term used for a type of Artificial Intelligence system that consists of an encoding method, a dynamical substrate (the reservoir), and a decoder (often called readout). The reservoir can be any dynamical substrate in which the states gradually contract over time (according to the echo state formulation [16]) and in which different inputs produce linearly separable states (the separation property according to the liquid state machine formulation [19]). Example substrates that have previously been used are: *in vivo* (in the living, meaning in the organism) cat visual cortex [22], bucket of water [11], artificial (spiking/non-spiking) neural networks [16,19] and *in vitro* neural networks [1,4], for a comprehensive review see [29].

Since data is often stored on computers, while reservoirs can perform their computations within a wide range of non-digital substrates, inputs must often be encoded into a format on which the reservoir can compute. Binary values have for example been encoded into the movement of a set of motors embedded in water [11], while input to *in vitro* neural networks has been encoded using electrical stimulation through: rate, spatial position, and timing [3,4,9,12–14,23,25,28,31,32] as well as light by means of optogenetics [10]. When constructing such encoders it is also important to determine the dynamic range of the reservoir, meaning the range of input values which the reservoir produces separable responses to. For example, we may attempt to encode a time series with state values defined over the range [0,1] as voltage variations over the same range. When these voltages are used as input to a biological network, we would need to make sure that the full range of voltages actually produce separable responses. Otherwise, we would loose all information that produces inseparable responses. It may for example be the case that the neurons do not produce spikes to voltage values below 0.5 V, meaning that the networks cannot compute on half the range of the time series.

If the reservoir is some form of unconventional substrate, i.e., not an artificial neural network, the decoder can be subdivided into a readout mechanism, which translates the states of the reservoir to digital data, and a linear machine learning component which is trained on the digitized states to produce the desired output. For biological networks, the readout would record the activity of the neurons, pre-process this activity, and translate it into vectors that can be used as input to the machine learning component. A schematic of a reservoir computing pipeline is shown in panel III in Appendix A.

The benefit of reservoir computing over the more traditional artificial neural network is that the reservoir, which performs the main computational work, is not trained and in some cases is implemented in a highly energy efficient substrate, such as biological neurons. Since only the linear decoder is trained, this grants energy efficiency over deep artificial neural networks which requires the full network to be trained.

In-Vitro Neural Networks. *In vitro* neural networks are networks of biological neurons kept alive outside of an animal, for example in a glass container. The neurons can be sourced from rodent brains, and specific regions like the cortex can be micro-dissected from the dissected brain to isolate specific cell types. The existing connections in the tissue can then be chemically cut such that the cells are separated (dissociated) from each other. When the dissociated neurons are added to their container the neurons will then self-organize over time into a 2D network covering the bottom surface. If the container is a Micro Electrode Array (MEA) which has an array of electrodes covering its bottom surface, the neurons that grow on top of the electrodes can be electrically stimulated, and their electrical response can be recorded. A microscopy image of one of the networks used in this study is shown in panel I in Appendix A. Depending on the pattern of activity, for example a voltage deflection of above 6 standard deviations of the voltage trace [2], spikes can be detected. These spikes are the output of biological neurons' computations. Electrical stimulation can induce spikes in the neurons adjacent to the electrode, which can spread throughout the network. The networks can thus potentially perform computations on input encoded through the electrical stimulation which is how they can be used as a reservoir in a reservoir computing system.

Stability Challenges in In-Vitro Neural Networks

Developmental Changes. As mentioned, *in vitro* neural networks go through developmental stages. Initially, the networks produce very few spikes, later they start producing network wide bursts of spikes [27] and sometimes they end in an avalanche critical phase [24], meaning that the size and length of chains of spiking activity follows a power law. These large changes in overall dynamics may have profound effects on the computational capacity of the neuronal network. Bursts have been shown to wipe hidden memories from networks [10], while avalanche critical dynamics have been shown to affect their dynamic range [28]. As networks develop and pass through these different dynamical phases the representation of input is likely to change, initially producing small responses, then large bursts, and then a wider range of burst sizes as the networks potentially reach the mature critical phase. An encoder calibrated for the dynamic range at one point in development may thus not be optimal for another as the bounds of the dynamical range changes. Furthermore a decoder trained on responses produced during one phase of development may not be capable of accurately decoding the same input at a different developmental phase due to these large scale changes in responses. The same input may for example elicit a small number of spikes at an early DIV and then a large network wide burst at a later, meaning that the features learned at the early DIV would not be the same as the ones the decoder would need to produce the correct output at the later.

Representational Drift. While network dynamics may appear stable when they reach maturity [24], representation of input may not. *In vivo* recordings of the neural response to certain stimuli has been shown to change gradually over time even though behavior stays constant [20], for review see [6]. Such representational drift would require constant re-training because the representation of input is always changing, thus limiting the energy benefits gained from the reservoir.

Medium Exchange Dependent Stability. An issue with the *in vitro* condition is that the neurons do not have an adaptive support system that ensure consistent nutrient conditions. This means that they instead are reliant solely on the nutrients available in the culture medium which is only changed intermittently (for example every third day). When the medium is exchanged, this can affect the dynamics of the networks [15] which could affect their computations. Indeed, in a previous study, [18], we found large variability in classifier performance on a similar experiment between two consecutive DIV suggesting that medium exchange and nutrient exhaustion can lead to unstable behavior.

Stimulation Dependent Changes. In vitro networks can also show plasticity in response to stimulation. However, eliciting changes in dynamics may only happen under certain conditions. Multiple early studies have reported plastic changes in spontaneous activity related to stimulation, but attempts at replicating these results have partly failed [33], with changes only appearing when bursts where suppressed. Similarly, evoked responses have been shown to change due to specific types of stimulation, but again dependent on network dynamics, specifically stable mean firing rate [5].

Biological networks can also show training effects. A training protocol based on the free energy principle has been shown to be capable of changing network behavior to improve performance on playing Pong [17] while passive learning also appears to be capable of improving network performance on chaotic time series prediction in 3D organoid cultures [4].

The full range of conditions under which plastic changes may occur, and how long the relevant effects may last are uncertain. Thus, when using multiple stimulation protocols on the same neural networks it becomes important to probe them for changes that may have compounding effects on downstream experiments. For example, if *in vitro* networks are to be used to solve multiple different tasks, learning effects from one task may affect performance on another. Similarly, we may need to conduct different stimulation experiments to calibrate the encoder, or to tune the networks to optimal dynamics. However, any one of such experiments could affect the outcomes of the other depending on the order in which they were done. This makes it important to probe the networks as they are assessed to examine how their behavior changes on relevant metrics like the dynamic range and stability of input representations.

In summary, biological neural networks are subject to a wide range of mechanisms that may make the interface with static *in silico* (in silicon, meaning in the computer) systems challenging. However, whether these mechanisms have an effect on relevant features for reservoir computing, like the dynamic range and representation of input encoded as stimulation timing, is not known. Because the presence of instability on either of these features will be detrimental both to research on optimal parameters for, and application of, *in vitro* biological neural reservoirs, assessment of their stability is essential.

2 Methods

Due to the wide range of mechanisms that may affect the performance of a reservoir system using *in vitro* neural networks, as described in the introduction section, we do not isolate and target any single effect in this study. Instead we probe the networks on different days *in vitro* while they undergo evaluation for encoding and phase tuning parameters:

First, at DIV 16, we identified electrodes viable for input encoding through a set of singular stimulations delivered to candidate electrodes. We also conducted a series of experiments consisting of long stimulation sequences at different frequencies meant to explore encoding and phase tuning parameters. Specifically, we aimed to identify the stimulation frequency that could optimally modulate the networks' phase to maximize their dynamic range. At DIV 22 we stimulated the networks at 1 Hz for 12 h, at DIV 25 and 28 we stimulated at 2 Hz for approximately 8.6 h and at DIV 31 at 4 Hz for 10.5 h. Finally, at DIV 34 we conducted an experiment in which we encoded a chaotic time series as stimulation timings, with an average frequency of 3.61 Hz, given over 0.18 h to test the networks' capabilities on a chaotic time series prediction task.

Thus the stimulations the networks were subjected to follows the pattern that may be necessary to utilize them as reservoirs as well as actual utilization on a reservoir computing task (chaotic time series prediction).

The experiments analyzed in this paper were conducted on DIV 19, 29 and 35. While we originally planned an initial experiment on DIV 19 as a baseline, then on each second day following nutrient media exchange, technical issues limited the days *in vitro* during which recording was possible.

Feeding and stimulation days are depicted in Appendix B.

All data and code used in this article as well as extended results are available from the digital repository at: https://osf.io/v6y2e/.

2.1 Culture Preparation and Maintenance

Six biological neural networks (labeled A1, A2, A3, B1, B2, B3) were established *in vitro* using primary rat cortical neurons (Thermo Fisher Scientific). Neurons were seeded on Cytoview 6-well plates (Axion Biosystems) at a density of 1,800 cells/mm2 for a total of approx. 190,000 cells per well (cell handling was conducted according to MAN0001574, Thermo-Fisher Scientific). A co-culture with rat primary cortical astrocytes was established using concurrent seeding of 180 cells/mm^2. Prior to seeding, the wells were coated with polyethyleneimine at 0.1% for 1 h, (Polysciences), washed with sterile deionized water, left to air dry over-night before a coating of natural mouse laminin (Thermo Fisher Scientific).

Following seeding, the cells were maintained using 95% Neurobasal Plus, 2% B27 Plus Supplement (50X), 1% GlutaMax Supplement, and 2% PenStrep (all from Thermo Fisher Scientific). The media were exchanged first 24 h post seeding, then every three days following. No experiments were conducted within 24 h following media changes.

2.2 MEA Recording System

We used the Maestro Pro system (Axion Biosystems) to record the neurons electrical activity through the MEA and to contain the neurons under appropriate conditions. The voltage trace was recorded at 12.5 khz, using a 3khz Kaiser window for low pass filtering and a 200 hz IIR for high pass filtering. Referencing method was set to "Median" and analog mode setting to "Neural Spikes". The software version for AxIS and Maestro Pro firmware was 3.6.2.2, Maestro Pro BioCore Version was "4B". CO_2 concentration was set to 5.0% while temperature was set to 37.0 C. Spikes were detected as voltage deflection of 7 standard deviations from the voltage trace at each electrode.

The microelectrode array type was CytoView MEA plate (M384-tMEA-6B, Axion Biosystems). Each Cytoview plate consists of 6 wells of 64 PEDOT electrodes with 100 um diameter and 300 um pitch arranged in a square 8×8 array. Networks were maintained in the Maestro Pro incubator continuously following the 13th DIV to keep recording conditions consistent, the MEA were only removed for 30 min during each media change.

2.3 Encoding Method

As mentioned in the introduction, information can be encoded to *in vitro* neural networks using multiple different methods. In this experiment we have chosen to encode input as the timing between stimulation pulses. This method allows relatively high input density of information as only two stimulation pulses are needed to encode an input value as opposed to frequency which would require multiple. It can also easily be combined with other encoding methods like spatial and ampere/voltage. Furthermore, the presence of spike time dependent plasticity in biological neural networks suggests that timing is a natural information code used by the brain.

An input thus consists of two electrical stimulations (0.4 ms bi-phasic stimulation) with varying delay between the first (cue) and second (probe), see Fig. 3. E.g. if the input value was 200, we input a cue and probe with a probe delay relative to the cue of 200 ms. The range of inputs examined, i.e. the delays between cue and probe were 0.2, 0.4, 0.6, 0.8, 1, 2, 3, 4, 5, 6, 7, 8, 9, 10, 11, 12, 13, 14, 15 s. A figure of a general experimental trial is given in panel II in Appendix A.

2.4 Electrode Selection

To ensure that our input resulted in network responses, we performed a series of analysis to identify optimal stimulation electrodes. First we recorded baseline activity for 1 h from each network starting at DIV 13 (see Appendix B for developmental trajectories based on spike and burst dynamics). Spiking and bursting activity over electrodes was then assessed to identify electrodes in proximity to neurons. We then selected a subset of the 5 most active electrodes for further testing. These electrodes were stimulated with a singular electrical pulses given at 30 s intervals. Spike counts were computed for 1024 ms time windows (epochs) following the stimulation and electrodes producing large responses within this epoch was selected for stimulation in the experiments.

2.5 Experimental Tasks

To assess the dynamic range and representational stability of the encoding method over time, the same input was given at DIV 19, 29 and 35.

Dynamic Range. The dynamic range was assessed using two separate classification tasks designed to identify the upper and lower bounds of the encoding method, as detailed in [18]. Specifically, these tasks are designed to identify the maximal timing between two stimulations that produce linearly separable responses (the upper bound) and how small timing difference can produce separable responses (the lower bound).

These tasks consisted of classifying whether spikes within a 1024 ms temporal window (an epoch) following the time of a probe stimulation were evoked either by an information encoding probe, or by a non-information carrying cue from a control trial consisting only of a cue stimulation. Schematics of the tasks are given in panel IV and V in Appendix A.

To identify the upper bound we attempted to classify whether epochs extracted directly after the probe stimulation of an experimental trial and directly after the cue stimulation of a control trial belonged to either the control cue or experimental probe. As in our previous study, we suggest that the evoked response from the cue and probe must interact to allow linear separability to occur, thus if the effects of the cue have dissipated by the time a probe stimulation is given, no interaction can take place and the probe response would be identical to that evoked by the control cue stimulation. Thus, we can identify the upper bound by finding the largest probe delays which can be correctly classified above a given threshold.

Identifying the lower bound similarly consisted of identifying if epoched, time binned spike counts belonged to control or experimental trials, however, the epoch extraction method was changed. Instead of extracting epochs directly following both control cue and experimental probe stimulations, we instead extracted the control cue epoch at the same temporal location as the experimental probe (computed relative to the cue stimulation in both control and experimental trials). For instance, for the 1000 ms probe condition we extracted experimental and control trial epochs at 1073 ms following the recorded cue stimulation times. Note here the addition of 73 ms, this is done to account for a set of components of a stimulation block that displaces the stimulation relative to values input in the Axion software and to account for stimulation artefacts, see further details in the digital repository. This ensures that performance of the classifier is solely based on the spike response of the network and is not confounded by effects of the electrical stimulation on the recording. The lower bound is here identified as the smallest probe delay that can be successfully classified above a given threshold. The lower bound task also functions to identify if the probe stimulation had an affect on the spike response of the network as successful classification in this task would require the probe stimulation to elicit spikes in the network, while the lower bound task could successfully classify an epoch solely based on a gradually decaying network response from the cue. Thus, for a probe delay to be included within the dynamic range of a network it has to be successfully classified in both tasks.

Representation Stability. To assess the stability of the networks' representation of the input, we tested the classifiers trained to assess the dynamic range on one DIV and tested them on spike responses elicited from both the same DIV as they were trained on and data from the two other days *in vitro*. If the representations were stable we would expect similar accuracy independent of whether the classifiers were tested on data from the same DIV or a different one. However, if the representations were unstable, we expect to see a decrease in accuracy.

2.6 Decoder

Readout. The decoder-readout extracted 1024 ms epochs. To test for representational stability and dynamic range within different spike code types (from 1 ms spike time to 1 s rate) we time-binned the 1024 ms epoch into time bins from 1 to 1024 ms in powers of 2. By increasing the time bin size we delete the information of the timing of spikes. If the information is contained in the specific timing of spikes, classification will only be possible using smaller time bins, however if the information is present in the overall rate of the response classification will be successful with larger time bins. The epoched and time-binned spike count vectors where then used to train and test the classifiers.

Classifier. We used Scikit-learn's LogisticRegression with settings: penalty = l2, solver = liblinear. Datasets were split into 100 random train test splits using scikit-learn's train_test_split with stratification, and a train-test ratio of 0.2. Average accuracy was used to assess model x time bin x condition performance.

Across both tasks, the logistic classifier was thus tasked with classifying if a given epoch belonged to a control cue trial or a experimental probe trial. We then trained the classifier on 80 % of trials from one DIV. The trained models were always tested on 20% of trials from a given DIV.

3 Results

The different networks all showed some degree of instability both on dynamic range and representation of input, but the level varied. In addition the specific changes were heterogenous. Aggregate results are given in the results section and extended results are given in the online repository at https://osf.io/v6y2e/.

3.1 Dynamic Range

To extract the dynamic range of the networks we found continuous sequences of probe delays in which both the upper bound and lower bound classifiers had an accuracy above 60, 70, 80 or 90% when the training and testing DIV were the same. At 80% and above accuracy threshold only half the networks had any valid dynamic ranges. Furthermore, the valid ranges were mainly present on DIV 19 with only one exception for network B3 which also had valid ranges on DIV 35. With a 70 % accuracy threshold all networks had at least one valid bound on one DIV. We therefore show the dynamic ranges for this threshold level in Fig. 1.

Most networks showed a high degree of instability in their dynamic range as can be seen in the different probe delays covered by the dynamic ranges over different days *in vitro*. A1 only had a single case of two adjacent probe delays performing at above threshold levels. B1 had a wide dynamic range for the smaller time bin sizes, but only on DIV 19 and had no valid bounds at the other days *in vitro*. B3 had the most stable range and showed similar bounds on each DIV for the smaller time bins. However, the bound range was relatively small starting at 200 ms and ending at 600–800 ms.

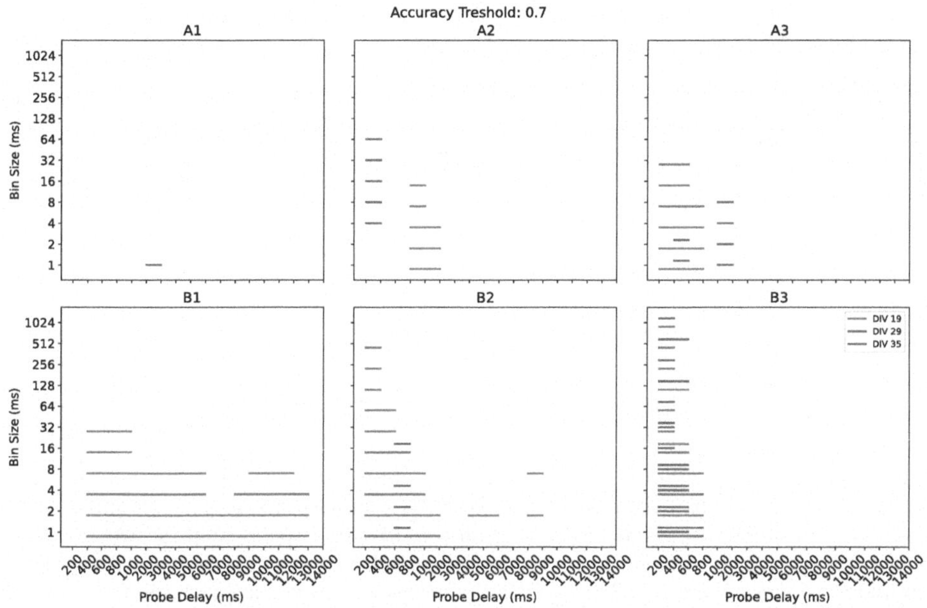

Fig. 1. Continuous ranges of probe delays where both the upper bound and lower bound classifiers could classify epochs e at 70% or higher accuracy when train == test DIV. The smallest and largest probe delay of a continuous sequence would indicate the upper and lower bounds of the dynamic range.

The dynamic ranges were also generally widest when smaller time bins were used which indicates that spike timing code has better capacity than rate code for encoding stimulation timing input.

3.2 Representation Stability

To isolate the effect of representational stability from changes in bounds, we selected upper and lower bound classifiers from overlapping dynamic ranges of at least three probe delays with a 70% accuracy threshold. This only amounted to the B3 network with time bin sizes 1,2,4,8 and probe delays 200, 400 and 600. The mean accuracy for the overlapping bounds across time bin sizes and probe delays are plotted for each training and testing day combination in Fig. 2 in black, with the means for each probe delay averaged over time bin size being shown in colored lines. The same procedure using all valid bounds and for all classifiers with higher than 70% accuracy when train and testing DIV were the same, is shown in the digital repository.

With two exceptions, the mean classifier performance decreased when the training and testing DIV was not the same indicating that the representations were not stable. The same general pattern was also the case for the averaged data when using all valid bounds and all classifiers with higher than 70% accuracy (data in digital repository).

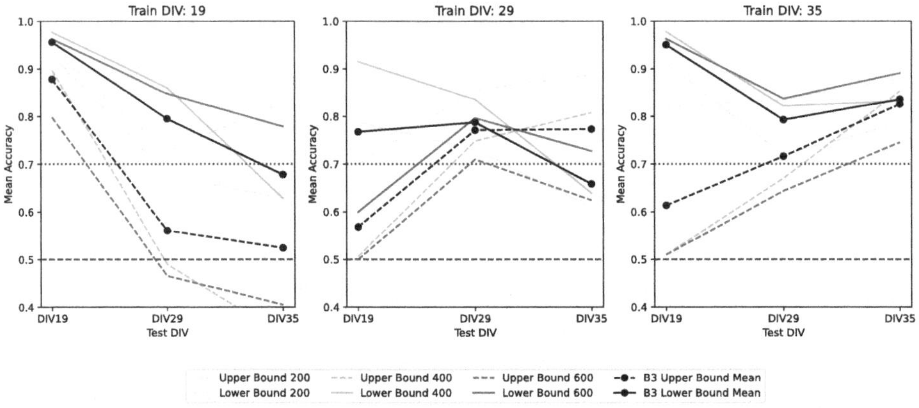

Fig. 2. Change in classifier accuracy when trained on one DIV and tested on all. Data is only shown for B3 as this network was the only one with overlapping bounds across all three experimental days at the 70% accuracy threshold (probe delays: 200, 400, 600 ms). In most cases the mean accuracy decreases when train does not equal test DIV compared to when train equals test DIV. Accuracies from the upper bound is given in dashed lines and from the lower bound is given in solid lines.

4 Discussion and Conclusion

Our main results demonstrate that the dynamic range and representations computed by our networks were unstable between experimental sessions separated by 6–16 days.

This instability poses multiple challenges to using *in vitro* neural networks as computational devices like reservoirs in reservoir computing systems, and raises issues in generalizing results between experiments on the same networks done on different days *in vitro* and of course between studies.

The instability of the dynamic range is particularly problematic. Establishing the dynamic range of a network to a specific encoding method would be necessary to use the networks for a computational task. That the dynamic ranges found in our experiments were so unstable that they generally cannot be generalized to other days *in vitro*, means that re-calibration of the encoder to the dynamic range at each specific DIV would be necessary. The total collapse of the dynamic range on certain days also means that the networks cannot be used to solve computational tasks on these days, at least using stimulation timing encoding. This also has consequences for comparing results from experiments conducted on the same networks but on different days *in vitro*. The instability of the network could easily mask the effects of experimental manipulations which may confound interpretation of the results.

However, tuning the networks to more optimal dynamics may help solve this issue as [28] found better dynamic ranges to input encoded as ampere variations when the networks were in an avalanche critical phase. The stability of the dynamic range while networks are in a critical phase would then need to be explored. Instability could also have been caused by variability in nutrient availability, but current technology exists that can use microfluidic pumps to allow continuous nutrient exchange with the culture medium which could remove this source of instability [15].

However, even if network dynamics can be stabilized, the instability of the input representations may still pose a challenge. How large this issue will be, will depend on the exact cause, whether this is representational drift, learning or development. If the cause is representational drift this will likely require constant re-training which may not be solvable. Learning and development may both stabilize, either when networks reach a fully mature state in the case of developmental effect, or through convergence to optimal solutions if learning caused the changes, meaning that re-training would only be temporarily necessary.

Given that our experiment was conducted together with multiple other experiments designed to assess encoding and phase tuning parameters of the networks we cannot pinpoint which specific mechanism caused the changes in accuracy we observed. Instead, our results suggests that caution should be taken when attempting to utilize parameters found to be optimal in one experiment conducted on one DIV in another experiment conducted at a different DIV. Thus, the next step would be to isolate each potential effect to identify their contribution. For example by inhibiting learning to control for drift versus changes in basal dynamics. However, in some cases, interaction effect may also need to be accounted for. Multiple species, including humans go through critical periods of development where their brains are particularly plastic, and if a specific skill is not learned during this period it may never be learned later in development [21]. Similarly, input stimulation may be necessary during the development of the networks to improve performance on a given task. Indeed, [31] have suggested that the super-critical pattern of activity characterized by large burst may represent a state of arrested development due to networks not receiving sensory input as would be the case in live animals. Thus learning and development may be intrinsically linked and not fully isolatable.

In conclusion we have shown that in vitro neural network demonstrate variable dynamic range and input representation over time. Absence of stable behaviour may thus make exploration of biological neural networks as computational systems challenging. However, we also highlight potential sources of this instability and methods that could partially alleviate them.

Acknowledgments. This work was partially financed by the Research Council of Norway's DeepCA project, grant agreement 286558.

Disclosure of Interests. The authors have no competing interests.

Appendix

A Methods Panel

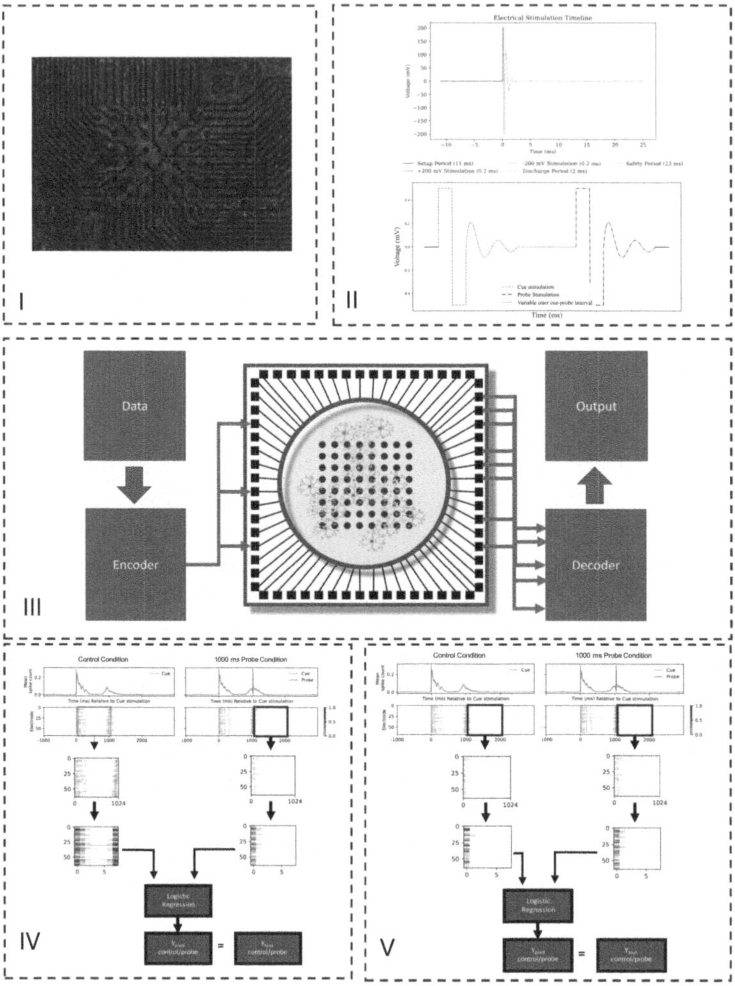

Fig. 3. I. Fluorescent microscopy of an *in vitro* neural network on an MEA interface after immunostaining: neurons (Neurofilament heavy; red), astrocytes (GFAP; green); cell nuclei (Hoechst; blue). II. Components of electrical stimulation and experimental trial design. III. Schematic reservoir system with *in vitro* neural network reservoir on MEA. IV. Upper bound analysis pipeline, raw spike times are epoched, time-binned, then passed to the logistic regression model. V. Lower bound analysis pipeline, following the same steps as the lower bound. (Color figure online)

B Developmental Trajectories

(See Fig. 4).

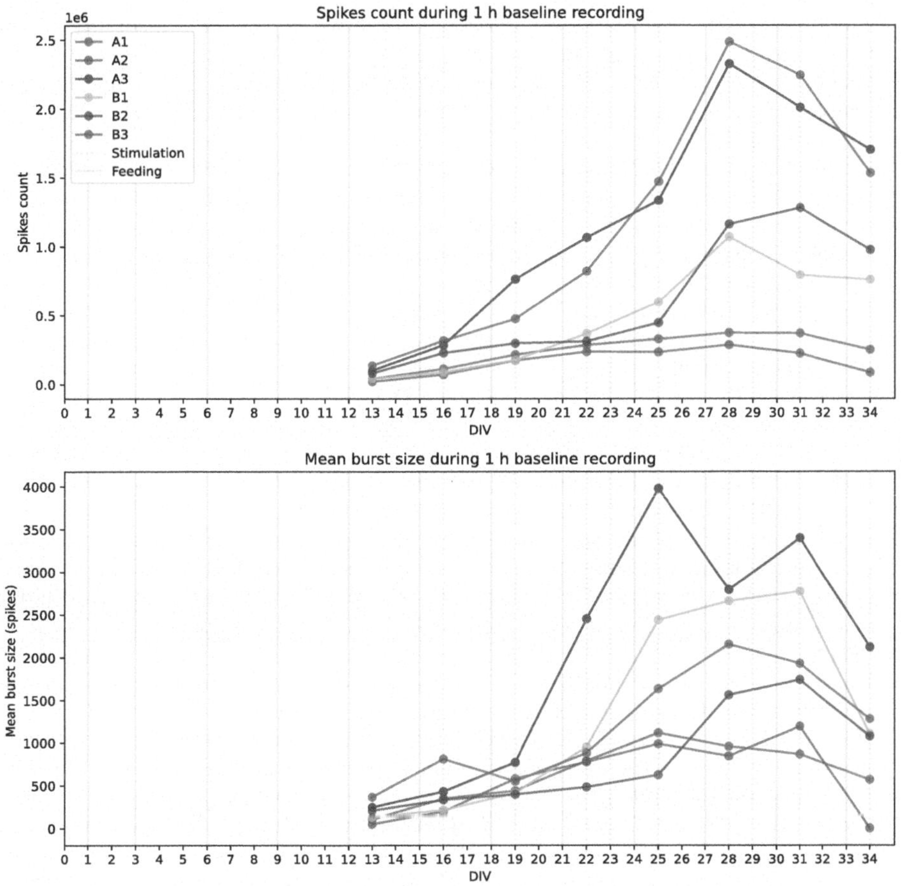

Fig. 4. Developmental trajectory of networks as measured through baseline spontaneous spikes (top) and bursts (bottom) during 1 h baseline recordings. Stimulation days are given as blue vertical lines, feeding days as grey and activity per well as horizontal.

References

1. Aaser, P., et al.: Towards making a cyborg: a closed-loop reservoir-neuro system. In: Artificial Life Conference Proceedings, pp. 430–437. MIT Press One Rogers Street, Cambridge, MA 02142-1209, USA Journals-Info ... (2017)
2. Axion Biosystems: Axis Navigator 2.0 User Guide. Axion Biosystems, 1819 Peachtree Rd, Atlanta, GA 30309, United States (2019)

3. Baljon, P.L., Chiappalone, M., Martinoia, S.: Interaction of electrically evoked responses in networks of dissociated cortical neurons. Phys. Rev. E **80**(3), 031906 (2009)
4. Cai, H., et al.: Brain organoid computing for artificial intelligence. bioRxiv, pp. 2023–02 (2023)
5. Chiappalone, M., Massobrio, P., Martinoia, S.: Network plasticity in cortical assemblies. Eur. J. Neurosci. **28**(1), 221–237 (2008)
6. Clopath, C., Bonhoeffer, T., Hübener, M., Rose, T.: Variance and invariance of neuronal long-term representations. Philos. Trans. R. Soc. B: Biol. Sci. **372**(1715), 20160161 (2017)
7. Cotterill, E., Hall, D., Wallace, K., Mundy, W.R., Eglen, S.J., Shafer, T.J.: Characterization of early cortical neural network development in multiwell microelectrode array plates. J. Biomol. Screen. **21**(5), 510–519 (2016)
8. Deitch, D., Rubin, A., Ziv, Y.: Representational drift in the mouse visual cortex. Curr. Biol. **31**(19), 4327–4339 (2021)
9. Dockendorf, K.P., Park, I., He, P., Príncipe, J.C., DeMarse, T.B.: Liquid state machines and cultured cortical networks: the separation property. Biosystems **95**(2), 90–97 (2009)
10. Dranias, M.R., Ju, H., Rajaram, E., VanDongen, A.M.: Short-term memory in networks of dissociated cortical neurons. J. Neurosci. **33**(5), 1940–1953 (2013)
11. Fernando, C., Sojakka, S.: Pattern recognition in a bucket. In: European Conference on Artificial Life, pp. 588–597. Springer, Cham (2003)
12. George, J.B., Abraham, G.M., Rashid, Z., Amrutur, B., Sikdar, S.K.: Random neuronal ensembles can inherently do context dependent coarse conjunctive encoding of input stimulus without any specific training. Sci. Rep. **8**(1), 1–10 (2018)
13. George, J.B., Abraham, G.M., Singh, K., Ankolekar, S.M., Amrutur, B., Sikdar, S.K.: Input coding for neuro-electronic hybrid systems. Biosystems **126**, 1–11 (2014)
14. Hafizovic, S., et al.: A CMOS-based microelectrode array for interaction with neuronal cultures. J. Neurosci. Methods **164**(1), 93–106 (2007)
15. Heo, R., Kim, H., Lee, K.J.: Dependence of synchronized bursting activity on medium stirring and the perfusion rate in a cultured network of neurons. J. Korean Phys. Soc. **68**(9), 1041–1048 (2016). https://doi.org/10.3938/jkps.68.1041
16. Jaeger, H.: The "echo state" approach to analysing and training recurrent neural networks-with an erratum note. Bonn, Germany: German National Research Center for Information Technology GMD Technical Report **148**(34), 13 (2001)
17. Kagan, B.J., et al.: In vitro neurons learn and exhibit sentience when embodied in a simulated game-world. Neuron **110**(23), 3952–3969 (2022)
18. Lindell, T.A.E., Ramstad, O.H., Sandvig, I., Sandvig, A., Nichele, S.: Information encoding and decoding in in-vitro neural networks on micro electrode arrays through stimulation timing (2024). arXiv:2404.10946 [q-bio.NC]
19. Maass, W., Natschläger, T., Markram, H.: Real-time computing without stable states: a new framework for neural computation based on perturbations. Neural Comput. **14**(11), 2531–2560 (2002)
20. Marks, T.D., Goard, M.J.: Stimulus-dependent representational drift in primary visual cortex. Nat. Commun. **12**(1), 5169 (2021)
21. Morishita, H., Hensch, T.K.: Critical period revisited: impact on vision. Curr. Opin. Neurobiol. **18**(1), 101–107 (2008)
22. Nikolić, D., Haeusler, S., Singer, W., Maass, W.: Temporal dynamics of information content carried by neurons in the primary visual cortex. In: Advances in Neural Information Processing Systems, vol. 19 (2006)

23. Ortman, R.L., Venayagamoorthy, K., Potter, S.M.: Input separability in living liquid state machines. In: International Conference on Adaptive and Natural Computing Algorithms, pp. 220–229. Springer, Cham (2011)
24. Pasquale, V., Massobrio, P., Bologna, L., Chiappalone, M., Martinoia, S.: Self-organization and neuronal avalanches in networks of dissociated cortical neurons. Neuroscience **153**(4), 1354–1369 (2008)
25. Pimashkin, A., Gladkov, A., Agrba, E., Mukhina, I., Kazantsev, V.: Selectivity of stimulus induced responses in cultured hippocampal networks on microelectrode arrays. Cogn. Neurodyn. **10**(4), 287–299 (2016). https://doi.org/10.1007/s11571-016-9380-6
26. Rule, M.E., O'Leary, T., Harvey, C.D.: Causes and consequences of representational drift. Curr. Opin. Neurobiol. **58**, 141–147 (2019)
27. Rutten, W.L., Van Pelt, J.: Activity patterns of cultured neural networks on micro electrode arrays. In: 2001 Conference Proceedings of the 23rd Annual International Conference of the IEEE Engineering in Medicine and Biology Society, vol. 1, pp. 737–740. IEEE (2001)
28. Shew, W.L., Yang, H., Petermann, T., Roy, R., Plenz, D.: Neuronal avalanches imply maximum dynamic range in cortical networks at criticality. J. Neurosci. **29**(49), 15595–15600 (2009)
29. Tanaka, G., et al.: Recent advances in physical reservoir computing: a review. Neural Netw. **115**, 100–123 (2019)
30. Tetzlaff, C., Okujeni, S., Egert, U., Wörgötter, F., Butz, M.: Self-organized criticality in developing neuronal networks. PLoS Comput. Biol. **6**(12), e1001013 (2010)
31. Wagenaar, D.A., Madhavan, R., Pine, J., Potter, S.M.: Controlling bursting in cortical cultures with closed-loop multi-electrode stimulation. J. Neurosci. **25**(3), 680–688 (2005)
32. Wagenaar, D.A., Pine, J., Potter, S.M.: Effective parameters for stimulation of dissociated cultures using multi-electrode arrays. J. Neurosci. Methods **138**(1–2), 27–37 (2004)
33. Wagenaar, D.A., Pine, J., Potter, S.M.: Searching for plasticity in dissociated cortical cultures on multi-electrode arrays. J. Negat. Results Biomed. **5**, 1–19 (2006)

Latent Agency: A Proto-Cognitive Property

Fernando Rodriguez(✉) and Phil Husbands

AI Research Group, Department of Informatics, University of Sussex, Brighton, UK
f.rodriguez-vergara@sussex.ac.uk

Abstract. Selectivity can be understood as a causal property that determines variable degrees of response specificity. In minimal autonomous systems, given their recursive dynamics, these responses are directed, not towards an hypothetical representation of the environment, but towards its own future states. In this sense, although co-specified by environmental circumstances, system state transitions can be characterized as dynamically more or less driven by an intrinsic 'steering force' underpinned by the degree of freedom that the organized states can provide under the influence of different environmental circumstances. In this work we present the idea of (latent) agency as a proto-cognitive property already present in autonomous systems and that can be generally measured through information metrics. We test these ideas and formulations through toy-experiments in the Game of Life cellular automaton.

Keywords: Agency · Proto-cognition · Autonomy · Information

1 Introduction

Intuitively, agency can be portrayed as the capacity of a given system to determine, at least to some extent, its own behavioral trajectory (to 'decide' what to 'do'). Or, in other words, to display some degree of a causal decoupling from the world that both surrounds it and realizes it. Also intuitively, we often think of agency as an evolutionary trait or capacity, that would have arisen at a relatively late stage, after complex adaptive intelligent systems were in place. It seems somehow –at least from our experience, to be a natural solution to a type of problem for which more archaic forms of adaptive behavior (like pure structural adaptivity or reinforcement learning) are insufficient or less well fit; roughly speaking, any situation requiring some form of improvised responses and/or truly goal-directed actions, such as, for instance, 'simple' tasks like driving a car. This stage-like-leap is appealing because agency can be conceived as a higher degree of intelligence, stemming from more complex cognitive structures. Unfortunately though, beyond intuitions of this kind, our actual understanding of the underlying mechanisms supporting any real form of agency are still very unclear and have been a matter of increasing debate [4,12,15,17,28,32,37].

T. Carletti et al. (Eds.): WIVACE 2024, CCIS 2532, pp. 292–306, 2025.
https://doi.org/10.1007/978-3-031-93631-9_23

Autonomous systems are organizationally closed, meaning that ongoing processes will recursively determine state transition from valid (i.e. viable) structural states into further valid states, where the whole set of viable states that a system can display without disintegrating, will be what defines its (autonomous) organization [23,24,35,42,45]. In this context, we propose to take a look back to the basics and start from the notion of selectivity. As we know, selectivity can be understood as a causal property that determines variable degrees of response specificity (depending on the complexity of the system, the environmental circumstances and the relation between them). Starting from this basic idea, we claim that, although physically co-specified by environmental circumstances, system state transitions in minimal autonomous (hence organizationally closed) systems can be characterized as dynamically more or less driven by an intrinsic 'steering force', underpinned by the degree of freedom that the organized sequential states can provide under the influence of different environmental circumstances. This is to say that, while the system's ongoing processes are unavoidably determined to an extent, a minimal form of agency will require a degree of under-determination in order to actively respond (as opposite to passively undergo viable transitions, hence depending only on an environmental "good-will"), this first condition arising from the structural mechanisms of the system will later enable enactions in a goal-directed fashion, therefore with an agential connotation. Simply put, while we know that both the states of the environment and the system will determine the upcoming state of the system, these contributions may not always be equal (less even an exact half-and-half), therefore naturally posing the question about the dynamics of this relation. Being thus, in the remainder of this manuscript we will explore this notion of a latent, proto-cognitive form of agency in terms of the asymmetrical causal contributions that the system and the environment exert (in spite of their inseparable nature) over the future states of the system and the possibility to formally approach a method for quantifying it.

2 Causality and Information

Building on Bateson's idea of "differences that make a difference" [6] and along similar conceptual lines about selectivity that we have examined until now, the Integrated Information Theory of Consciousness (IIT) [2,3,30,39] introduces the formal notion of intrinsic information, in order to quantify the causal effect a system, given its structure, has over itself. Roughly speaking, the IIT claims that integrated information, a measure of emergent causality (often also expressed in terms of causal power), reflects the degree to which a given system intrinsically exists as such (so independent of external observation), hence, as a causally decoupled entity that becomes a phenomenological observer itself [31,40]. Technically, the goal of the IIT is to provide a formal framework for quantifying consciousness from physical mechanisms, under the premise that the ontological existence of the system entails a phenomenological identity [2,40], under the broader paradigm of causal emergence [52].

Basically, intrinsic information is the idea that the causal links of an irreducible system have to be self-contained. In turn, cause-effect information is the purported measure for the degree of causal-power exerted, *within* and over itself, by the very existence of a system/mechanism (and a hint of the implied intrinsic subjective observer), given the state transition probabilities from past and to future states, applied to the minimal mechanisms (units) of a candidate system [2]. On these grounds, causality becomes fundamentally linked to selectivity, which is understood as informative (with a connotation of specificity), insofar as the state of a system and its mechanisms logically specify a finite set of possible past and future states while discarding others [2,22,26,30]. From there, integrated (intrinsic) information (ϕ) is a second step; somehow an attempt to formally solve the combination problem [13] through the notion of a causally emergent unity, assuming that each of the elements of the system may have some degree of phenomenological experience on its own, but that they *integrate* into an irreducible physical/phenomenological mechanism. Simply put, integrated information is the aggregation of what intrinsic information reveals.

Although the theory has gathered a lot of interest, it has also received plenty of criticism; pointing to its lack of a principled justification for the leap from maximally integrated information to consciousness [16,27,41]; inconsistencies among different measures for integrated information [26]; unconvincing mechanism accounting for temporality, especially regarding the exclusion principle [29,34,38]; and the apparent impossibility of consistently apply its methods to more complex systems [1,25], among others. In this respect, while we don't commit to the axiomatic postulates of the IIT [2,30,40], nor believe there is a strong enough reason to accept the proposed causality-consciousness identity, we do praise the attempt of orienting the discussion into a scientific domain. Moreover, we believe that, as others have done by building on some of these ideas to approach particular problems, usually with respect to the linked notion of emergence [14,46], the application of the specific notion of intrinsic information may be quite fruitful in the context of our present investigation about agency, under a strict selectivity-causality interpretation. We will henceforth develop this idea.

To this end, we will make use of the cause and effect repertoires introduced in [30]. IIT repertoires assign a markovian probability to every possible state transition, thereby producing two probability distributions; a cause (past-to-present) repertoire (ci) and an effect (present-to-future) repertoire ei.

$$ci = EMD\,(p\,(S^p|S^c)\,||\,p^{uc}(S^p) \tag{1}$$

$$ei = EMD\,(p\,(S^f|S^c)\,||\,p^{uc}(S^f) \tag{2}$$

where the terms S^p (past), S^c (current) and S^f (future) refer to the system states, considered in the context of subsequent timesteps (hence, equivalent to S^{t-1}, S^t and S^{t+1}). ci and ei stand for cause and effect information respectively, and cei for cause-effect information quantifying the causal power measurable from the system. EMD stands for the Earth Mover's Distance [36,51] which is applied to compare cause/effect repertoires against non-intrinsically causal, unconstrained conditions (denoted by the superscript uc) to provides a measure of how much the system determines its own behavior (state transitions) away from random changes or pure external determination, such as ordinary entropy decay. From these, cause-effect (intrinsic) information is obtained by taking the minimum between them: $cei = min(ci, ei)$. The minimum reflects the shared degree of causality that we can ascribe to its structure in both directions.

Roughly speaking, the notion of cause and effect repertoires can be simplified and illustrated as in Fig. 1, where the sets S^{t-1} and S^{t+1}, both contain all the possible states of the system, so that:

$$S^{t-1} = S^{t+1}$$

$$\{x_1, ..., x_n\} = \{z_1, ..., z_n\}$$

And where the multiple arrows connect each of these states to the current state of the system (represented by y). Hence, in this case, IIT repertoires assign a probability to each of these arrows, thereby producing two weighted mappings; a cause repertoire, which assigns a probability related to every possible (past) state of the system at $t-1$, and an effect repertoire which does the same for the upcoming (future) states at $t+1$.

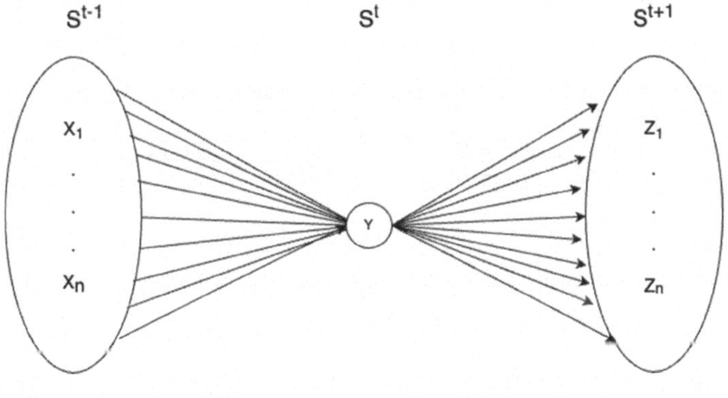

Fig. 1. A simplified illustration of cause and effect repertoires. The sets S^{t-1}, S^t, S^{t+1} represent past, current and future states respectively; repertoires assign probabilities to each of the arrows. See main text for further details.

3 A Toy Case Scenario from the Game of Life

Since cybernetics, toy models have often been used as proof-of-concept in cognitive science and artificial life [18,20,43,44]. In this same vein, we will first test whether intrinsic information could be applied to approach a measure of causality (or agency), by comparing and interpreting some information measures for the simplest case to be found in the Game of Life; namely, a single cell with only two states and its environment, made of the surrounding 8 cells also known as

Moore neighborhood [50], that can display 256 different environmental configurations (512 overall). From the rules of the GoL [8,11,19] we know that a cell can only be 'alive' (active) if it is already active and there are 2 or 3 active cells in its neighborhood, or, if not currently active, only if the sum of active cells in its neighborhood is exactly 3. This can also be expressed as:

$$C_y = \begin{cases} 1 & \text{if } C_x = 1 \text{ and } \sum M(C_x) = 2 \vee \sum M(C_x) = 3 \\ 1 & \text{if } C_x = 0 \text{ and } \sum M(C_x) = 3 \\ 0 & \text{otherwise} \end{cases} \tag{3}$$

where $M(C_x)$ stands for the Moore neighborhood of the cell $C_x \rightarrow C_y$. Note that the neighborhood in this case, corresponds to E_x (environmental category) instead of e_x (environmental instance), because for the cell-system the specific configurations do not make any difference, only their sum is important. And since there are: 28 combinations for $E_x = 2$, 56 combinations for $E_x = 3$ and 172 combinations for the remaining non viable (i.e. deactivation) alternatives, this give us the following counts:

$$C_x \rightarrow C_y = \begin{matrix} \\ 0 \\ 1 \end{matrix} \begin{Vmatrix} 0 & 1 \\ 200 & 56 \\ 172 & 84 \end{Vmatrix} \tag{4}$$

From where we can derive the probability matrices:

$$T(C_x) = \begin{vmatrix} 0.78 & 0.22 \\ 0.67 & 0.33 \end{vmatrix} \quad ; \quad T(C_y) = \begin{vmatrix} 0.54 & 0.46 \\ 0.4 & 0.6 \end{vmatrix}$$

Now then, we can follow the IIT formulation (from Eqs. 1–2) and calculate information, by first computing cause and effect repertoires based on the current state of the cell. For the sake of the example, we will start with $cell = 1$:

$$crep(cell = 1) = p(cell^p \mid cell = 1) = (\frac{56}{140}, \frac{84}{140}) = (0.4, 0.6)$$

$$erep(cell = 1) = p(cell^f \mid cell = 1) = (\frac{172}{256}, \frac{84}{256}) = (0.67, 0.33)$$

Where *crep* and *erep* stand for cause and effect repertoires and the super-indices p and f refer to past and future states respectively.

We can also obtain the unconstrained (past) UC^p and (future) probability distributions UC^f through direct counting. First, by considering that UC^p represents the probabilities of the past state of the system without any knowledge of its current state (transition into C_y if looking at Eq. 4), so unconstrained by it, which logically imply a uniform distribution. Similarly, UC^f represents the probabilities of the future state of the system without any causal input from (or, again, unconstrained by) S_x, which is the same as not having any knowledge about the current state of the system, hence a vertical sum of the elements in the transition matrix from Eq. 4.

Put differently, whereas the unconstrained past is formalized as a simple homogeneous distribution (assuming unconstrained outputs), the unconstrained future distribution is taken as independent of the current state of the system, although still dependent on its causal structure (i.e. as a system with unconstrained inputs). Formally, this can be concretely expressed as:

$$UC^{p} = p(\sum_{i=1}^{n}(C_{y_i}|C_x = 0), \sum_{i=1}^{n}(C_{y_i}|C_x = 1)) = (0.5, 0.5)$$

$$UC^{f} = p(\sum_{i=1}^{n}(C_{x_i}|C_y = 0), \sum_{i=1}^{n}(C_{x_i}|C_y = 1)) = (0.73, 0.27)$$

Then, information is computed by comparing cause/effect repertoires against their unconstrained reciprocal distributions:

$$ci = EMD(crep(cell = 1) \mid\mid UC^{p}\ ) = EMD((0.4, 0.6) \mid\mid (0.5, 0.5)) = 0.1$$

$$ei = EMD(erep(cell = 1) \mid\mid UC^{f}\ ) = EMD((0.67, 0.33) \mid\mid (0.73, 0.27)) = 0.0547$$

Thus $cei = min(ci, ei) = 0.0547$. Which, leaving aside hypothetical connection to phenomenological properties of any kind, is basically an expression of the fact that our knowledge about the ON state of a single cell is informative to the extent it give us an insight into its past and future states. From this it follows that, when put in terms of causality, may be interpreted as the constraints that the state of the ON cell places upon its state transitions, hence the degree of self-determination with respect to its environment (i.e. the Moore neighborhood within the context of our example). Then, by repeating the process for $cell = 0$ we obtain:

$$ci = EMD((0.54, 0.46) \mid\mid (0.5, 0.5)) = 0.0376$$

$$ei = EMD((0.78, 0.22) \mid\mid (0.73, 0.27)) = 0.0547$$

Then making $cei = min(ci, ei) = 0.0376$. This, as it could be expected, entails that active cells on the grid has a slight higher causal power (0.017) than non-active cells, which is reasonable considering the amount of transitions leading to ON and OFF states by the dynamics of the Game of Life.

As a final note may be important to mention that there is no need for ulterior integration calculations or analysis, as presented within the formal framework of the IIT because the object of study we have selected (a single cell on the grid) is the minimal possible one.

4 Some Considerations

Aside from the general criticisms to some aspects of the IIT that we introduced above, there are a few specific aspects that require further examination. In concrete, we will move away from the past-current-future view by examining only the two states of the system involved in any system transition. By recalling the formulation of repertoires from Eqs. 1 and 2, we can express a single causal direction as:

$$rep_{xy}(S_x) = p(S_y \mid S_x = X) \tag{5}$$

Here, rep_{xy} corresponds to the effect repertoire, which, while conceptually different, is formally the same (X represent the specific state of S_x). The notation has been modified to avoid unnecessary confusions further on.

Tentative starting points could have been the enactivist notion of intentional (in the sense of directedness) action [21,49] or interactional asymmetry [5,15]. However, as we discussed above, we'd like to avoid cognitive properties of higher order than minimal autonomous systems. Being thus, we will leave aside from our constructs the idea of the future-environment element and incorporate the notion of asymmetry only in terms of the influence that the environment has over the system, in order to enable a formal method to later examine it in opposition to that of the system upon itself. To this end, we will introduce a second repertoire:

$$rep_{ey}(E_{xy}) = p(S_y \mid E_x(S_x = X)) \tag{6}$$

where the term E_x refers to the equivalent categories available to the system, S_x to the state of the same system and X to some specific state. The argument is E_{xy} because rep_{ey} is a weighted mapping of the equivalent category defined by S_x and S_y. Thus, it expresses the probability of the next state of the system, considering the interpretations that the system can make from the conditions of the environment.

5 Coming Back to GoL

As mentioned, the objective environmental instances is the set all possible configurations that the 8 cells in the neighborhood can produce, whereas the environmental sets are only 3 and are a product of their sums, so that: $C(8,2) = 28$, $C(8,3) = 56$ and $C(8,q) = 172$. The symbol q refers to the rest of the sums: $E_{xy} \neq 2 \vee 3$, whereby only non-active states are generated. From this, we can systematize transitions as enactions (see Table 1).

We have previously examined this system on its own, then, we now need to derive the possible transitions only as a function of the environment (i.e., the environment-system component of the coupling). This is presented in Table 2.

After this, we finally are in position to derive the environmental causal components as probability distributions (repertoires) from Table 2. As it can be inferred, the only environmental category that enables some degree of causal freedom to the system is $E_{xy} = 2$, because for all cases in which the sum of the surroundings cells is 3, the subsequent state of the single cell in the centre will unavoidably be ON ($C_y = 1$), independently of its own state. Conversely, for any other case, the subsequent state of the single cell at hand will be OFF ($C_y = 0$). This exemplifies different degrees of causal environment-state determination and can be related to the entropy of these distributions (in bits):

$$H(rep_{ey}(E_{xy} = 2)) = H(0.5, 0.5) = 1$$

Table 1. Table displaying the possible system-environment combinations and their resulting state transitions, for a single cell in the GoL.

$(S_x, E_{xy})/C_y$	0	1	Total
$(0, 2)$	28	0	28
$(0, 3)$	0	56	56
$(0, q)$	172	0	172
$(1, 2)$	0	28	28
$(1, 3)$	0	56	56
$(1, q)$	172	0	172
Total	372	140	512

Table 2. Table displaying only environmental states encountered before transitioning and their correspondent resulting states. Hence, as environment-to-system influence

E_{xy} / C_y	0	1	Total
2	28	28	56
3	0	112	112
q	344	0	344
Total	372	140	512

$$H(rep_{ey}(E_{xy} = 3)) = H((rep_{ey}(E_{xy} = q)) = 0$$

Therefore, illustrating the two extreme cases; the former shows how for some environmental categories, the selectivity of the system allows a degree of under-determination that may be exploited by the system. Conversely, the other two cases totally preclude any kind of under-determination. In both cases the system is self-determined, the key difference is that, for E_{xy} what determines the state transition (so the enaction) is the state of the system (C_x). The entropy of the correspondent distributions given by rep_{xy} can be calculated accordingly:

$$H(rep_{xy}(C_x = 0)) = H(0.78, 0.22) = 0.76$$

$$H(rep_{xy}(C_x = 1)) = H(0.67, 0.33) = 0.914$$

Regarding the environment, as we have seen, entropy tell us the degree to which a system could have the possibility to exert causal effects in its own future states, if any at all. This seems intuitively easier to understand it the other way around; a probability distribution with very uncertain outcomes (like one close to a uniform distribution), would probably provide a smaller possibility for the system to influence its future states. Because in order for the system to self-determine its states it needs to exploit the environmental under-determination, for which the more focused tendencies will be represented by lower entropy values. Therefore, likely ideal conditions would be, at least generally speaking, high environmental entropy along with low system entropy levels.

Given that totally determined and uniform probability distributions are the extreme cases for lowest and highest entropy values respectively, and considering that we'd like for the system-system and the environment-system cases to be as close as possible to said type of distributions; we can apply a distance information measure to see how far are the actual repertoires from the ideal ones. For this, as we have mentioned above, we will make use of the Earth Mover's Distance. Hence, for the first environmental case ($E_{xy} = 2$) we obtain:

$$\delta_U = EMD(\, rep_{ey}(E_{xy} = 2) \,||\, pU \,) = 0$$

where pU stands for the discrete uniform distribution and δ_U for the difference to the ideal case (how far from total under-determination). Then, for the totally determined cases ($E_{xy} = 3$ and $E_{xy} = q$) we get:

$$\delta_U = EMD(\, rep_{ey}(E_{xy} = q) \,||\, pU \,) = 0.5$$

Which, given the simplicity of the example at hand, can be interpreted directly like the ideal environment-system condition (insofar as eventually allowing some form of agency) and the total opposite scenario, respectively. As it can be seen, the lower the entropy value, the higher the chances for agency. Put another way, δ_U captures the lack of under-determination, or how (potentially) entangled the response of the system is with respect to the interpretation that it does from the environment.

We believe, however, that it is not necessary to incur an unfeasible number of calculations (like power-set permutations or something alike), but only to compare the repertoire against the totally determined case in which the highest probability within the repertoire is taken as the only possible (i.e.; $max(rep_{xy}(C_x = X)) = 1$, while all remaining elements are made zero), as this would represent the minimum value for such change to be possible and any other case will be therefore, less likely.

Accordingly, we will denote this construction as pF (focused) and δ_F as the difference (in work, or energy) required to transform the $rep_{xy}(C_x)$ into pF. For the system repertoires we obtain:

$$\delta_F = EMD(\, rep_{xy}(C_x = 0) \,||\, pF \,) = 0.22$$

$$\delta_F = EMD(\, rep_{xy}(C_x = 1) \,||\, pF \,) = 0.33$$

Which basically indicate the distances (i.e., amount of energy/work) that would be required to transform the state of the system into the ideal (the most focused) case possible. Unlike the previous case (δ_U), for δ_F the interpretation is a bit less straightforward, because it conceptually requires us to embrace the impossibility of mind-like phenomena. Otherwise, we would probably wish to compare also against a uniform distribution, along the lines of δ_U, expecting the *agent* to be capable of meaningfully acting by exerting as low causal power as possible (so to look for possible sparks of agency). This probably requires further elaboration that is out of the scope of the current work, however, we will briefly expand on this on the Discussion section.

Finally, as we anticipated, we can also apply EMD to compare between the repertoires. Given that we know that for $E_{xy} = 3$ and $E_{xy} = q$ there's no possible under-determination, we will just examine the $E_{xy} = 2$ case:

$$\delta_{xe}(C_x = 0) = EMD(\,rep_{xy}(C_x = 0) \;||\; rep_{ey}(E_{xy} = 2)\,) = 0.28$$

$$\delta_{xe}(C_x = 1) = EMD(\,rep_{xy}(C_x = 1) \;||\; rep_{ey}(E_{xy} = 2)\,) = 0.17$$

Which, again, is simple to relate to previous results and our knowledge of the minimal dynamics of the isolated cell in the GoL. Here, as we should have expected from the above EMD comparisons, we get that the distance between the repertoire for $C_x = 1$ and $E_x = 2$ is less than for $C_x = 0$. Certainly, the numerical results we have seen until now are quite evident themselves, or at least seem to be easy to follow; this, however, will not always be the case as the complexity of the systems being examined increases. In this sense, our goal has been to demonstrate the minimal possible case in the most intuitive fashion, so that we could develop an intuition about these methods. In the following section, we will explore a more complex case, from dynamics patterns in the Game of Life.

6 A More Complex Case from the GoL

Patterns in *Life* have different probabilities of maintaining their structural states, transitioning into new structures (i.e., instances of the same organization), or to disintegrate, as a consequence of their interaction with the environment, which has proven to be a good ground for toy models of autopoiesis and autonomy [7–10, 33]. Being thus, and in order to further test our metrics, we analyzed transitions among sets of possible patterns and applied the same measurements that we used for a single cell (the patterns we will discussed are presented in Fig. 2).

Fig. 2. Some of the patterns of the GoL we investigated. Green represents active cells, black non-active cells acting as the 'membrane' of the system, whereas blue represent the environmental cells surrounding the system, which may be active or not. (Color figure online)

To this end, we simulated all the possible transitions all of these (and other) patterns could undergo (2^20 to 2^24, depending on the number of environmental cells) and searched for the number of occurrences of the same patterns in the grid domain after transitions, then made the system and environment repertoires by computing the probability distributions from the counts and, finally, measured the entropy of each individual distribution, while making comparisons between them using the EMD. While for reasons of space we cannot display full tables or results from these experiments, we will provide some examples to illustrate the feasibility of these methods. First of all, given our knowledge of the transition counts, to obtain the system-system repertoires and entropy is a very much straightforward process (results are displayed in Table 3).

Table 3. Entropy values obtained from the $S_x \rightarrow S_y$ repertoires, measuring system-system influence. H_x stands for $H(rep_{xy}(S_x))$, so for the entropy of such distributions.

S_x	blinker	pb0	block	gliderA	gliderB	flag	tetrisT	tetrisL	worm	boat
H_x	2.672	2.666	1.852	2.853	1.676	3.053	1.329	2.623	2.934	2.398

After this, in order to obtain a measure of the effect of the environment over the system, we first, following the steps from the previous section, separated the possible state-environmental categories combinations to then, look for the subsets of these environmental categories that (the elements e_i within each E_{xy}) could have been interpreted differently if the state of the system were to be different. Thus, given some enaction $(S_x, e_i) \rightarrow S_y$, where $e_i \in E_{xy}$, we want to find at least a set $K_x \subset E_{xy}$, for which an arbitrary (viable) state S_u, under the same environmental circumstances e_i, could have produced another enaction: $(S_u, e_i) \rightarrow S_z$, where $S_u \neq S_z$ and $S_z \neq S_y$, and where the set $Alt(S_x) = \{K_x^1, ..., K_x^n\}$ describes all the alternative interpretations for each of the n states of the system. In other words; because we know that, if there are no alternative transitions to some $S_x \rightarrow S_y$ given some external conditions, then S_y is being determined by the environmental perturbation on the system, inasmuch as it (the system) does not have any interpretation availability except for the one at hand.

For instance, if we start by looking at the transitions between the 2 canonical states of the glider (denoted as 'gliderA' and 'gliderB' in Fig. 2, gA and gB respectively from now on), we find that, for $gA \rightarrow gB$ there is just one alternative ($worm \rightarrow gB$) accounting only for 32 cases of a total of 32,800. Leading to $H(rep_{ey}(gA, alt(gB)) = 0.0112$. Similarly, for the opposite state transition $gB \rightarrow gA$, we get: $H(rep_{ey}(gB, alt(gA)) = 0.0651$, which although higher due to more structural alternatives (including a recurrent case to gA ($n = 32$)), is still quite low compared to the entropy values of xy repertoires presented in Table 3. This, of course, is to be expected from a toy scenario such as GoL, especially taking into the account the low level of complexity of the patterns at hand.

Ideally, as we have discussed above, we would like to encounter an entropy value as closest to zero as possible for the $S_x \rightarrow S_y$ repertoire (i.e., strongly determined) and, conversely, the highest possible for the entropy of the $E_x \rightarrow S_y$ repertoire, so close to a uniform distribution, which would imply that the system has as many responses for a given environmental case, as viable states. With this in mind, we would examine so cases along these lines.

Table 4. Comparison of entropy values obtained from the repertoires correspondent to the GoL patterns from Fig. 2. S_x stands for the state of the system, while $HE(S_u)$ is a shortage for $H(rep_{ey})(S_x, alt(S_u)))$, hence for the environment-system repertoires. The subindices $ifmax$ and min refer to the next maximum entropy value for each S_x (or the next after blinker and pb0, if lower) and minimum respectively (further details in the main text).

S_x	$HE_{alt}(blinker)$	$HE_{alt}(pb0)$	HE_{ifmax}	HE_{min}
blinker	0.633	1.322	tetrisL = 0.381	$gA = 0.030$
pb0	nf	1.583	block = 0.389	$gB = 0.089$
block	nf	0.570	flag = 0.687	block = 0.133
gliderA	0.235	1.251	tetrisL = 0.151	$gB = 0.011$
gliderB	0.767	1.340	tetrisL = 0.266	$gA = 0.065$
flag	0.644	0.949	$gB = 0.352$	$gA = 0$
tetrisT	2.060	nf	flag = 0.591	tetrisT = 0.525
tetrisL	1.352	1.853	flag = 1.139	$gA = 0.120$
worm	0.986	0.929	worm = 0.524	$gB = 0.095$
boat	1.445	1.314	flag = 0.740	tetrisL = 0.373

As it is visible from the results in Table 4, there is a general tendency towards lower values from environment-system entropy measures, along the lines we have discussed until now. As a matter of fact, the only case in which $H(rep_{ey}) > H(rep_{xy})$ (i.e., that the environment-system potential influence is lower than that of the reciprocal system-system influence), is for the *tetrisL* pattern, when transitioning into a blinker.

Clearly this does not entail that this pattern has an agency in the cognitive connotation of the term, nor that it is choosing to undergo that specific state-transition, which would require representational and probably also phenomenological properties which are alien to minimal autonomous systems. Notwithstanding, the point that we would like to make here is that the possibility of choosing seems to exist, or at least, the formal development we have presented suggests so. And being so, we reckon that future work should be focused on understanding how could cognitive systems exploit this feature.

Previously, in more autopoietic views of cognition [47,48], in the case of a coupled system-environment pair, it was asserted that physically there is a symmetry that autonomous systems are not capable of breaking (unless we invoke higher order cognitive processes). We have described here, though, how another kind of asymmetry may be viable. Namely, the determination of the future state of the system itself, which does not require any form of mental representation whatsoever and for which we believe it is possible to characterize a non homogeneity of causal contributions. For this purpose, we have developed some formulations that can be applied in discrete cases, albeit we acknowledge that a more direct formulation instead of a series of comparisons would be a great advance. In this sense, the key difference between environment and (autonomous) system, is that the latter remains as such, instead of basically dissolving into something new. In particular, we suggest the origin of agency, insofar as a minimal property enabling under-determination, can be traced (logically and evolutionarily) to the mechanistic dynamics of autonomous organizations, hence prior to biological or mental phenomena.

References

1. Aguilera, M., Di Paolo, E.: Integrated information in the thermodynamic limit. Neural Netw. **114**, 136–149 (2019)
2. Albantakis, L., et al.: Integrated information theory (IIT) 4.0: formulating the properties of phenomenal existence in physical terms. PLoS Comput. Biol. **19**(10), e1011465 (2023). https://doi.org/10.1371/journal.pcbi.1011465
3. Balduzzi, D., Tononi, G.: Integrated information in discrete dynamical systems: motivation and theoretical framework. PLoS Comput. Biol. **4**(6), e1000091 (2008). https://doi.org/10.1371/journal.pcbi.1000091
4. Baltieri, M., Iizuka, H., Witkowsi, O., Sinapayen, L., Suzuki, K.: Hybrid life: integrating biological, artificial, and cognitive systems. WIREs Cogn. Sci. e1662 (2023). https://doi.org/10.1002/wcs.1662
5. Barandiaran, X., Di Paolo, E., Rohde, M.: Defining agency: individuality, normativity, asymmetry, and spatio-temporality in action. Adapt. Behav. **17**(5), 367–386 (2009). https://doi.org/10.1177/1059712309343819
6. Bateson, G.: Steps to an echology of mind: collected essays in anthropology, psychiatry, evolution and epistemology. Jason Aronson (1972)
7. Beer, R.: Autopoiesis and cognition in the game of life. Artif. Life **10**, 309–326 (2004)
8. Beer, R.: The cognitive domain of glider in the game of life. Artif. Life **20**, 183–206 (2014)
9. Beer, R.: Characterizing autopoiesis in the game of life. Artif. Life **21**, 1–19 (2015)
10. Beer, R.: On the origins of gliders. In: Proceedings of the ALIFE 2018: The 2018 Conference on Artificial Life. ALIFE2018: The 2018 Conference on Artificial Life. Tokyo, Japan, pp. 67–74 (2018). https://doi.org/10.1162/isal_a_00019
11. Berlekamp, E., Conway, J., Guy, R.: Winning Ways for Your Mathematical Plays, vol. 2. Academic Press, New York (1982)
12. Biehl, M., Virgo, N.: Interpreting systems as solving pomdps: a step towards a formal understanding of agency. In: Buckley, C.L., et al. (eds.) Active Inference, IWAI 2022. Communications in Computer and Information Science, vol. 1721. Springer, Cham (2023). https://doi.org/10.1007/978-3-031-28719-0_2
13. Chalmers, D.: The combination problem for panpsychism. In: Godehard, B., Ludwig, J. (eds.) Panpsychism. Oxford University Press (2017)
14. De Rosas, F., Mediano, P., Jensen, H., Seth, A., Barret, A., Carthart-Harris, R.: Reconciling emergences: an information-theoretic approach to identify causal emergence in multivariate data. PLOS. Computat. Biol. **16**(12) (2020)

15. Di Paolo, E., Burhmann, T., Barandarian, X.: Sensorimotor Life: An Enactive Proposal. Oxford University Press (2017)
16. Doerig, A., Schurger, A., Hess, K., Herzog, M.: The unfolding argument: why IIT and other causal structure theories cannot explain consciousness. Conscious. Cogn. **72**, 49–59 (2019)
17. Froese, T.: Irruption theory: a novel conceptualization of the enactive account of motivated activity. Entropy **25**(5), 748 (2023)
18. Froese, T., Stewart, J.: Life after Ashby: ultrastability and the autopoietic foundations of biological individuality. Cybern. Hum. Knowing **17**(4), 83–106 (2010)
19. Gardner, M.: Mathematical games: the fantastic combinations of john Conway's new solitaire game "life." Sci. Am. **223**, 120–123 (1970)
20. Husbands, P.: Never mind the iguana, what about the tortoise? Models in adaptive behavior. Adapt. Behav. **17**(4), 320–324 (2009)
21. Kirchhoff, M.: Autopoiesis, free energy, and the life-mind continuity thesis. Synthese **195**, 2519–2540 (2018)
22. Lombardi, O., López, C.: What does "information" mean in integrated information theory. Entropy **20**, 894 (2018)
23. Maturana, H.: The organization of the living: a theory of the living organization. Int. J. Man Mach. Stud. **7**(3), 313–332 (1975). https://doi.org/10.1016/S0020-7373(75)80015-0
24. Maturana, H., Varela, F.: Autopoiesis: the organization of the living. [De maquinas y seres vivos. Autopoiesis: la organizacion de lo vivo]. 7th edn. from 1994. Editorial Universitaria (1973)
25. Mediano, P., Rosas, F., Bor, D., Seth, A., Barret, A.: The strength of weak integrated information theory. Trends Cogn. Sci. **26**(8), 646–655 (2022)
26. Mediano, P., Seth, A., Barret, A.: Measuring integrated information: comparison of candidate measures in theory and simulation. Entropy **21**(1), 17 (2018)
27. Merker, B., Williford, K., Rudrauf, D.: The integrated information theory of consciousness: a case of mistaken identity. Behav. Brain Sci. **May**(19), 1–72 (2021)
28. Moreno, A., Etxeberria, A.: Agency in natural and artificial systems. Artif. Life **11**, 161–175 (2005)
29. Northoff, G., Zilio, F.: From shorter to longer timescales: converging integrated information theory (IIT) with the temporo-spatial theory of consciousness (TTC). Entropy **24**(270) (2022). https://doi.org/10.3390/e24020270
30. Oizumi, M., Albantakis, L., Tononi, G.: From the phenomenology to the mechanisms of consciousness: integrated information theory 3.0. PLOS Comput. Biol. **10**(5) (2014). https://doi.org/10.1371/journal.pcbi.1003588
31. Pautz, A.: What is the integrated information theory of consciousness. a catalogue of questions. J. Conscious. Stud. **26**(1-2), 188–215 (2019)
32. Potter, H., Mitchell, K.: Naturalasing agent causation. Entropy **24**(472) (2022). https://doi.org/10.3390/e24040472
33. Rodriguez, F., Husbands, P.: A saucerful of secrets: open-ended organizational closure in the game of life. In: ALIFE 2024: Proceedings of the 2024 Artificial Life Conference, p. 4. MIT Press (2024). https://doi.org/10.1162/isal_a_00712
34. Rodriguez, F., Husbands, P., Ghosh, A., White, B.: Frame by frame? A contrasting research framework for time experience. In: ALIFE 2023: Ghost in the Machine: Proceedings of the 2023 Artificial Life Conference, p. 75. MIT Press (2023). https://doi.org/10.1162/isal_a_00688
35. Rubin, S.: Cartography of the multiple formal systems of molecular autopoiesis: from the biology of cognition and enaction to anticipation and active inference. BioSystems **230**(104955) (2023)

36. Rubner, Y., Tomasi, C., Guibas, J.: A metric for distributions with applications to image databases. In: Proceedings of the 1998 IEEE International Conference on Computer Vision, Bombay, India (1998)
37. Seifert, G., Sealander, A., Marzen, S., Levin, M.: From reinforcement learning to agency: frameworks for understanding basal cognition. Biosystems **235**, 105107 (2024). https://doi.org/10.1016/j.biosystems.2023.105107
38. Singhal, I., Mudumba, R., Srinivasan, N.: In search of lost time: integrated information theory needs constraints from temporal phenomenology. Philos. Mind Sci. **3**(13) (2022). https://doi.org/10.33735/phimisci.2022.9438
39. Tononi, G.: An information integration theory of consciousness. BMC Neurosci. **5**(42) (2004)
40. Tononi, G., Koch, C.: Consciousness: here, there and everywhere? Philos. Trans. R. Soc. B **370**(1668) (2015). https://doi.org/10.1098/rstb.2014.0167
41. Tsuchiya, N., Taguchi, S., Saigo, H.: Using category theory to asses the relationship between consciousness and integrated information theory. Neurosci. Res. **107**, 1–7 (2016)
42. Varela, F.: Principles of Biological Autonomy. North Holland (1979)
43. Varela, F., Maturana, H., Uribe, R.: Autopoiesis: the organization of the living systems, its characterization and a model. Biosystems **5**(4), 187–196 (1974)
44. Varela, F., Thompson, E., Rosch, E.: The Embodied Mind: Cognitive Science and Human Experience. The MIT Press, Cambridge (1991)
45. Varela, F.: Patterns of life: intertwining identity and cognition. Brain Cogn. **34**(1), 72–87 (1997)
46. Varley, T.F.: Flickering emergences: the question of locality in information-theoretic approaches to emergence. Entropy **25**(54) (2022). https://doi.org/10.3390/e25010054
47. Villalobos, M., Silverman, D.: Extended functionalism, radical enactivism and the autopoietic theory of cognition: prospects for a full revolution in cognitive science. Phenomenol. Cogn. Sci. **17**, 719–739 (2018)
48. Villalobos, M., Palacios, S.: Autopoietic theory, enactivism, and their incommensurable marks of the cognitive. Synthese **198**(1), 71–87 (2019). https://doi.org/10.1007/s11229-019-02376-6
49. Weber, A., Varela, F.: Life after kant: natural purposes and the autopoietic foundations of biological individuality. Phenomenol. Cogn. Sci. **1**, 97–125 (2002)
50. Weisstein, E.: Moore neighborhood (2024), from MathWorld–A Wolfram Web Resource. https://mathworld.wolfram.com/MooreNeighborhood.html
51. Weng, L.: What is Wasserstein distance? (2017). https://lilianweng.github.io/posts/2017-08-20-gan/what-is-wasserstein-distance
52. Yuan, B., et al.: Emergence and causality in complex systems: a survey of causal emergence and related quantitative studies. Entropy **26**(108) (2024). https://doi.org/10.3390/e26020108

Author Index

A
Alahvirdi, Davoud 16
Alkilabi, Muhanad 98
Almansoori, Ahmed 98
Antonic, Nemanja 56

B
Baldini, Paolo 1, 160
Beauthier, Charlotte 125
Benvenuti, Filippo 160
Braccini, Michele 1, 160

C
Cambier, Nicolas 111
Carletti, Timoteo 208
Cerdeira, Hilda A. 223
Chinzer, Erica 250
Collinson, Ethan 236
Crovetti, Carlo 137

D
D'Addese, Gianluca 137
Damiano, Luisa 84

E
El Saliby, Michel 70

F
Fellermann, Harold 236
Filomeno Coelho, Rajan 125
Fischer, A. 198
Frasca, Mattia 208

G
Gambuzza, Lucia Valentina 208
Gigliotta, Onofrio 250
Giorgini, Diana 250
Glover, Tom Eivind 42
Goddard, Benjamin D. 28
Gratia, Antoine 173

H
Hubermont, Antoine 186
Husbands, Phil 292

I
Ingold, R. 198
Ioannou, Eirini 28

J
Jain, Sanyam 1

K
Kongni, Steve J. 223

L
Lindell, Trym A. E. 276
Louodop, Patrick 223

M
Marcantonio, Matteo 148
Masier, Stefano 148
Mauroy, Alexandre 16
Medvet, Eric 70, 160
Michaelidou, Eleni 28

N
Nadizar, Giorgia 70
Nguyen Tien Hoang, Dat 111
Nichele, Stefano 1, 42, 276
Njougouo, Thierry 223
Nobile, Clara 250

O
Osipov, Evgeny 42

P
Perrouin, Gilles 173
Pietquin, Julien 16

T. Carletti et al. (Eds.): WIVACE 2024, CCIS 2532, pp. 307–308, 2025.
https://doi.org/10.1007/978-3-031-93631-9

Ponticorvo, Michela 250
Pool Marquez, Sylverio 125

R
Ramírez-Ávila, Gonzalo Marcelo 208
Ramstad, Ola H. 276
Reimers, Felix 1
Rodriguez, Fernando 292
Roli, Andrea 1, 84, 160
Romero Moreno, Guillermo 28
Rubin, Sergio 84
Rusin, Francesco 160

S
Sainvitu, Caroline 125
Salvato, Erica 70
Sandvig, Axel 276
Sandvig, Ioanna 276
Sartenaer, Annick 125
Schobbens, Pierre-Yves 173
Serra, Roberto 137
Shrestha, Aarati 1
Simo, Gaël R. 223
Sion, Antoine 148
Stano, Pasquale 84

T
Tapper, Keren 28
Tchitnga, Robert 223
Temple, Paul 173
Thomson, Sarah L. 70
Tomaselli, Cinzia 208
Trendafilov, D. 198
Trendafilov, Dari 98
Tuci, Elio 16, 56, 98, 148, 186, 208
Tzafestas, Elpida 261

V
Vellinger, Aymeric 56, 186
Villani, Marco 137
Vitanza, Alessandra 250

Z
Zaugg, Ch. 198

The manufacturer's authorised representative in the EU is Springer Nature Customer Service Centre GmbH, Europaplatz 3, 69115 Heidelberg, Germany. If you have any concerns regarding our products, please contact ProductSafety@springernature.com

Printed and bound by CPI Group (UK) Ltd, Croydon, CR0 4YY
30/09/2026
02225560-0009